WAIT FIVE MINUTES

WAIT FIVE MINUTES

Weatherlore in the Twenty-First Century

Edited by
Shelley Ingram
and Willow G. Mullins

University Press of Mississippi / Jackson

The University Press of Mississippi is the scholarly publishing agency of
the Mississippi Institutions of Higher Learning: Alcorn State University,
Delta State University, Jackson State University, Mississippi State University,
Mississippi University for Women, Mississippi Valley State University,
University of Mississippi, and University of Southern Mississippi.

www.upress.state.ms.us

The University Press of Mississippi is a member
of the Association of University Presses.

Previous iterations of John Laudun's chapter "Weathering the Storm: Folk Ideas about Character" have appeared in *The Louisiana Folklore Miscellany*, vols. 28–29. The University Press of Mississippi has the journal's permission to publish the revised chapter in *Wait Five Minutes: Weatherlore in the Twenty-First Century*.

Copyright © 2023 by University Press of Mississippi
All rights reserved
Manufactured in the United States of America

First printing 2023
∞

Library of Congress Control Number: 2022061300

Hardback ISBN: 978-1-4968-4435-4
Trade paperback ISBN: 978-1-4968-4436-1
E-pub single ISBN: 978-1-4968-4437-8
E-pub institutional ISBN: 978-1-4968-4438-5
PDF single ISBN: 978-1-4968-4439-2
PDF institutional ISBN: 978-1-4968-4440-8

British Library Cataloging-in-Publication Data available

To the folks at NOAA and all those who forecast the never-ending weather.

—WGM

To my dad, Mark, and my paw paw, Donald, two men who love to sit and talk about the weather.

—SI

CONTENTS

Acknowledgments xi

Introduction: And Now, the Weather xiii

Section I: Belief

Introduction: The Sky Is Telling It 3

Chapter 1. Divergent Weatherlore in Christian Hermeneutics:
Climate Change and Vernacular Rhetoric in Our Current
Environmental Crisis 10
Emma Frances Bloomfield and Sheila Bock

Chapter 2. "Of Biblical Proportions": Flood Motifs
in Personal Narratives of Katrina Survivors 28
Kate Parker Horigan

Chapter 3. In the Bones: Prognostication and Weather
in the Twenty-First Century 43
Willow G. Mullins

Chapter 4. Contrails to Chemtrails: Atmospheric Scientists
Respond to Challenging Belief Narratives 59
Anne Pryor

Chapter 5. From Clockwork Weatherman to
Atomic Environmentalist 78
Máirt Hanley

Section II: Text

Introduction: The Romance of the Weather. 99

Chapter 6. "The World of Sensible Seasons Had Come Undone":
Climate Change and Regional Folklore in Barbara Kingsolver's
Flight Behavior 104
Hannah Chapple

Chapter 7. Early Modern Special Snowflakes 122
Christine Hoffmann

Chapter 8. Mothering the Storm: Black Girlhood and
Communal Care in Literature of Katrina 144
Jennifer Morrison and Shelley Ingram

Chapter 9. "You Don't Need a Weatherman": Bob Dylan's Windlore 160
James I. Deutsch

Chapter 10. "I'll Never Forget the Thunderstorm of 1960,
I Think It Was": Storm Stories. 176
Lena Marander-Eklund

Section III: Tradition

Introduction: Feeding the Storm 199

Chapter 11. Framing the Flood: Strategic Environmental Storytelling
in Appalachia. 205
Jordan Lovejoy

Chapter 12. Weathering the Storm: Folk Ideas about Character 226
John Laudun

Chapter 13. It Always Rains on a Picnic: Weatherlore and Community
Narrative at St. Patrick's Irish Picnic and Homecoming 248
Kristen Bradley

Chapter 14. The Folk Wisdom of Lawns 261
Todd Richardson

Chapter 15. Canning for the Apocalypse: Climate Change, Zombies, and the Early Twenty-First-Century Canning Renaissance 276
Claire Schmidt

About the Contributors 291

Index 297

ACKNOWLEDGMENTS

Thanks to my family, like my mama, who said, "I think I'll actually read this one," and my friends, who always send me weather memes. Thanks especially to my friend and colleague Clancy Ratliff, who encouraged this book from the time it was a little baby tweet, and to Constance Bailey, just because. And, of course, to Willow. We bonded at the University of Missouri over our shared secret desire to go back to college and become meteorologists—putting together this collection has been the next best thing.

—SI

Of course, thanks to my family for all the extreme weather adventures, the NOAA weather radio, and the support along the way. I learned my color wheel in a hurricane hole during David and Frederic, listening to the rain on the deck and the waves and my dad on the ham radio, crackling with reports from windward and from those whose anchors had slipped. And to the rest of my family for putting up with my hunt for the perfect weather app and my tendency to stand outside and size up the sky when the tornado sirens go off. Thanks to Cora Willard for the cake and veg, and the students in the fall 2021 Cultural Sustainability course at Washington University for talking more about the weather than they signed up for. Even though sometimes we appear to share a brain, I will say that I could ask for no better collaborator than Shelley, who taught me what revisions really look like. If this whole academic thing doesn't work out, we should still consider becoming meteorologists.

—WGM

Introduction

AND NOW, THE WEATHER

Since we began planning this volume, the United States has experienced at least four major extreme weather events, and there have been multiple editorial meetings that we had to postpone because of the weather. Thunderstorms, hurricanes, tornado warnings, snowstorms, and flash floods—and the power outages and road and school closures they bring with them—have shaped both the writing and the process of writing this volume. You can't escape the weather, and the weather, it feels like, has become more insistent, more fierce, more present.

Blown in on the winds of all this weather comes weatherlore: the folk beliefs and traditions about the weather that are passed down casually among groups of people. Weatherlore includes memes marking the 2021 Texas ice storm that quip, "The only thing Texas knows about ice and salt is a margarita"; advice shared among neighbors on how to wrap pipes against freezing; and conspiracy theories about whether snow is really snow or whether it is plastic manufactured and dispersed by a cabalistic government instead. Weatherlore can be predictive, such as the belief that more black than brown fuzz on a woolly bear caterpillar, *Pyrrharctia isabella*, signals a harsh winter. It can be commentary to ease daily social interactions, like asking whether it is "hot (or cold) enough for you," or to remark on politically inflected climate anxiety. Or it can be simply ubiquitous, much like the saying "if you don't like the weather, wait five minutes, and it will change." The weather governs our lives. It fills the gaps in conversations, determines our dress, and influences our architecture. No matter how much our lives may have moved indoors, no matter how much we may rely on technology, we still watch the weather. We still engage in weatherlore.

Weatherlore is one of the absolute truths of folklore. It has featured in foundational documents since the field was conceived: William Thoms seems to suggest the inclusion of weatherlore in his definition of the term "folk-lore" (see Roper 2007); Richard Inwards ([1869] 2013) published a collection of weatherlore texts in 1869; Fanny D. Bergen and W. W. Newell wrote a brief account of weatherlore in the *Journal of American Folklore* in 1889, one year after the first issue of the journal itself; the topic was revisited by H. A. Hazen in 1900. Wayland Hand's 1964 collection *Popular Beliefs and Superstitions of North Carolina* includes some weather-related beliefs. In the same period, one of the more comprehensive of these collections is Douglas Ward's (1968) article "Weather Signs and Weather Magic." Weatherlore is not only the purview of folklorists: meteorologist Edward B. Garriott produced a similar text in 1903 under the auspices of the US Weather Bureau in an attempt to separate fact from fiction. These texts had a feel of play about them—they recounted the charming verses and spoke in terms of historical agricultural practices.

Then, as the collecting of texts gave way to context and performance, the weather faded from the scholarly view of folklorists. The field seems to take for granted that weatherlore *is* folklore, but it has rarely been treated to extensive critical examination. The collections of weatherlore that are published—and they are largely collections with little analysis—appear in *The Farmers' Almanac* and in light-hearted, human-interest pieces in local papers and meteorological newsletters. Eric Sloane's 1963 *Folklore of American Weather* makes some attempt to contextualize beliefs about the weather, but its primary goal is to list popular weather sayings and label them true, false, or merely possible. Perhaps weatherlore is like the weather itself: it is so much about everyday life that most don't feel the need to write about it. Folklore studies has not spilt a lot of ink on the weather.

The playful approach that marks early texts is still meaningful,[1] but alongside it has risen an urgent need to examine vernacular understandings of weather and climate. There are invaluable contributions by folklorists to discussions of folk responses to disaster, like Carl Lindahl's work with Katrina and Rita survivors, particularly his essay "Legends of Hurricane Katrina: The Right to Be Wrong, Survivor-to-Survivor Storytelling, and Healing" and *Second Line Rescue: Improvised Responses to Katrina and Rita*, his collection with Barry Jean Ancelet and Marcia Gaudet (see, respectively, Lindahl 2012; Ancelet, Gaudet, and Lindahl 2013). Also influential for us is Kate Parker Horigan's (2018) masterful *Consuming Katrina: Public Disaster and Personal Narrative*, whose critique of "resilience" has become central to our own understandings of weather and climate disasters. And in *When They Blew the Levee: Race, Politics, and Community in Pinhook, Missouri*, David Todd Lawrence and Elaine J. Lawless (2018) explore the real, concrete impacts that political machinations concerning the effects of weather have on communities. Because, like it or not, weather is political, and our future depends upon it.

As we move further into the twenty-first century, though, an increasing awareness of climate change and its impacts on daily life, like in the kinds of weather events that marked the progress of this very book, calls for additional folkloristic reckonings with the weather. Weatherlore helps us know and shape global political conversations about climate change and biopolitics at the same time as it influences individual, group, and regional lives and identities. We use weather, and thus its folklore, to make meaning of ourselves, our groups, and, quite literally, our world. *Wait Five Minutes* brings together essays that look in some way to the weather: from personal experiences of picnics and suburban lawns to critical analyses of storm stories, novels, and flood legends, the chapters in this volume draw from folkloric, literary, and scientific theories to offer up new ways of thinking about this most ancient of phenomena.

Spurious Tornadoes

Wait Five Minutes tracks two trajectories we have identified in contemporary weatherlore: first is a desire for our experiences of weather to match our beliefs *about* weather, and second, should that weather pose a threat, is a belief that we will survive it. When we set out to produce a book about the weather, we anticipated a number of essay submissions that would document updated versions of the kinds of collections that marked the study of weatherlore from the nineteenth and early-twentieth centuries: legends of weather, weather sayings, and maybe some weather superstitions. What we received suggested a more complicated relationship between people and the weather, a relationship shaped by larger systems, seen through personal encounters, and in conversation with deeply held beliefs. Much weatherlore expresses a longing for better weather in the present, for better knowledge of the weather to come, and for better narrative cohesion of past weather events. As a result, much of our weatherlore reveals a negotiation between belief and experience, a negotiation that is on full display in one of the earliest recorded photographs of a tornado.

The photograph appeared in a number of newspapers at the time and purports to show a tornado that struck Waynoka, Oklahoma, in 1898. In the photo, two men stand beside an orchard and watch along the road as an ominously dark funnel cloud fills the sky. This tornado is not, however, the one that struck Waynoka. While it might be a photograph of *a* tornado, the men in the image were not watching *that* tornado; they were probably not watching *any* tornado.

The photograph is an early photo manipulation—a composite of several (possibly fake) images. The "original" image of the tornado (without the people or the orchard) was sent to *Weather Monthly Review* by a Mr. Connor, who claimed that it depicted the Waynoka tornado. However, the editors suspected a fake, if a very good fake, and chose not to publish it. A year later, the editors received another photo showing the same twister, now with the new setting and purporting to be from Kirksville, Missouri. The editors commented: "When we first saw it our funnel cloud was stirring up the dust and incidentally frightening the inhabitants of Waynoka, in far-off Oklahoma, and this was more than a year ago. The scene has now changed to a quiet road in Missouri across which our Oklahoma tornado cloud appears to be crossing, while a couple of artistic Rubens watch its progress in wonder and amazement" (Henry 1899, 204). An article on "spurious tornado photographs" in the Weather Bureau's *Monthly Weather Review* went further, claiming that the tornado itself was likely not a tornado at all, but rather an ink drawing on glass printed onto the negative (Henry 1899, passim). The image of the road and men was likely taken at sunset to produce the lowering clouds and the cast lighting.

So, there was no tornado, road, or men, and it was neither Oklahoma nor Missouri. The author ends his frustrated rant with both a warning that this photo will likely appear again labeled as another actual tornado and a bit of vernacular thinking about the weather. He writes, "The argument seems to be: if there was a disaster, it must have been a tornado; if a tornado, it must have been a funnel; if a funnel, there must be a picture; this is a photograph, therefore it will do" (Henry 1899, 204). This belief about extreme weather at the time—that it must be a tornado that looks like a tornado should look—drove the hoax. This is not the first ever photograph of a tornado, though it sometimes gets labeled as such. And despite earlier authenticated photos, there is something iconic about this one with its perfect funnel hovering menacingly above the pastoral scene that gives it an afterlife on the internet.

The story of this photo captures much about weatherlore in general and weather legendry in particular. It features a text passed around, linking itself to one place or another through local points of reference. It *could* be true and builds off of verifiable events, actual tornadoes that hit Waynoka and Kirksville. More to the point for this volume, however, it shows the desire of a population to know what a tornado looks like, of scientists and photographers to meet that need, and especially for the weather itself to resemble its lore. Alfred J. Henry, the Weather Bureau meteorologist, alludes to this last point in his final line, discussing the wonder and amazement of facing down such a perfect funnel. The lore tells of funnel clouds that stretch from pillowy clouds to the earth that cause horrible destruction. The photograph flips the deductive reasoning of meteorology to inductive, producing the case to prove the theory. Tornadoes are dramatic—they make good pictures, and they make good stories.

This photo and this story of perhaps the world's first weather meme illustrate the twinned themes of belief and survival that are threaded together throughout this volume. Folks wanted to believe in the tornado and its appearance in a Missouri field. The photograph summed up the mythology of the American heartland, a land of plenty for self-starting young Americans, a land ravaged occasionally by God. But equally compelling is the story of the men in the photo, for we must assume, given the length of exposure and the lack of evidence claiming otherwise, that the men survived the storm. This photo isn't just a photo about the romance of funnel clouds: it's a photo of survival.

Storm Survivals

By contrast, a popular contemporary internet meme about the weather features a scene of a seemingly suburban backyard. The image centers on a thin plastic patio table and chairs with one of the chairs laying on its side. The name of a weather nonevent appears at the top or bottom, sometimes accompanied by the phrase "We Will Rebuild" or "Never Forget." The meme first appeared tagged to an earthquake in Virginia in 2011, produced by FunnyJunk (KnowYourMeme 2011), but it has reappeared for a number of events, including Hurricane Irene in 2011, the Los Angeles earthquake of 2014, the Houston ice storm of the same year, the Melbourne earthquake of 2021, storms in Britain generally, and hurricanes Iselle, Isaac, and Zeta. Variations of the meme show other outdoor objects slightly askew. The website KnowYourMeme connects this particular meme to the kind of overblown language about rebuilding and remembrance that came out of the 2001 terrorist attacks, but it expresses something fundamental about how people talk about the weather. Much of the weatherlore that we see is about adaptation, survival, and how we negotiate or mitigate the power that weather has on our everyday lives.

It is interesting to think about what "survival" means in terms of weatherlore. Survivals, Edward Tylor suggests, are remnants of earlier culture that have lasted into the current age in the form of folklore (Dorson 1968, 182). Garriott (1903, 5), writing for the Weather Bureau, believes that "our first parents" gained their "weather wisdom" from observation and experience, passed down through descendants and to us "in the form of trite sayings or proverbs," sayings that future generations held on to even as they failed to hold up in new locations. Such trite sayings and proverbs about the weather are the very model of cultural survivals. And perhaps the patina of ancient wisdom that covers a claim like "Red sky at night, sailors delight" helps link the person who holds this knowledge to some far-flung, sea-faring ancestor. In most of the early collections of weatherlore, some acknowledgment is made of its seeming timelessness, its direct connection to earlier humans and earlier ways.

But weatherlore is not timeless; as our weather has changed, so has our lore. Inwards recognized this in 1869. He notes that some of his collected weatherlore was "giving evidence of the slowly changing climate of this country," and he forecasts that "it is not unlikely that at some distant date most of the predictions will be found inapplicable" (Inwards [1869] 2013, 13). The changing nature of weather and weatherlore, as Hannah Chapple discusses in her chapter in this volume, means, for example, that a predictive belief about the weather could become conditional: instead of "when," it's "if." A belief that had been stable—take "April showers bring May flowers," for instance—becomes destabilized, so that May flowers may come about *only if* there are showers in April. In order to survive, our lore must adapt because, as both the natural cycles of the Earth and anthropogenic climate change show us, things do not always stay the same, even the weather.

We humans must, therefore, accept certain contradictory things about our natural world: we are having a direct negative impact on the climate and must work to stop the devastation such impact is bringing about, while, at the same time, there is very little we can do about the weather. No matter how many Sharpies a former president may use to change a weather map, we cannot force a hurricane to hit Alabama. In the face of such daunting odds, our weatherlore often becomes a matter of *survival*, of negotiating the relationship between humans and weather. There is another popular weather meme that closes in on a panicked man, his arms filled with loaves of bread and containers of milk, screaming as he runs through the grocery store. It is most often tagged "Just seen a snowflake," though variations include "I just felt the wind blow" and "Don't forget the bread and milk." It crosses over with other popular memes, such as those featuring Jon Snow from *Game of Thrones* or the abominable snowman from the 1984 animated *Rudolph the Red-Nosed Reindeer* film. These "bread-and-milk" memes play on the American vernacular tradition of stocking up on bread, milk, and eggs before a snowstorm, sometimes referred to as the "French-toast run." They acknowledge the extent to which people engage in this pantry-stocking survivalist tradition, which we discuss later in the book, while also poking fun at the accompanying weather-induced panic. They even provide a sense of boastful pride for the people in areas that *do* experience extreme weather and manage to keep their cool. Both the "bread-and-milk" and the "we-will-rebuild" memes function as a way to release some of the tension of an extreme weather event through humor while also reminding us in that moment of how little there is that we can do to make the bad weather go away. It is ultimately out of our control whether a storm hits a locale or misses it, drops inches of precipitation or none at all.

These memes tell us something about how we frame encounters with the weather in terms of both individual and community survival on a tangible, immediate level. Like so much other weatherlore, they communicate a message that people survive—and while the loss of life in extreme weather events may heighten the stakes, the general message stays the same. When the sun comes up, people will be able to look out at their upright patio furniture as they eat their delicious French toast. But we are intrigued by the idea that such lore may have primed many of us to think of climate change in the same way—as an inevitable upheaval of nature that will nonetheless be survived. Narratives about the weather, from Noah until now, have shaped political responses to climate change by reinforcing the belief that a miracle will happen or a last-minute reprieve in the form of natural or technological intervention will be granted. The hurricane will turn, the tornado lift, the sun will come out tomorrow, and the climate will right itself. There seems to be a deeply embedded vernacular belief that all will, eventually, be well.

If You Don't Like the Weather

Thus, weatherlore is important. As the authors in this volume show us, individual experiences with and expressions of the weather in our world are deeply connected to our beliefs about community, culture, politics, and identity. The undeniable impact of climate change means that we need to pay deep attention to such expressions; if there is any hope of addressing the perils of our changing environment head-on, we must seek to understand why people feel the way that they do about the weather. Sometimes that means looking back at how notions of selfhood, preservation, and the natural world were embedded within the literature of early modern Europe, as Christine Hoffmann does in her chapter for this volume, "Early Modern Special Snowflakes." At other times, it requires us to consider the beliefs people hold about their place within the universe, as we see in Reverend Máirt Hanley's personal reflection "From Clockwork Weatherman to Atomic Environmentalist." As such, this collection represents a wide range of viewpoints considering a wide range of topics and texts. In it are chapters that find new ways to think about traditional expressions of weatherlore and others that present contemporary complications of what we might mean by "weather" and "lore." Taken as a whole, they offer up a picture of our weatherlore as integral to the relationship we humans have with the many different facets—cultural, physical, spiritual—of the world around us.

This volume is divided into three sections, with each section prefaced by a short exploratory introduction. The introductions address an element of weatherlore related to the central tenet of the section before giving a brief overview of the section's chapters. The first section, "Belief," includes a group of chapters that push beyond the collecting of weather sayings and rituals to consider the many ways that weatherlore and belief converge. From religion to prophecy to conspiracy, we find that belief is perhaps the most fundamental element of contemporary weatherlore. The second section, "Text," examines personal narratives, music, and literature that draw meaning, in some way, from the weather. These chapters consider different ways we tell stories of the weather, how those stories make their way into our art and our metaphors, and how we use them for ideological and political purposes in the inscription of culture. We call the final section "Tradition," though it may be more accurate to describe the chapters here as explorations into the array of vernacular responses people and communities have to the weather. Included in this section are chapters that often take a personal approach to discussing the value and meaning of weatherlore in our communities and in our domestic and personal lives. Of course, these divisions are somewhat arbitrary as all the chapters are in conversation with each other—and we find this fact incredibly exciting since it suggests that weatherlore is a foundational component of human society. More importantly, perhaps, it suggests a way forward, a way to think about weatherlore as we move further and further into the twenty-first century.

Finally, we'd like to say a word about our title, *Wait Five Minutes*. We draw this title from the folk expression "You know what they say about Missouri: if you don't like the weather, wait five minutes, and it will change." This expression is unique to Missouri—and Ohio, Minnesota, and Texas. And Indiana, Colorado, North Carolina, and Rhode Island. Mississippi and Louisiana say it too. Mark Twain said it about Hannibal or San Francisco or Hartford. In fact, we have found evidence of this vernacular expression being used from one end of the United States to the other. It speaks to a desire to believe in the weather and to claim a weather identity. It gives us a way to predict the unpredictable, even in the face of decades-long climate change. And it tells us that we believe we can survive weather both ordinary and extraordinary if we just wait five minutes.

Notes

1. Antone Minard's (2010) "'Like a Dying Duck in a Thunderstorm': Complex Weather Systems through the Lens of Folk Belief and Language" is a particular favorite of the authors.

References

Bergen, Fanny D., and W. W. Newell. 1889. "Weather-Lore." *Journal of American Folklore* 2, no. 6: 203–8. https://doi.org/10.2307/534149.

Dorson, Richard, ed. 1968. *Peasant Customs and Savage Myths*. Chicago: University of Chicago Press.

Garriott, Edward B. 1903. *Weather Folk-Lore and Local Weather Signs*. Honolulu: University of the Pacific Press.

Hand, Wayland. 1964. *Popular Beliefs and Superstitions of North Carolina*. Durham, NC: Duke University Press.

Hazen, H. A. 1900. "The Origin and Value of Weather Lore." *Journal of American Folklore* 13, no. 50: 191–98. https://doi.org/10.2307/533883.

Henry, Alfred J. 1899. "Spurious Tornado Photographs." *Monthly Weather Review* 27, no. 5 (May): 203–4. https://journals.ametsoc.org/view/journals/mwre/27/5/1520-0493_1899 _27_203b_stp_2_0_co_2.xml.

Horigan, Kate Parker. 2018. *Consuming Katrina: Public Disaster and Personal Narrative*. Jackson: University Press of Mississippi.

Inwards, Richard. [1869] 2013. *Weather Lore: A Collection of Proverbs, Sayings, and Rules Concerning the Weather*. Cambridge, UK: Cambridge University Press.

KnowYourMeme. 2011. "We Will Rebuild." https://knowyourmeme.com/memes/we-will -rebuild.

Lawrence, David Todd, and Elaine J. Lawless. 2018. *When They Blew the Levee: Race, Politics, and Community in Pinhook, Missouri*. Jackson: University Press of Mississippi.

Lindahl, Carl. 2012. "Legends of Hurricane Katrina: The Right to Be Wrong, Survivor-to-Survivor Storytelling, and Healing." *Journal of American Folklore* 125, no. 496: 139–76.

Minard, Antone. 2010. "'Like a Dying Duck in a Thunderstorm': Complex Weather Systems through the Lens of Folk Belief and Language." *Western Folklore* 69, no. 1: 109–19.

Roper, Jonathan. 2007. "Thoms and the Unachieved 'Folk-Lore of England.'" *Folklore* 118, no. 2: 203–16.

Sloane, Eric. 1963. *Folklore of American Weather*. New York: Hawthorne Books.

Ward, Donald J. 1968. "Weather Signs and Weather Magic: Some Ideas on Causality in Popular Belief." *Pacific Coast Philology* 3: 67–72. https://doi.org/10.2307/1316674.

Section I

BELIEF

Introduction

THE SKY IS TELLING IT

One morning in southwestern Louisiana, I stepped outside to drink my first cup of coffee and noticed something ominous. "Uh oh," I thought. "The sky is green." I hadn't seen a green sky since I'd moved out of the Midwest, so I hurried to check the local weather. Sure enough, we were under a tornado watch. "Did y'all see the green sky?" I asked excitedly as I walked into my introduction to folklore class an hour later, thinking we could take a few minutes to discuss weatherlore. Their response was immediate and unequivocal: "What's a green sky?"

We argued over this for a while, and I even made them get out of their seats to look out the window. None of the students in the class who were from Louisiana had really heard of, much less recognized, that tint of sickly yellow green that lays heavy in the air when a storm is coming. While I could see that the sky was so clearly giving us a warning of what might lie ahead, all my students saw was a typically humid south Louisiana spring day.

In "Tornado Stories in the Breadbasket," Larry Danielson (1990, 39) talks about the "boiling, greenish clouds" that would churn across his Kansas prairie, a sure sign that a storm was rolling in. The popular science education YouTube series *SciShow* calls the green sky warning "a common piece of celestial fortune-telling" (2018). One frustrated person complained to a group of climate researchers that their neighbors in Mississippi kept saying, "We don't see no green." "The sky is telling it!" this interviewee said. "If it's green, there's a tornado coming!" (Klockow, Peppler, and McPherson 2014, 802). Meteorologist Frank W. Gallagher III, William H. Beasley, and Craig F. Bohren (1996) wanted proof of the green sky phenomenon, and they got it—a series of experiments they performed in the 1990s showed that the sky *can*, in fact, turn green. So how could my students and I look up at the same sky and not see it the same way? This seemed to be a moment when folklore shaped our physical perception of the world—did my students not see the color because their lore did not ask them to? Did they simply *not believe* that a sky could be any color but blue?

We believe a lot of things about the weather. As Edward B. Garriott (1903, 5) said, humans have "ever employed inherited and acquired weather wisdom in the daily affairs of life." When bones and bunions begin to ache, expect the clouds to fill the lake. When the dragonflies swarm, expect a storm. An orange sky comes before a bad wind. We believe in the predicative power of the natural world, so that holly bushes foretell the severity of winter, cats bathing behind their ears or running around like wild predict rain or wind, and a ring around the moon means snow. We engage in ritual or ritualistic behavior, from the mundane activity of opening all the windows during a tornado warning to the sacred burning of a blessed palm frond to ward off a hurricane. Some of our rituals are intensely private, like the Cajun MawMaw who buried the umbilical stumps of her children under the eaves of her home to guard against bad weather, while others are idiosyncratic behaviors that are too complicated to even explain (one story I heard involved a woman slipping on her granddaughter's discarded shoes during a thunderstorm).[1] And yes, a lot of folks believe that the sky tells us when a tornado is coming.

Many of the early works on folklore and the natural world were collections of just such folk beliefs, with some of them focused exclusively on separating "true" folk wisdom from false. But there are other ways to think about belief and weather. Sabina Magliocco (2004, 14) says of the "socially constructed nature of reality" that "even when we are within our own culture, our experience of reality is context-dependent." Environmental researchers, perhaps surprisingly, have drawn similar conclusions about the individual experience of local weather. Trevor A. Harley (2003, 115) finds that nostalgia drives people's memories of local weather in Britain and that this nostalgia can be based on incorrect facts about actual weather so that "we might for example remember the weather as being hot and sunny at a time when we were particularly happy, or pouring with rain when we were unhappy." He argues that we also tend to let exceptional events—like a very cold winter—become typical events in our memories so that all winters in our past become like that one particularly chilly January. Weather thus becomes a metacognitive tool for shaping autobiographical memory even when that memory recalls factually false data. Furthermore, that false data could be dependent not just on memory but on beliefs about culture as well. White Christmases, for example, are relatively rare events in lowland England, but beliefs about what makes the perfect Christmas—in this case, some kind of extreme winter weather event—means that white Christmases live large in people's memories of their childhoods (Harley 2003). A similar phenomenon was reported in connection with the 1971 Los Angeles earthquake. People remembered unusual, and contradictory, weather in the days before the earthquake: from an unnatural stillness to big winds, from an extreme cold to an uncommon heat. Weather reports, however, showed no strange weather—it was a perfectly ordinary day (Anderson 1974, 335). What matters in these instances is the belief that extraordinary events warrant extraordinary weather or that our perfect childhood Christmases were always covered in snow.

It is not just the memory of past weather experiences that can depend upon belief. There is mounting evidence that folks' experiences of the weather in the relative present can be influenced by cultural and social factors so that "worldviews can alter the subjective experience of certain [weather] events," revealing a "gap between the perceived weather and real weather" (see, respectively, Lyons, Hasell, and Stroud 2018, 877; Shao 2016, 736). This is particularly important when it comes to climate change. A significant number of studies have begun to show that people experience the weather differently depending on whether or not they believe in anthropogenic climate change. When asked in one study to comment on the temperature of the past summer, respondents who dismiss concerns about climate change report average temperatures when, in fact, they experienced *above* average temperatures (Howe and Leiserowitz 2013). Wanyun Shao (2016, 738) finds that the "perception of local weather is subject to the process of motivated reasoning" so that perceptions of reality are driven by a desire to support one's own beliefs about climate and climate change. And another group of researchers finds that respondents' personal experiences of extreme draught and "polar vortexes" fell along partisan lines, with Republicans less likely to report that they experienced these extreme weather events even if the climate data said they did. These researchers conclude that actual experience of some weather events can be predicted by political partisanship (Lyons, Hasell, and Stroud 2018, 886).

So, if we can shape the weather to match beliefs about past events and if we can shape the weather to match ideological or political beliefs, then it stands to reason that beliefs about tornadoes and, perhaps more importantly, about *place* means that my students and I really could see two different skies that morning. People often use their own "folk science" to understand or predict the weather, a "place-based and culturally situated environmental knowledge" that comes from living somewhere—what geographers might call "landscape consciousness" (Klockow, Peppler, and McPherson 2014, 796). However, things like place attachment, the optimism bias, and "local (vernacular) knowledge" all help dictate why people believe the things about the weather that they do (Peppler, Klockow, and Smith 2017, 33). For example, researchers have repeatedly found that people make pretty bad forecasters when it comes to believing that they are at risk of a tornado hitting them directly. Folk beliefs about the land and their own place relative to it have led to the development of "attitudes and perceptions that are at odds with existing scientific knowledge" (Peppler, Klockow, and Smith 2017, 34).

But rather than simply listing out the ways in which people get things "wrong," as scientists so often do, some researchers are interested in the *ways of knowing* a place, about how "beliefs about tornado risk are rooted in local places and embody experiences in those places" (Klockow, Peppler, and McPherson 2014, 803). So, instead of pointing out, for example, that tornadoes can indeed climb up hills, they listen to the man who says that "they just don't do things like that—or didn't . . ." during his recollection of the 2011 Alabama tornado outbreak (Klockow, Peppler, and McPherson 2014, 791). The man knows this place, his place, and no tornadoes had ventured up that hill in his lifetime. Like that one frigid January standing as the ghost of winters past, the tornado-free hill had helped this man develop his personal 'climatology' of tornado paths "based on generalized observations as part of the lived experience" (Peppler, Klockow, and Smith 2017, 35). Does this mean, perhaps, that my students had simply not had enough experiences with tornados to incorporate a green sky into their folk science, into their own climatologies?

A quick internet search shows that the number of popular nonfiction books about the tornadoes of Oklahoma, Kansas, and the rest of the Great Plains—traditional "Tornado Alley"—vastly outnumber the number of books written about tornadoes in Louisiana or Mississippi. And yet, the yearly average number of tornado days is highest in these two Deep South states (Cappucci 2020). In fact, stretches of the deep south, long nicknamed "Dixie Alley" by weather scientists, are more prone to deadly, long-tracking tornadoes than anywhere else in the country. In 2020, there were sixty-eight deaths from tornadoes in this southern region.[2] There were eight deaths in *all other states combined*, including Oklahoma and Texas (NOAA 2021, n.p.). But belief in the traditional boundaries of Tornado Alley persists, and meteorologist Matthew Cappucci (2020, n.p.) argues that it all "boils down to public perception, rooted in years of storm chasing, cinematography and geography."

The wide-open spaces of the plains make for more frequent sightings, their geography more hospitable to breathtaking vistas of these beautiful but violent twisters. Tornadoes are less likely to be "rain wrapped" in the plains, which allows them to be seen taking that iconic funnel shape. And they have become part of the identity of people who live there, as Danielson (1990) memorably writes in his essay. The life of a tornado story "is a vigorous one in midwestern talk," serving as a "means of expressing a distinctive identification with one's regional home" (Danielson 1990, 30 and 39). One popular internet meme shows a photo of a tornado siren with the words "Let me play for you the song of my people" superimposed on it, while another suggests that a tornado is nothing more than an Oklahoma woman's "perfect cuddle weather." Perhaps, too, the myth of westward expansion, with its homesteaders fighting to survive on a harsh new land, elevates the tornado to the realm of the sublime. Other extreme weather events happen in this part of the country, like droughts and blizzards and ice storms, but the tornado reigns supreme in the crafting of a regional identity.

If belief can shape our perception of the weather and if regional and cultural identity help shape our beliefs, then I can begin to make sense of my class's disagreement about what the sky was telling us. For the Gulf South states of Louisiana, Mississippi, and Alabama, and South Atlantic states, like Georgia and South Carolina, their identity is perhaps tied to a different extreme weather event. When I search for internet memes about tornadoes in the Gulf South, I got hurricanes instead. Hurricanes and humidity seem to be the dominant weather traits down there, natural elements that the region uses to help make sense of itself and claim a geographical identity, despite the prevalence of killer tornadoes. Maybe each region gets to claim one extreme weather event for itself: the plains get tornadoes, California gets earthquakes or, increasingly, fires, the Great Lakes get snow. With five years in a row of busier-than-average Atlantic hurricane seasons and with the memories of Camille and Katrina and Harvey and Florence always just a stone's throw away, perhaps there is simply not much need in the South for belief in green skies.

The chapters that follow in this section continually push beyond the collection of eccentric weather rituals to consider the many ways that weatherlore and belief converge, diverge, interact, and inform. In the first chapter, "Divergent Weatherlore in Christian Hermeneutics: Climate Change and Vernacular Rhetoric in Our Current Environmental Crisis," Emma Frances Bloomfield and Sheila Bock call for a resistance to neat categorizations of believer/nonbeliever when thinking about Christianity and its relationship to climate change. Through a combination of close textual analysis and interviews, they highlight the variety of ways and the varying degrees to which Christians engage in ecological thinking and behaviors. Kate Parker Horigan, in her chapter "'Of Biblical Proportions': Flood Motifs in Personal Narratives of Katrina Survivors," also examines an engagement with biblical language and discourse as she examines how survivors of Hurricane Katrina in New Orleans engage biblical motifs of punishment and survival to draw local meaning from the flood.

Moving from the ecclesiastical to the mystical, Willow G. Mullins asks us to consider the role of magic and prophecy in the knowing of weather. Her chapter, "In the Bones: Prognostication and Weather in the Twenty-First Century," explores how the history of weather forecasting, juxtaposed as it is against personal, embodied experiences of the weather, has resulted in an understanding of weather and climate that depends on both faith and feeling. We end this section with two chapters that deal directly with public belief about the weather. Anne Pryor, in "Contrails to Chemtrails: Atmospheric Scientists Respond to Challenging Belief Narratives," introduces us to Steve Ackerman and Jonathan

Martin's monthly radio show *The Weather Guys*, broadcast on Wisconsin Public Radio. She details the sorts of speculative conspiracies atmospheric scientists bump up against all the time, situating the ideas of both the scientists and the skeptics within the wider world of belief. Finally, Máirt Hanley, a Church of Ireland priest, was once told by his bishop that he did not approve of Hanley's praying for the weather. In a chapter that includes interviews and correspondences, Hanley's "From Clockwork Weatherman to Atomic Environmentalist" reflects on this disapproval in a consideration of the faith, theology, and folk belief of the people in his pews.

Notes

1. Thank you to Laura Lege-McGovern, Rhonda Robison Berkeley, and Paddy Bowman, among others, for sharing some of their weatherlore with us.

2. States included in this count: Louisiana, Mississippi, Alabama, Tennessee, Georgia, South Carolina, and North Carolina.

References

Anderson, David M. 1974. "The Los Angeles Earthquake and the Folklore of Disaster." *Western Folklore* 33, no. 4: 331–36.

Cappucci, Matthew. 2020. "*Tornado Alley in the Plains* Is an Outdated Concept: The South Is Even More Vulnerable, Research Shows." *Washington Post*, May 16, 2020. https://www.washingtonpost.com/weather/2020/05/16/tornado-alley-flawed-concept/.

Danielson, Larry. 1990. "Tornado Stories in the Breadbasket: Weather and Regional Identity." In *Sense of Place*, edited by Barbara Allen and Thomas J. Schelereth, 28–39. Lexington: University of Kentucky Press.

Gallagher, Frank W., III, William H. Beasley, and Craig F. Bohren. 1996. "Green Thunderstorms Observed." *Bulletin of the American Meteorological Society* 77, no. 12: 2889–98.

Garriott, Edward B. 1903. *Weather Folk-Lore and Local Weather Signs*. Honolulu: University of the Pacific Press.

Harley, Trevor A. 2003. "Nice Weather for the Time of Year: The British Obsession with the Weather." In *Weather, Climate, Culture*, edited by Sarah Strauss and Benjamin S. Orlove, 103–18. Oxford: Berg Publishers.

Howe, Peter D., and Anthony Leiserowitz. 2013. "Who Remembers a Hot Summer or a Cold Winter? The Asymmetric Effect of Beliefs about Global Warming on Perceptions of Local Seasonal Climate Conditions in the U. S." *Global Environmental Change* 23, no. 6: 1488–1500.

Klockow, Kimberly E., Randy A. Peppler, and Renee A. McPherson. 2014. "Tornado Folk Science in Alabama and Mississippi in the 27 April 2011 Tornado Outbreak." *GeoJournal* 79, no. 6: 791–804.

Lyons, Benjamin, Ariel Hasell, and Natalie Jomini Stroud. 2018. "Enduring Extremes? Polar Vortex, Drought, and Climate Change Beliefs." *Environmental Communication* 12, no. 7: 876–94.

Magliocco, Sabina. 2004. *Witching Culture: Folklore and Neo-Paganism in America*. Philadelphia: University of Pennsylvania.

NOAA. 2021. "2020 Preliminary Killer Tornadoes." January 26, 2021. https://www.spc.noaa.gov/climo/torn/STATIJ20.txt.

Peppler, Randy A., Kimberly E. Klockow, and Ricard D. Smith. 2017. "Hazardscapes: Perceptions of Tornado Risk and the Role of Place Attachment in Central Oklahoma." In *Explorations in Place Attachment*, edited by Jeffrey S. Smith, 33–45. London: Routledge.

SciShow. 2018. "Why Does the Sky Turn Green before Tornadoes?" May 29, 2018. YouTube video, 2:18. https://youtu.be/ro3fIrbyXqM.

Shao, Wanyun. 2016. "Are Actual Weather and Perceived Weather the Same? Understanding Perceptions of Local Weather and Their Effects on Risk Perceptions of Global Warming." *Journal of Risk Research* 19, no. 6: 722–42.

Chapter 1

DIVERGENT WEATHERLORE IN CHRISTIAN HERMENEUTICS

Climate Change and Vernacular Rhetoric in Our Current Environmental Crisis

Emma Frances Bloomfield and Sheila Bock

In 1967, historian Lynn White Jr. (1967, 1203) argued that Christian teachings about the environment are at least partly to blame for our current "ecologic crisis." Even as early as the 1960s, White (1967, 1203) was concerned about how "changes in human ways often affect nonhuman nature," such as animal extinction and habitat destruction. He notes that the drive for human expansion and dominion can be traced to biblical teachings that "no item in the physical creation had any purpose save to serve man's purposes," so he characterized Christianity as "the most anthropocentric religion the world has seen" (White 1967, 1205). One of the verses White refers to is the dominion verse, Gen. 1:26, which can be interpreted as granting humans dominion, power, and control over the natural world. These teachings reify an anthropocentric perspective where nonhuman interests are subordinate to human needs (Milstein 2011). The pervasiveness of this perspective can be seen in various polls that show Christians' sustained skepticism toward the science of climate change (e.g., Gauchat 2012; Pew Research Center 2017) and even their welcoming of climate change as a sign of the apocalypse (Barker and Bearce 2013). But the relationship between Christianity and climate-change skepticism is far from deterministic (Bloomfield 2019; Veldman 2019). Indeed, as Erika Brady (2001), David J. Hufford (1982; 1994), Bonnie Blair O'Connor (1995), and Leonard Norman Primiano (1995; 2012) remind us, a belief system is never all-encompassing since its interpretation and enactment is always shaped by the lived experiences of individuals. Despite prominent associations between Christianity and climate-change skepticism, there are a variety of different ways and degrees to which Christians engage in ecological thinking and behaviors that are hard to neatly categorize.

One prominent example that illustrates the varying perspectives of Christianity is the Creation Care movement, which is an iteration of Evangelical Christianity that unites faith with environmentalism. While small and still growing, the Creation Care movement has emerged as a powerful and prominent social movement that creates space for an eco-conscious identity in Christianity (Bloomfield 2020). Using Julia Corbett's (2006) spectrum of environmental ideologies, we can label dominion Christian interpretations of the relationship between humans and nature as anthropocentric, or human centered, while Creation Care members' perspectives are far more eco-centric, or nature centered. The presence and rhetoric of Creation Care members reveal the hermeneutical latitude of Bible verses, offering an illustration of "the religious vernacular," an approach to the study of religion that "highlights the power of the individual and communities of individuals to create and re-create their own religion" (Primiano 2012, 383) and "understands religion as the continuous art of individual interpretation and negotiation of any number of influential sources" (Primiano 2012, 384).

While biblical text serves as a source of authority, *how* Christians interpret biblical teachings on the relationship between humans and nature creates resources from which people can argue for divergent perspectives on human-nature relationships. In other words, the story of Genesis offers polysemous verses that chain out into various interpretations about proper, God-ordained environmental practices that manifest differently among Christians. These interpretations are often paired with or complemented by observations and understandings of the natural world, such as the weather, as a source of evidence for human-nature relationships. This chapter explores the different interpretations that emerge from the same text, the Bible, and how those interpretations are articulated in ways that help to both construct a group identity (i.e., Christian climate skeptic or Creation Care advocate) and create an explanatory framework for making sense of extreme weather observed in the world.

First, we explore the two poles of Corbett's environmental ideology spectrum, anthropocentric and eco-centric, using case studies of established Christian climate groups. Then, we wish to complicate the notion that these two "official" readings of Christianity in regard to the environment are the only interpretative options. In the spirit of Primiano's (1995, 45) warning against separating popular religion from "institutionalized" religion, we also attend to how individuals express not only anthropocentric and eco-centric perspectives but also a diversity of "vernacular" religious interpretations in between. We thus combine close readings of discourse published online by various Christian groups about the environment with qualitative interviews in order to analyze "the way people actually live their religious lives" in complicated, nuanced, and individual ways not always represented in public discourse or by established groups (Primiano 1995, 45).[1]

An Overview of Christianity and Climate Change in Public Discourse

Before beginning our analysis, it is important to contextualize our discussion by exploring the relationships between Christianity and climate change and how they both influence and are influenced by public discourse. There is increasing evidence and consensus that anthropogenic climate change is both urgent and severe. Despite scientific certainty on the matter, climate change still rages on as a persistent and prominent controversy such that even top officials in governments around the world espouse climate-change skepticism. Media coverage of climate change contributes, in part, to this persistence by showering attention onto polarized voices (Bolsen and Shapiro 2018) and through the tendency to frame topics in journalism as balanced between two sides of a story (Dixon and Clarke 2013). Such representations give highly motivated and polarized voices, who are not representative of the public's climate attitudes,[2] an even playing field to grab headlines and media attention. In the case of climate change, this preference for balance (Patterson and Lee 1997) means that the media may cover issues of the environment from two perspectives—climate-change believers and climate-change skeptics—creating a "false balance" between the two sides that does not reflect scientific knowledge and understanding (Dixon and Clarke 2013).

If we associate climate-change skeptics with the anthropocentric view that "humans are superior to and dominate the rest of creation," then environmentalists and climate-change believers may be classified as eco-centric, holding perspectives that acknowledge that "humans are an interdependent, integral part of the biological world but no more or less important than other portions of it" (Corbett 2006, 27). As two broad categories, these are "extreme" end points that only partially represent the full spectrum of environmental ideologies (Corbett 2006, 29). While Corbett (2006) does not seek to quantify the amount of people who might fall into these categories, other groups have developed their own heuristics for making similar estimations. For example, the Yale Project on Climate Change (YPCC) groups people in the United States into six groups on a scale between "alarmed," or holding eco-centric beliefs, to "dismissive," or holding anthropocentric beliefs (Goldberg et al. 2021). While these are not perfect analogues, the alignment provides a rough estimate of eco-centric and anthropocentric beliefs in the US population. Despite what mediated and political messages might convey, 73 percent of the population is at least "cautious" about climate change, while the remaining 27 percent are "disengaged," "doubtful," and "dismissive" (Goldberg et al. 2021). The two extremes of the poles (alarmed and dismissive) make up 34 percent of the scale, meaning that blends and mixtures of anthropocentric and eco-centric beliefs constitute roughly 66 percent of climate attitudes. We contend that this blending represents a molding of official statements from anthropocentric and eco-centric representatives to people's everyday lived experiences with the environment.

Furthermore, we seek to unpack this middle ground as it relates to Christians because 70.6 percent of the US population use the Bible as an authoritative resource to different extents, and interpretations of it craft their identity and ecological attitudes (Pew Research Center n.d.). As with the variety of beliefs expressed in the YPCC's Six Americas project, we will analyze how Christians negotiate a variety of different environmental attitudes using the same text. As previously mentioned, Christianity and the environment have had a complicated relationship; some, like White, view them as oppositional, while others recognize their potential compatibility and alignment. Despite the presence of Christian environmentalists, public discourse about and media coverage of Christianity and climate change often pit them against one another. Common understandings of Christianity frame it as antienvironmental, drawing upon White's associations, the dominion verse, and perceived hierarchies of humanity ruling over nature. In order to uncover this perspective in more detail, we will highlight the discourse of the Cornwall Alliance as a prominent anthropocentric voice in contemporary discourse (Bloomfield 2019; Evans 2018).

To represent an eco-centric perspective, we will analyze the discourse of the Evangelical Environmental Network (EEN), arguably the first Creation Care group in the US that combined specifically Evangelical Christianity with proenvironmental attitudes (Bloomfield 2019; Bloomfield 2020). While the Creation Care movement still contains markers of anthropocentrism and thus has not fully abandoned human-nature binaries, it is also true that they represent what amounts to an eco-centric view under the framework of Christianity (Bloomfield 2020). For example, to see humans and nature as fully equal would likely be so distinct from Christianity as to be unrecognizable as such. Thus, the EEN represents an eco-centric identity as performed within the Christian faith.

In what follows, we perform a close reading of texts from online publications from the Cornwall Alliance and the EEN in addition to interviews with nearly one hundred Christians about their environmental beliefs (originally interviewed for Bloomfield 2019). The close reading is informed by a combination of rhetorical analysis and folkloristic study of belief that attends to patterns and recurring themes in how the environment is constructed in relation to religion, with particular attention to the trope of weather. By attending to official discourses and vernacular constructions, we seek to complicate a simple, deterministic reading of Christianity and the environment and explore ways that religion, faith, and the weather can be "inventional" resources for environmental advocacy.

Anthropocentric Weatherlore

To represent an anthropocentric environmental ideology, we analyze the Cornwall Alliance (CA), a fundamentalist Christian group started by Dr. E. Calvin Beisner in 2005. The CA receives media attention for their prominent denial of climate change, polarized statements, and deployment of literal readings of the Bible to defend their preference of humanity over nature. The CA thus represents an anthropocentric view that prioritizes human comfort, wellbeing, and profit over nonhuman interests. In our analysis of the CA's discourse, we focus on their expression of the human-nature relationship and how specific Bible verses and interpretations related to weather lead them to these conclusions.

One of the discursive patterns easily identifiable in the CA's rhetoric is the separation they create between being pro-Christian and proenvironment. For the CA, these two ideologies, Christianity and environmentalism, are staunchly opposed. In its "What We Do" description, the CA (n.d.c, n.p.) overtly states that the environmental movement is "overwhelmingly anti-Christian" and its teachings "undermine the fundamental Christian doctrines of God, creation, humanity, sin, and salvation." The separation between the two is in part motivated by interpretations of certain Bible verses. For example, the CA (2009, n.p.) draws on Gen. 1:1–31, the first verses in the Bible, and Gen. 8:21–22 as evidence that the "Earth is robust, resilient, self-regulating, and self-correcting." The first verses of Genesis describe God's process of creating the heavens and the Earth. While these verses do not explicitly say that the Earth is resilient, they can be interpreted as meaning that God is in control and has an influence on all parts of it. The verses of Gen. 8:21–22 provide more explicit support for the Earth's resilience. These verses note: "The Lord smelled the pleasing aroma and said in his heart: 'Never again will I curse the ground because of humans, even though every inclination of the human heart is evil from childhood. And never again will I destroy all living creatures, as I have done. As long as the earth endures, seedtime and harvest, cold and heat, summer and winter, day and night will never cease'" (New International Version). These verses are thought to contain the words of God, claiming that the Earth will never again be "cursed" after the Great Flood, which can be interpreted to mean that the Earth is protected by God from future disasters. These verses also explicitly mention the changing seasons and temperatures as under God's control and ultimately outside of the realm of human influence. If the CA interprets the Bible as saying that the Earth will recover and remain stable even when faced with "some damage" from humans, then the group will likely view statements of the Earth's fragility and susceptibility to human influence as contrary to the Bible's teachings.

Another reason the CA provides for separating Christianity and the environment is their interpretation of what the Bible says about poverty. The CA argues that Christians should "take seriously the Bible's emphasis on protecting the vulnerable from harm," referencing Pss. 12:5, 35:10, 41:1, 72:4, 72:12; Prov. 31:9; and Gal. 2:10 (Legates and van Kooten 2014, 7). All these verses provide statements of protection and respect for the poor, the needy, the weak, and the oppressed. For example, Ps. 12:5 quotes God saying, "I will protect [the poor] from those who malign them." Ps. 72:4 makes a hortatory claim that people should "defend the afflicted among the people and save the children of the needy; may he crush the oppressor." The CA (n.d.a, n.p.) frames supporting and protecting the poor as matters of economics, noting that "to protect the poor, we must . . . make energy and all its benefits more affordable, and so free the poor to rise out of poverty." In an open letter to politicians, the CA (2009, n.p.) describes environmental policies as "fruitless, indeed harmful policies to control global temperature." Instead of being distracted by the environment, the CA (n.d.b, n.p.) argues that good Christians should prioritize "enabling more and more of our neighbors to enjoy good health, long life, freedom, justice, and spiritual and material abundance." The CA (n.d.b, n.p.) thus creates a hierarchy where the primary goals of Christianity are to lift "the needy out of poverty" and spread "the gospel of Jesus Christ." Indeed, the CA (n.d.a, n.p.) notes that environmentalism promotes "far-reaching, costly policies" that will damage the economy and, thus, "the poor."

In pitting developing countries against the environment, the CA sets up a false dichotomy whereby people must choose between helping humans and helping nonhuman nature. Environmentalists, however, view helping the environment and developing countries as simultaneous goals under "climate justice." The CA (n.d.a, n.p.) distorts climate justice by arguing that environmental policies—as opposed to neoliberalism and colonialism, as climate justice activists assert—are the primary creators of "more generations of poverty and the high rates of disease and premature death that accompany it." Interestingly, the CA (n.d.a, n.p.) does argue that it invokes "environmental stewardship." But, in using the modifier "responsible" (i.e., responsible environmental stewardship), the CA (n.d.a, n.p.) creates its own brand of stewardship. Counter to scientific consensus, the CA argues that "obtaining energy from fossil fuels and so adding CO_2 to the atmosphere" is "not an abuse of the earth [sic] but is instead a vitally important way of improving human wellbeing" (Legates and van Kooten 2014, 7). In addition to going against mainstream knowledge about the role of carbon in climate change, the CA also frames "not an abuse of the earth" as equivalent to "improving human wellbeing." In other words, the CA seems to conflate what is good for the poor as a performance of environmental stewardship, which is an anthropocentric transformation of the meaning of "stewardship" as related to the benefit of humanity as opposed to nonhuman nature or human-nature writ large.

In addition to pulling from Bible verses as inventional resources, the CA also directly implicates daily weather patterns as reasons to deny climate science. Weather becomes a useful, empirical tool that they can point to in order to create antienvironmental arguments. For example, in an article about Tropical Depression Imelda, the CA uses historical references to claim that extreme weather events "have wreaked havoc upon coastal populations since before the beginning of written history" (Balgord 2019, n.p.). This claim supports the idea that there is nothing unique or distinct about contemporary weather patterns, which many researchers point to as examples of how warming causes more frequent and serious weather events. Instead, the CA notes: "Severe hurricanes and typhoons will continue to occur and to plague mankind from time to time during the remainder of this century. They are not newly invented, man-caused extreme weather events under the sun and should never be portrayed as such" (Balgord 2019, n.p.). Instead of pointing to extreme weather events as evidence of climate change and the need to mitigate it, the CA uses the same evidence to invert the argument and to frame these events as normal occurrences. Emphasizing extreme weather as standard occurrences through history, the CA draws from the maxim often attributed to Mark Twain, "Climate is what we expect, weather is what we get," to signify how "extreme weather does not necessarily indicate a drastic change in climate" (Sadar 2019, n.p.). This argument emphasizes the difference between weather and climate, that one is day-to-day and the other based on long-term patterns, to discredit arguments that pull from weather events as evidence of a changing climate. This maxim may also be interpreted as casting doubt on climate science, framed as "predictive" and often incorrect, whereas weather is tangible and felt in the present.

The CA's anthropocentrism is reflected in its rhetoric regarding the hierarchy of human life above nonhuman life and how weather patterns are used as evidence of the Earth's stable, God-controlled climate. The CA's use of Bible passages and appeals to weather as an argumentative resource support their underlying anthropocentric and conservative beliefs whereby human needs, especially economic ones, outweigh nonhuman needs in all scenarios. To use the weather or Bible passages as evidence for climate change and thus the need for proenvironmental action would be to sacrifice the CA's ideological commitments. For the CA, there is a clear separation between Christianity and the environment whereby environmental concerns are not only contrary to its interpretations of the Bible but also to the "reality" of weather that the CA and its members observe. Both hermeneutic processes, interpreting Bible verses and interpreting the clouds, lead to an anthropocentric faith reflected in the CA's rhetoric.

Eco-centric Weatherlore

Opposed to the anthropocentrism of the CA, the EEN, a group started in 1993 by Reverend Mitch Hescox, represents a Christian, eco-centric environmental ideology. Although we might not consider the EEN fully eco-centric in terms of Corbett's spectrum, it is an eco-centric iteration of the Christian identity (Bloomfield 2020) that is often in direct conversation with the CA. For example, the CA overtly refers to the EEN as betraying Christian values and ethics and hopes that the CA's discourse expresses "another side of the story" about Christianity and the environment (Beisner 2015, n.p.). The EEN thus represents an alternative to the CA's environmental interpretations that tempers human needs with the recognition of the interconnectedness between human and nonhuman life. In our analysis of the EEN's discourse, we focus on their expression of the human-nature relationship and how specific Bible verses and interpretations lead to these conclusions.

Although pulling from the same primary text, the EEN draws very different conclusions about the negotiation between Christianity and the environment. One of the first statements the EEN (n.d.a) makes on its website is "Biblically understood, 'the environment' is actually part of God's creation, of which human beings are also a part." This statement emphasizes a nonhierarchical, more interconnected relationship between all life as equal parts of God's creation instead of humans being superior to nature as in the CA's hierarchy. To solidify this idea of interconnectedness and the importance of both human life and nonhuman life, the EEN (n.d.c) references Col. 1:16, which reads, "For by him all things were created, in heaven and on earth, visible and invisible, whether thrones or dominions or rulers or authorities—all things were created through him and for him." By emphasizing the common ancestry of humans and nonhuman nature, the EEN uses the Bible to support the idea that the two are equally important. The EEN (n.d.c) also references Matt. 25:37–40, which emphasizes the interconnectedness of God and the weakest and most vulnerable of society. The passage reads:

> Then the righteous will answer him, "Lord, when did we see you hungry and feed you, or thirsty and give you something to drink? When did we see you a stranger and invite you in, or needing clothes and

clothe you? When did we see you sick or in prison and go to visit you?" The King will reply, "Truly I tell you, whatever you did for one of the least of these brothers and sisters of mine, you did for me." (New International Version)

While the CA might interpret this same quotation as meaning that we should prioritize the needs of humans, the EEN views nonhuman nature and humans as siblings, both being created by God the Father, and so would include in its interpretation the need to also protect and care for nonhuman nature. To put the finest point on its belief of interconnectedness, the EEN firmly states: "Indeed, one cannot fully worship the Creator and at the same time destroy His creation, which was brought into being to glorify him. Worshiping the Creator and caring for creation is all part of loving God. They are mutually reinforcing activities. It is actually unbiblical to set one against the other" (EEN, n.d.c, n.p.). Unlike the separation that the CA creates between following Christianity and advocating for nonhuman nature, the EEN views them as interdependent behaviors that are articulated as such in the Bible.

Instead of emphasizing the dominion verse of Gen. 1:26, the EEN (n.d.c, n.p.) draws from other Bible verses that show "God's relationship to [the rest of creation] and how God wants us to behave towards it." In addition to referencing verses we have already discussed, the EEN (n.d.c, n.p.) also cites extensive passages from Genesis that detail how God created not only humans but also plant and animal life, and then "called it good." For example, the EEN (n.d.c, n.p.) cites Gen. 1:31 that in part states, "God saw all that he had made, and it was very good." If God created something and called it good, then the EEN interprets this as meaning that it certainly has value, a purpose, and is intimately related to the other acts of creation, such as humanity.

In addition to interpretations of Bible verses as inventional resources for its eco-centric performance of Christianity, the EEN also makes use of weather as evidence. For example, as part of the EEN's (n.d.b, n.p.) founding doctrine, the "Caring for Creation Pledge," it notes that "the weather is biting back with increasing frequency and ferocity." The EEN (n.d.b, n.p.) attributes this increase of extreme weather events to the "experiment" that humans are running "on the only home we have: God's home." This statement asserts not only a need for care for human life harmed in extreme weather events but also that these events are damaging the home/the Earth that God provided to humans. In an article describing the damages of Hurricane Harvey, the EEN characterizes the storm as a "meteorological monster" that had "no historical precedent" (Douglas 2017, n.p.). Unlike the CA's attempts to characterize extreme weather events as nothing new, the EEN emphasizes their severity as anomalous and due to human actions. The article about Harvey continues: "It was the most extreme rainfall event ever observed in the United States; a 1 in 1,000-year-flood" (Douglas 2017, n.p.). While the CA interprets this weather event in a context of a natural timeline, the EEN sees Harvey as an aberration that points to how "the warming we're witnessing is turning up the volume on weather extremes, especially hotter heat waves, drier droughts and heavier rainfall amounts" (Douglas 2017, n.p.). Making a direct connection to Christianity, the EEN describes extreme weather events and especially hurricanes as causing "Old Testament flooding," thereby elevating the importance and resonance of the extreme weather events for their audience (Douglas 2017, n.p.).

The EEN also uses weather as a poetic resource, whereby the "fog" of Pittsburgh's polluted skies mirrors Eph. 5:11, which states, "Have nothing to do with the fruitless deeds of darkness, but rather expose them" (Krost 2017). This connection links the figurative darkness and blindness of those who would deny the reality of the weather to the physical fog that clouds physical sight. The fog, of course, comes from industrial pollution, which is a main contributor to climate change and the Earth's warming. The EEN further connects the harms of industrial pollution to the need to care for the poor and vulnerable. In one article, the EEN quotes from the *Fourth National Climate Assessment* (2017): "People who are already vulnerable, including lower-income and other marginalized communities, have lower capacity to prepare for and cope with extreme weather and climate-related events and are expected to experience greater impacts" (Ball 2018, n.p.). The EEN then continues, "Followers of Christ, who commanded us to care for the poor, should be especially concerned about this finding" (Ball 2018, n.p.). Similar to the CA, the EEN is motivated in part by biblical teachings on charity and the poor. But, instead of viewing proenvironmental actions as damaging to the poor, the EEN views environmental advocacy as forms of climate justice that help protect vulnerable populations from further exploitation and damage caused by environmental collapse.

Due to the damage caused by extreme weather events and industrial pollution, the EEN calls for these events to inspire people to make "everything we do more flood-resistant and storm-tolerant" and for people to see climate change as "an opportunity for innovation and reinvention" (Douglas 2017, n.p.). The EEN also discusses how individuals can help others who are suffering during heat waves. This provides evidence that the EEN is still very concerned with human life and well-being while seeing human life as interconnected with the environment's well-being. In responding to heat waves, the EEN notes, "while still working on the awareness and prevention of further global warming for the long-term, small acts of kindness can act to cool others in the short-term" (Goebel 2019, n.p.). The EEN uses a both/and temporal framing by which we can work now to help humanity while also working for long-term environmental benefits. Conversely, the CA frames the short term and long term as a zero-sum game and frames environmentalism as sacrificing short-term comfort and well-being for far-off and uncertain environmental benefits.

In discussing the CA's anthropocentrism and the EEN's eco-centrism, we have explored the poles of the environmental spectrum of which there are various gradations and iterations. In what follows, we turn to the voices of individual Christians in how they negotiate their faith in relation to the environment. While we may categorize some as more anthropocentric or eco-centric than others, we see a wider variety of viewpoints and opinions articulated, which point to the variety of ways that vernacular religions and the relationship between faith and the environment are performed.

Disrupting the Eco/Anthro Binary

Beliefs as complicated as faith and environmentalism rarely fit into a simple binary (Bloomfield 2019). In line with folkloristic scholarship attentive to the complex dynamics of belief, we seek to complicate the strict determinism of seeing Christians as always climate-change deniers and also the idea that Christians only occupy the poles of anthropocentric or eco-centric beliefs. Because each Christian's interpretations of Bible verses are influenced by a variety of personal and social factors, it is important to focus "on the believers themselves, the variety of their life experiences, and the general capacity for variation as individual systems of belief incorporate and integrate other systems" (Green 1997, 713). In other words, as we interrogate the relationship between Christianity and the environment, we must also attend to the ways that weatherlore is created, interpreted, and made personal to individuals. We thus complement our previous discussions of organized interpretations of religion with the analysis of interviews with Christians who reflect various perspectives along the spectrum of environmental ideologies. By unpacking the voices of Christians and examining their individual perspectives, we can dig deeper into the anthropocentric/eco-centric binary and view Christian environmental attitudes in relation to lived experiences. In what follows, we present examples from interviews with individuals who articulated a variety of perspectives that roughly track Corbett's spectrum from anthropocentrism to eco-centrism.[3] The spectrum provides a starting point for examining the variety of ways that Christians adapt their faith and personal experiences to form attitudes about human-nature relationships.

The anthropocentric hierarchy in which human needs supersede the needs of nonhuman nature is a common trope in conversations. For example, Liam (2018, n.p.) argues that the proper relationship between humans and nature would not require "sacrifice or inconvenience" because of the priority for human "well-being." These beliefs reflect "unrestrained instrumentalism," which means that resource consumption "need not be restrained or limited in any way" (Corbett 2006, 28). Many interviewees express some form of hierarchy in which humans outranked nonhuman nature. But even within the shared idea that nonhuman nature is to be used or consumed, various individuals still express different perspectives on the environmental spectrum. For example, Charles (2018, n.p.) notes that the Earth and nonhuman animals have no value beyond their function as a "food source, entertainment, and curiosity." In acknowledging nonhuman animals can be used for reasons "beyond their purely instrumental value to include" other purposes such as "aesthetic" ones, these views represent what Corbett (2006, 28) calls "preservationism," which is a blend of anthropocentrism and eco-centrism. Benjamin (2018, n.p.) also expresses preservationist views when he notes, "environmental care is good" as long as it does not overtake other priorities "the Lord commanded." Benjamin (2018, n.p.) also notes that it is "right in a certain perspective" to care for the environment because it can be a form of expressing the Christian value of "love to their neighbor." Benjamin's beliefs thus deviate from a purely anthropocentric view that would not acknowledge value in environmental care.

Evoking more eco-centric ideas, Savannah (2016, n.p.) points to the interrelationship between humans and nonhuman nature by calling humans and animals "brothers and sisters" because they are both created by God the "Father." This idea is shared by Denise (2016, n.p.), who notes that "God made man [sic] from the same materials he used to create animals." The idea of shared substance between humanity and nonhuman nature expresses eco-centric ideas of "transformative ideologies" that question humanity's dominion and power over nature (Corbett 2006, 28).

Although there are eco-centric ideas present in Christian discourse, they are more often closer to values-driven ideologies, which Corbett (2006) places between preservationism and transformative ideologies. This perspective recognizes the intrinsic value of nonhuman life and views all life as part of a "biotic community" but also appeals to "traditional hierarch[ies]" (Corbett 2006, 28). This middle space along the spectrum is perhaps best reflected by Denise (2016, n.p.), who says that while humans "may be the crown of creation, we are closer to the rest of creation than to God." This comment expresses a values-driven ideology of the close relationship between humans and nonhuman nature but also reifies humanity's position as the crown, the top, and the pinnacle of creation above all forms of life. Therefore, even the discourse of people who consider themselves part of the Creation Care movement contain references to human-nonhuman nature hierarchies. This pervasiveness indicates that even the most eco-centric Christians must negotiate a primary tenet of the faith as humans made in God's image.

Among the interviews performed, some respondents reference the weather as evidence for both anthropocentric and eco-centric beliefs. For example, Abigail (2018, n.p.) drew on the idea of the Earth being self-correcting and not needing human action as she explains the following: "The 'Save the planet' mantra has been going on for a long time, with the emotional reaction related to cooling, heating, acid rain, diseases, etc. Each time it was resolved without any intervention." For Abigail, seeing the weather's flexibility and variability place it beyond the realm of human influence and control and thus not a concerning part of everyday life as it was resolved without the need for people to act. Furthermore, Abigail (2018, n.p.) uses "extra snow and colder temps" as reasons that climate-change advocates have "lost" their "credibility" and thus have the "burden of proof" to argue for why climate change is real.

Unlike Abigail, Riley (2016, n.p.) thinks that extreme "weather" events that can be "irrefutably connected to climate change" can serve as catalysts to correct climate denial. With evidence of climate change causing these "major events," Riley is hopeful that attitudes will begin to shift. There is survey evidence of this possibility that shows that a majority of Texas residents believe in climate change and want officials to act on it, which are increases from prior surveys (Treviso 2019). This change is attributed in part to Texans "see[ing] the real impacts of extreme weather," such as Hurricane Harvey (Treviso 2019, n.p.). Weather, for Riley, serves as empirical evidence of the Earth and its weather systems changing beyond normal limits. For Sean (2016, n.p.), "changing weather patterns" are "what worries me most" about climate change. Sean does not see extreme weather patterns as argumentative resources, as Riley does, but as harbingers of global destruction. Laurel (2016, n.p.) shares these fears, noting "If we do not solve our climate change issues, increasing weather extremes will lead to widespread starvation and migration in various parts of the world." Sean and Laurel both express concern about climate change, which reflects eco-centric attitudes, but Sean (2016, n.p.) emphasizes the "effects [weather has] on the land's ability to feed us [and] give us clean water," which reflects an instrumentalist perspective that the environment is valuable because of the resources it provides to people. Interestingly, both Sean and Laurel's comments were given in response to a survey question that asked them to "Please describe your 'end of the world' or apocalyptic beliefs. When will the world end? How will it end?" That is, weather was brought up in direct responses to a question about the apocalypse, indicating that weather, at least in these two cases, serves as a tangible and observable indicator of the coming apocalypse, which, to Sean and Laurel at least, is a very real but nevertheless intangible tenet of their faith. Weather is thus invoked by interviewees for a variety of purposes: as a reason to doubt climate change, as evidence for climate change, as a source of fear, and as a source of hope.

The Importance of Weatherlore in Our Climate Crisis

Our analysis shows that both the Bible and observations of weather are important sources of evidence that reflect individual and community beliefs about climate change and its implications. Although people may draw from the same sources of information (i.e., the Bible and observations of the weather), they interpret spiritual and material evidence differently to reach divergent conclusions about the proper, Christian relationship between humans and the environment. In our analysis, we focused not on how Christian beliefs shape understandings of weather, assuming a one-way flow of influence. Rather, we brought attention to *how* weather is interpreted in relation to Biblical text, grounding our analysis within a framework that foregrounds the ways in which meaning is constructed and negotiated at both community and individual levels.

One of the most important lessons to be learned by exploring Christian weatherlore in this way is how the relationship between Christianity and environmentalism, however deterministic it may appear or has been interpreted as being, is rather complicated and layered. This relationship is far from deterministic, and even noting that someone largely follows anthropocentric or eco-centric environmental beliefs based on their faith is an incomplete picture. Studying the variety of ways that environmental beliefs manifest in discourse can ground our discussions of the nuanced interpretations that people hold toward the environment.

Acknowledging that as many people (and perhaps even more; see Goldberg et al. 2021) fall in between as on the extremes of the spectrum of environmental beliefs should call us to recognize the commonalities that different iterations of weatherlore share and how those might even serve as inventional resources for collaboration and cooperation around environmental protection (Bloomfield 2019; Bloomfield 2020). In a world where there is scientific consensus and yet many controversies and public concerns around a range of topics, such as climate change, evolution, and vaccines (Ceccarelli 2011; Kitta 2012; Kitta 2015), it is crucial to look beyond binaries and strict dichotomies to explore the variety of beliefs and frameworks that guide official and vernacular understandings. The case of climate-change debates and other scientific controversies make clear that framing the people who hold beliefs that do not align precisely with scientific consensus as somehow "less 'enlightened'" (Mullen 2000) or willfully ignorant detracts from any meaningful dialogue that can, and indeed must, take place to enact any significant change.

Notes

1. This strategy has been highlighted as specifically important for environmental issues but has been understudied. For example, Lauren E. Cagle and Denise Tillery (2015, 159) argue that scholars should investigate "how personal experiences shape perceptions of climate change." Furthermore, the role of the natural environment and, specifically, weather in the study of Christian beliefs have received little attention in folklore studies. One notable exception includes Daniel N. Wojcik's (1997) study of apocalyptic beliefs in which he briefly addresses understandings of weather within Evangelical Christian belief about prophecies of the apocalypse.

2. Goldberg et al. (2021) note that a consistent 10 percent of the population is "dismissive" toward climate change and are likely unreachable, indicating that 90 percent are already alarmed, concerned, or otherwise reachable.

3. This data was originally collected for Bloomfield (2019; 2020) and consists of fifty-four survey responses, twelve phone interviews, and thirty-six online chat exchanges that spanned 2016 and 2018. Phone interviews and online chat exchanges were arranged through solicitations for conversations posted to various Christian and climate-change subreddits on Reddit.com. Interviews were semi-structured and began with questions such as "What are your views on the environment?" and "How has your faith influenced these views?" Respondents agreed to have their comments referenced in the research anonymously. Demographic information from the survey is available in Bloomfield (2020). No demographic information was collected in interviews to retain anonymity. Interviewees all identified as Christians with the vast majority identifying as Evangelical. When referencing conversations, randomly assigned pseudonyms are used that are not representative of gender, race, or age.

References

Abigail [pseud.]. 2018. Online direct message. January 23, 2018.

Balgord, William. 2019. "Under-Reported Tropical Storm Imelda Nevertheless Makes Herself Known to Residents of Southeastern Texas." Cornwall Alliance, September 20, 2019. https://cornwallalliance.org/2019/09/under-reported-tropical-storm-imelda-nevertheless-makes-herself-known-to-residents-of-southeastern-texas/.

Ball, Jim. 2018. "Climate 'Zombie' Reports that Just Won't Stay Buried." Evangelical Environmental Network, November 29, 2018. https://www.creationcare.org/climate_zombie_reports_that_just_won_t_stay_buried.

Barker, David C., and David H. Bearce. 2013. "End-Times Theology, the Shadow of the Future, and Public Resistance to Addressing Global Climate Change." *Political Research Quarterly* 66, no. 2: 267–79.

Beisner, E. Calvin. 2015. "The Radical Religion of Environmentalism." Cornwall Alliance, October 8, 2015. https://cornwallalliance.org/2015/10/the-radical-religion-of-environmentalism/.

Benjamin [pseud.]. 2018. Phone call with author. January 20, 2018.

Bloomfield, Emma Frances. 2019. *Communication Strategies for Engaging Climate Skeptics: Religion and the Environment.* New York: Routledge.

Bloomfield, Emma Frances. 2020. "The Reworking of Evangelical Christian Ecocultural Identity in the Creation Care Movement." In *The Routledge Handbook of Ecocultural Identity*, edited by Tema Milstein and José Castro-Sotomayor, 195–207. New York: Routledge.

Bolsen, Toby, and Matthew A. Shapiro. 2018. "The US News Media, Polarization on Climate Change, and Pathways to Effective Communication." *Environmental Communication* 12, no. 2: 149–63.

Brady, Erika. 2001. "Introduction." In *Healing Logics: Culture and Medicine in Modern Health Belief Systems*, edited by Erika Brady, 3–12. Logan: Utah State University Press.

Cagle, Lauren E., and Denise Tillery. 2015. "Climate Change Research Across Disciplines: The Value and Uses of Multidisciplinary Research Reviews for Technical Communication." *Technical Communication Quarterly* 24, no. 2: 147–63.

Ceccarelli, Leah. 2011. "Manufactured Scientific Controversy: Science, Rhetoric, and Public Debate." *Rhetoric & Public Affairs* 14, no. 2: 195–228.

Charles [pseud.]. 2018. Online direct message. January 19, 2018.

Corbett, Julia B. 2006. *Communicating Nature: How We Create and Understand Environmental Messages*. Washington, DC: Island Press.

[CA] Cornwall Alliance. n.d.a. "Protect the Poor: Ten Reasons to Oppose Harmful Climate Change Policies." Accessed April 6, 2018. https://cornwallalliance.org/landmark-documents/protect-the-poor-ten-reasons-to-oppose-harmful-climate-change-policies/.

[CA] Cornwall Alliance. n.d.b. "What Drives Us." Accessed October 1, 2019. https://cornwallalliance.org/about/what-drives-us/.

[CA] Cornwall Alliance. n.d.c. "What We Do." Accessed April 24, 2018. https://cornwallalliance.org/about/what-we-do/.

[CA] Cornwall Alliance. 2009. "An Evangelical Declaration on Global Warming." May 1, 2009. http://www.cornwallalliance.org/articles/read/an-evangelical-declaration-on-global-warming/.

Denise [pseud.]. 2016. Survey response. April 1, 2016.

Dixon, Graham N., and Christopher E. Clarke. 2013. "Heightening Uncertainty around Certain Science: Media Coverage, False Balance, and the Autism-Vaccine Controversy." *Science Communication* 35, no. 3: 358–82.

Douglas, Paul. 2017. "Harvey: A Storm Like No Other." Evangelical Environmental Network, September 5, 2017. https://www.creationcare.org/harvey_a_storm_like_no_other.

[EEN] Evangelical Environmental Network. n.d.a. "Beliefs." Accessed April 2, 2018. https://www.creationcare.org/why_creation_care_matters.

[EEN] Evangelical Environmental Network. n.d.b. "The Caring for Creation Pledge." Accessed April 2, 2018. https://www.creationcare.org/the_caring_for_creation_pledge.

[EEN] Evangelical Environmental Network. n.d.c. "Why Creation Care Matters." Accessed April 2, 2018. https://www.creationcare.org/why_creation_care_matters.

Evans, John H. 2018. *Morals Not Knowledge: Recasting the Contemporary U.S. Conflict between Religion and Science*. Oakland: University of California Press.

Gauchat, Gordon. 2012. "Politicization of Science in the Public Sphere: A Study of Public Trust in the United States, 1974 to 2010." *American Sociological Review* 77, no. 22: 167–87.

Goebel, Tori. 2019. "Helping Our Neighbors During Heatwaves." Evangelical Environmental Network, August 15, 2019. https://www.creationcare.org/heatwavehelpers.

Goldberg, Matthew, Xinran Wang, Jennifer Marlon, Jennifer Carman, Karine Lacroix, John Kotcher, Seth Rosenthal, Edward Maibach and Anthony Leiserowitz. 2021. "Segmenting the Climate Change Alarmed: Active, Willing, and Inactive." Yale Program on Climate Change Communication, July 27, 2021. https://climatecommunication.yale.edu/publications/segmenting-the-climate-change-alarmed-active-willing-and-inactive/.

Green, Thomas A. 1997. *Folklore: An Encyclopedia of Beliefs, Customs, Tales, Music, and Art.* Santa Barbara, CA: ABC-CLIO.

Hufford, David J. 1982. "Traditions of Disbelief." *New York Folklore* 8, no. 3: 47–55.

Hufford, David J. 1994. "Folklore and Medicine." In *Putting Folklore to Use*, edited by Michael Owen Jones, 117–35. Lexington: University Press of Kentucky.

Kitta, Andrea. 2012. *Vaccinations and Public Concern in History: Legend, Rumor, and Risk Perception.* New York: Routledge.

Kitta, Andrea. 2019. *Kiss of Death: Contagion, Contamination, and Folklore.* Logan: Utah State University Press.

Krost, Christina. 2017. "Lifting the Fog." Evangelical Environmental Network, December 8, 2017. https://www.creationcare.org/christinakrost/moms_fracking_pennsylvania.

Laurel [pseud.]. 2016. Survey response. July 9, 2016.

Legates, David R., and G. Cornelis van Kooten. 2014. "A Call to Truth, Prudence, and Protection of the Poor 2014: The Case against Harmful Climate Policies Gets Stronger." Cornwall Alliance, September 2014. https://cornwallalliance.org/wp-content/uploads/2019/07/2014-Call-to-Truth-full.pdf.

Liam [pseud.]. 2018. Online direct message. April 27, 2018.

Milstein, Tema. 2011. "Nature Identification: The Power of Pointing and Naming." *Environmental Communication* 5, no. 1: 3–24.

Mullen, Patrick B. 2000. "Belief and the American Folk." *Journal of American Folklore* 113, no. 448: 119–43.

O'Connor, Bonnie Blair. 1995. *Healing Traditions: Alternative Medicine and the Health Professions.* Philadelphia: University of Pennsylvania Press.

Patterson, Robert, and Ronald Lee. 1997. "The Environmental Rhetoric of 'Balance': A Case Study of Regulatory Discourse and the Colonization of the Public." *Technical Communication Quarterly* 6, no. 1: 25–40.

Pew Research Center. n.d. "Religious Landscape Study." Accessed September 30, 2019. https://www.pewforum.org/religious-landscape-study/.

Pew Research Center. 2017. "Mixed Messages about Public Trust in Science." December 8, 2017. http://www.pewresearch.org/science/2017/12/08/mixed-messages-about-public-trust-in-science/.

Primiano, Leonard Norman. 1995. "Vernacular Religion and the Search for Method in Religious Folklife." *Western Folklore* 54, no. 1: 37–56.

Primiano, Leonard Norman. 2012. "Afterward: Manifestations of the Religious Vernacular. Ambiguity, Power, and Creativity." In *Vernacular Religion in Everyday Life: Expressions of Belief*, edited by Marion Bowman and Ulo Valk, 382–94. New York: Routledge.

Riley [pseud.]. 2016. Survey response. March 27, 2016.

Sadar, Anthony J. 2019. "Climate Is What We Expect, Weather Is What We Get." Cornwall Alliance, June 10, 2019. https://cornwallalliance.org/2019/06/climate-is-what-we-expect-weather-is-what-we-get/.

Savannah [pseud.]. 2016. Survey response. March 26, 2016.
Sean [pseud.]. 2016. Survey response. March 26, 2016.
Treviso, P. 2019. "Climate Change an Increasing Concern for Texas Voters, Poll Finds." *Houston Chronicle*, September 11, 2019. https://www.houstonchronicle.com/news/houston-texas/houston/article/Climate-change-an-increasing-concern-for-Texas-14429888.php.
Veldman, Robin Globus. 2019. *The Gospel of Climate Skepticism: Why Evangelical Christians Oppose Action on Climate Change*. Oakland: University of California Press.
White, Lynn, Jr. 1967. "The Historical Roots of Our Ecologic Crisis." *Science* 155, no. 3767: 1203–7.
Wojcik, Daniel N. 1997. *The End of the World as We Know It: Faith, Fatalism, and Apocalypse in America*. New York: New York University Press.

Chapter 2

"OF BIBLICAL PROPORTIONS"

Flood Motifs in Personal Narratives of Katrina Survivors

Kate Parker Horigan

> I know there is a lesson in the storm—
> ha! ha! you idiot, what could that be?
> . . .
> I'm here, the light says, breaking in the East.
> You never asked for me, but I have come
> just as I came, slowly that first morning.
> I've never given up on mortal man.
> **—PETER COOLEY**, "Adam after the Hurricanes"

A decade ago, I was in the early stages of analyzing interviews that I had transcribed for the Surviving Katrina and Rita in Houston (SKRH) project.[1] Reading through hurricane survivors' dramatic and moving narratives, I was struck by parallels between heartfelt statements from survivors and hateful commentary from fundamentalist public figures like John Hagee and Pat Robertson (Gross 2006; Smith 2010). From these divergent camps came the same story: Hurricane Katrina was a punishment from God against sinners in New Orleans. At the time, I presented a paper at the 2010 annual meeting of the American Folklore Society on the topic. But there are many layers to this phenomenon, and it has continued to prod me into deeper thinking over the years; this chapter shows how survivors' uses of the traditional punishment story are dynamic expressions of local concerns. As statements about divine retribution continue to surface with every disaster and as scholars like Dorothy Noyes (2021) and the editors of this volume invite us to reconsider how we talk about the weather, Katrina survivors' explanations of the storm in terms of God's wrath continue to be significant and instructive. As in the lines quoted above from New Orleans poet Peter Cooley (2006), these survivors draw on familiar biblical motifs and, in so doing, juxtapose the promises and limitations of making meaning from the flood.

Deluge, Punishment, Escape: Flood Motifs

Hurricane Katrina was, like other modern climate catastrophes, more than a weather event. The storm itself caused damage in New Orleans, but the true devastation in that area came in the form of subsequent levee failures and flooding of densely populated areas.[2] Due to centuries of systemic racism, working-class African Americans were disproportionately affected by the flood. By most estimates, about 80 percent of the low-lying city was under water, with over one thousand people dead, and thousands more left stranded during a slow and inadequate official response. Thus, it seems apt that when Charles A. Darenbourg was interviewed for the SKRH, he proclaimed: "When I look at the city of New Orleans since the storm, I say this is something of biblical proportions. It was like reading the Old Testament" (Ancelet, Gaudet, and Lindahl 2013, 240).[3] "Biblical proportions" brings to mind, most basically, the dramatic scale of Katrina's events and suffering, but, more specifically, this statement evokes the flood described in Gen. 6–9. Though not all the SKRH interviewees make a direct connection to the Old Testament, many make direct or indirect references to flood motifs present in biblical and other mythological narratives.

In his *Motif-Index of Folk-Literature*, Stith Thompson (1955–58) categorizes "world calamities" under A1000–1099 in mythological motifs. Several motifs applicable to Katrina survivors' narratives are found within this section, including Motif A1010, Deluge; A1018, Flood as punishment; and A1020, Escape from deluge. Historical and contemporary folklore scholarship indicates the prevalence of these motifs in myths as well as other folk-narrative genres. Notably, Alan Dundes's (1988, 1–2) edited volume *The Flood Myth* compiles research on what he and others have identified as "one of the most widely distributed narratives known," including historical, scientific, psychological, and comparative approaches to "the story of a cataclysmic deluge" (see also Kluckhohn 1959).[4] Though the volume's starting point and primary focus is the flood as presented in Genesis, it includes a discussion of flood myths from many parts of the world. Selections from the work of Louis Ginzberg (1909) and Francis Lee Utley (1961) also demonstrate how flood motifs are not constrained to the genre of myth, as they explore, respectively, "Noah and the Flood in Jewish Legend" and variants of the folktale "The Devil in the Ark (AaTh 825)."

From Dundes's collection, perhaps the most relevant piece for understanding Katrina stories is Hans Kelsen's "The Principle of Retribution in the Flood and Catastrophe Myths." Kelsen's comparative study notes the wide range of actions for which retribution might be delivered in myth, sometimes general wickedness but more often a particular offense against a deity or violation of a prohibition. Some Katrina survivors refer to the storm as "wrath" or "the Lord's work," evoking retribution in a general sense, and others explicitly identify the flood as divine punishment. The survivors' personal narratives express ambiguity, though, whereas the myths Kelsen ([1943] 1988, 129) presents tend to deal in moral absolutes, with "catastrophe employed as a factor in the victorious contest of the good principle against the evil." In a more recent investigation of retribution, William Hansen (2009, 245) explores what he terms "poverty of cause," referring to the pattern in myth where "a substantial cosmic loss is brought about by a single instance of minor misbehaviour." For instance, as punishment for one person's misstep, all humans lose eternal life. Hansen updates scholarship such as Kelsen's ([1943] 1988, 126) that theorizes this seemingly disproportionate scale between cause and effect as a form of "primitive" thinking. Hansen (2009, 246) notes that, despite the pervasiveness of this pattern, it is not prevalent as a worldview: "as a theme, poverty of cause appears to belong to the realm of folk narrative rather than to the realm of folk philosophy." In his analysis, the etiological nature of such narratives creates the apparent imbalance between offense and punishment. He explains that many etiological narratives describe a relationship between a local cause and a local effect. It is only when, at the end of some narratives, this effect is applied on a larger scale to all humans (or bears, serpents, etc.) that it becomes disproportionate. In Hansen's (2009, 247) words, "The cause is local, whether the effect is local or cosmic." In the case of Katrina, in which the stories are not etiological in a cosmic sense, the punishment of the flood is generally presented as proportionate to the preceding offense: even while drawing on the universality of the mythic motif, these narratives construct both offense and effect in local terms. The proportions of the flood may be biblical, but its cause and its consequences are very close to home.

Katrina survivors reveal in their narratives how they view the storm's consequences and which pre-storm behaviors they see as having merited punishment in their comparisons of conditions before and after the storm. Attending to these comparisons in some ways follows Tikva Frymer-Kensky, who compares the flood described in the Atrahasis Epic with that of Genesis. In Atrahasis, the gods are disturbed by humans' noise, and they send various catastrophes to reduce the human population. In contrast, God's postdiluvian command to Noah to "be fruitful and multiply" reveals that "the flood story in Genesis is emphatically not about overpopulation" (Frymer-Kensky [1978] 1988, 66). Frymer-Kensky ([1978] 1988, 66) suggests the following: "A study of the changes that God made in the world after the flood gives a clearer picture of the conditions prevailing in the world before the flood, of the ultimate reason that necessitated the flood which almost caused the destruction of man, of the essential differences between the world before the flood and the world after it." Likewise, focusing on how Katrina survivors compare New Orleans before and after Katrina lends insight into which aspects of their communities these narrators found reprehensible.

Katrina survivors' dynamic use of the divine-retribution motif enables them to make claims to both local and larger-than-local concerns (Shuman 1993). As Amy Shuman (2005, 60) notes, "Rather than see the act of narrating as an act of filling in already-familiar slots in an already-familiar sequence, we will find it more often to be an act of searching for a way to integrate received motifs, seemingly uncategorizable experiences, and events that do not immediately make sense." When faced with a catastrophic flood, New Orleans residents search for familiar ways to narrate the experience. Many, for instance, note that what they saw looked like a movie because fabricated, popular-culture images corresponded more closely with what they observed than their own past experiences did (Horigan 2018, 30–32). For others, as with the survivors described in this chapter, the best "available narratives" for this uncategorizable experience are flood myth motifs (Shuman 2005). Shuman (2005, 60–61) cautions us that using allegory to narrate personal experience does not result in a direct correspondence:

> In [Walter] Benjamin's terms, the mythologizing or allegorizing of experience points not to coherence but to fragmentation. The seemingly fixed and coherent allegorical text is an illusion in the modern world; it is a sign of what is not possible; it is a record of catastrophe. . . . As a response to catastrophe, it offers redemption not by creating coherence but by observing the hope and promise still evident in the fragmented ruins of the past.

Examining the relationship between the personal and allegorical in the narratives below reveals the "fragmented ruins" of Katrina, and it also suggests the "hope and promise"—especially in the escape motifs—that these narrators hold out for their flooded city and upended lives. Like other folk narratives, personal narratives about Katrina use traditional elements in creative ways to speak to the narrators' present moments; survivors use flood motifs to indicate what is deserving of punishment and who is deserving of rescue—to critique and reflect on their social world.

"Man Put His Hand in God's Plan and Just Destroyed It": Survivor Narratives

The SKRH database housed at the University of Houston contains over four hundred interviews with survivors of hurricanes Katrina and Rita. Of this extensive collection, I have worked most closely with a group of thirty-five interviews (with twenty-nine different people) that I transcribed for the survivor-led documentation project. Within this group, all the interviewees lived through Katrina, and most were from the New Orleans area.[5] To study how survivors incorporate flood motifs in their narratives—particularly Motif A1018, Flood as punishment, I began with a broad search of these thirty-five transcripts for certain keywords, noting their frequency and their context. Subsequent close readings of the transcripts helped me locate less direct references. Although I searched for terms from Thompson's motif index, such as "punishment" (of which there were no relevant uses) and "sin" (of which there were a couple), I also searched for language more commonly used by interviewees such as "reason" and "crime." Search terms also included many variations of references to God (such as "Lord," "Almighty," and "He"). Almost none of the interviewees referred to deities from outside of Christianity.

The search terms with the highest yield were "God" and related appellations, which occurred in thirty-nine relevant instances across fourteen interviews (with twelve different people). I considered an instance relevant if it appeared in the context of explaining why the storm happened and God's role in it. There were many other references to God that I did not consider directly relevant and therefore did not tally, for example, as exclamations (as in "oh my Lord") and expressions of gratitude (such as "thank God"). Of the interviewees in this sample, only two had no references to God in their narrative. So, in sum, 40 percent of these interviewees had some relevant reference to God as they talked about their interpretations of the hurricane. Carl Lindahl (2012, 168) writes of the SKRH interviews: "Given three general choices of where to lay the blame—on people, nature, or God—the human factor is by far the most commonly reported, and God comes second. The idea that Katrina was exclusively or principally a natural disaster seldom surfaces. The speakers strive to find a conscious purpose in an event that has utterly altered their lives." Lindahl's larger point is to emphasize the credibility with which survivors approach the explanation that the city's levees were intentionally destroyed to flood African American neighborhoods and protect wealthy white neighborhoods and tourist areas, and my analysis supports his claims. However, it is not simple to separate attributions of blame in these interviews; even in those instances of which survivors appear to be putting God "first" in terms of who was responsible for the deluge, the "human factor" is integrated in the narrative management of divine retribution. After all, who is God punishing but humans? Part of the power of the mythological motif is that it places particular people and behaviors within that punishable category.

In the excerpts from survivor interviews that follow, each narrator incorporates the idea of punishment in a unique way, underscoring their individual concerns through their use of this traditional motif. In one of his interviews for SKRH, Bernie Porché answers a question about what New Orleans was like prior to Katrina:

> They had a lot of crime rate in New Orleans . . . [T]hey had a lot of corruption going on in New Orleans. And truly, to be honest, I really do believe that the Lord was tired. I mean, New Orleans was having too much sinful stuff going on, New Orleans as a whole. I mean, they had corrupt cops, and they had corrupt politics that's still going on . . . I really do believe that if they had taken the money that was given to them years ago to fix the levee systems, we would have never been in Houston, Texas today. (SKRH 2006a)

Porché echoes similar sentiments throughout his interview. In the first part of his description here, he references the human wickedness that he believes preceded and likely provoked the flood. He evokes the theme common in many flood myths that humans essentially wore out their deities' patience, noting that "the Lord was tired." Other interviewees make similar observations, expressing the conviction that New Orleans residents had been given multiple chances and were due for retribution; these remarks also resonate with secular observations about weather patterns, for example, that New Orleans had been lucky to be spared direct hits from hurricanes in the past and that it was bound to happen at some point.

However, Porché does not end with his general statement about the "sinful stuff" in New Orleans; he goes on to specify the sinful behavior. He mentions the high rate of crime, a common theme in survivors' interviews, but he quickly shifts to describing corruption among police and politicians. Then he names a particularly egregious offense of these corrupt officials—not strengthening the city's levee system prior to the storm—and cites this as the cause for the flood and his forced departure. Criticism regarding the levees is a common thread in other survivors' narratives.[6] Most notably, though, Porché's account offers a remarkable negotiation of the punishment motif: the very same behavior that is deemed sinful (corruption), which exhausted the Lord and prompted the flood, is the human behavior that failed to protect people from the flood. Thus, he is able to adapt the allegorical narrative in a way that both reflects the magnitude of his experience—this was a world-changing catastrophe—and pinpoints the specificity of his social critique—this was negligence by our corrupt public officials. Looking at how this narrative describes conditions before and after the flood also sheds light on the condemnable actions. Porché indicates that corruption in the city is "still going on." Later in his interview, he again describes sin in terms of officials' shortcomings, in particular the failures to evacuate vulnerable elderly people prior to the storm, and he expresses his wish that the experiences of Katrina will bring about change. In Porché's engagement with allegory, he shows that the ruins of the past and his hope for the future are directly related: city officials have put their self-interest ahead of care for their constituents, and this is both the cause and the lesson of Katrina's deluge.

Larry X, another survivor, also uses flood motifs to comment on human behavior. When he narrates the storm itself, he evokes the aura of renewal associated with Genesis: "The wind and everything had stopped, the sun came out, it looked—it was unbelievable. It was something like in the Bible . . . The whole sky lit up bright after, something been washed away" (SKRH 2006c). He describes how his neighbors came out and cooked food together in a peaceful and almost celebratory scene. Shortly afterward, though, his story turns to the levee failures and subsequent flood and even shifts from Old Testament imagery of remaking the world to a New Testament vision of apocalyptic destruction. He begins talking about Revelation and the parallels to it that he witnessed following the flood of New Orleans. He remarks that Revelation says "the world is going to be destroyed by water and fire. We had water and fire. We had fire in the water" (Lindahl 2012, 170). Though he is still drawing on allegory, this moment bears none of the redemptive potential of the pre-flood scene. He concludes: "We were pushed away from our home by man not by God or . . . nature. . . . Man did this here, busted that levee open" (Lindahl 2012, 170). X's juxtaposition of flood motifs offers another sort of comparison: rather than reflecting on conditions before and after Katrina, he contrasts the possibilities the storm first presented with the destruction it ultimately delivered, especially in his working-class, Black neighborhood. Here again, though in a different progression of allegorical motifs, we see both hope and ruins: this could be a chance for renewal, and then, this is apocalyptic. In terms of local concerns, this means a shift from a community coming together—in a glimpse of what Rebecca Solnit (2009, passim) terms "a paradise built in hell"—to people being "pushed away" from their homes by the destruction of the levees.

Both X and his fellow SKRH interviewee Lester Harris frame the punishment motif not so much as retribution directly from God but as a tragically botched opportunity. Harris offers a similar contrast between the redemptive possibilities of a heaven-sent flood and the failure of certain people to recognize that potential: he reflects that Katrina "was supposed to change people on a better way," but he does not see evidence of a positive change in the aftermath (SKRH

2007). Whereas Porché finds that local authorities are themselves the source of sin, in X's and Lester Harris's accounts, public officials are not directly named as the cause of punishment, but they are still responsible for its effects. For example, Harris assesses the levee failures as follows: "To me it was something that was premeditated for it to happen and a lot of people, thousands of people are dead because of a bad mistake" (Lindahl 2012, 164). Harris reveals here his unique engagement with the motif of divine retribution. As he puts it: "The way I look at it, the reason for Katrina—it was supposed to have been a blessing. But man put his hand in God's plan and just destroyed it. And that's why it's the way it is now" (SKRH 2007). In Harris's narrative negotiation of mythological motifs, the promise inherent in the allegory of the flood is shattered by the "bad mistake" of those who destroyed (or, minimally, neglected) the levees.

Some survivors use flood motifs in ways that foreground hope in the wreckage, but even these are ambiguous in their assessment of the cause and consequences of the New Orleans deluge. For example, Patrice M describes, at one point, the water covering the city as appearing calm and "awesome" (in the sense of "awe-inspiring") and shares her contemplative realization: "You know, the storm had to happen, because of history. It had to happen . . . God allow things to happen, you know, for us to either go this way, or go that way"; she goes on to explain that God does not "want us to be on both sides of the fence" (SKRH 2006d). In this explanation, she offers not only the weight of allegory as evidence that humans were destined for a flood but also her interpretation—in agreement with many mythological uses of the motif—that such a flood is a test of faith. Elsewhere in her narrative of Katrina, M has made clear that she remained strong in her faith throughout the course of the flood, trusting in what she believes to be a divine plan (Horigan 2018, 24–30). She also reflects that the trial has strengthened her in some ways, gesturing toward the regenerative potential of the flood allegory: "Katrina, to me, it makes you appreciate things more" (SKRH2006d). And yet, in a second interview months later, M has not moved any closer to mythologizing her experience in a positive manner: "It was a natural disaster, you know something that God had sent to us, for what reason we don't know now, but I guarantee you, by the time this . . . where is it going, to the Library of Congress? We may not be here but the people [after] us will know why that storm came" (SKRH 2006e).[7] In this analysis, she remains assured that the scale and source of the experience were both biblical, "something that God had sent," but she declines to name a reason for it, punishment or otherwise. Instead, she defers interpretation about what it means and what will come of it to future generations (and, by referencing the archive in which her interview will be deposited, invites future listeners to participate in that meaning making, presumably by surveying the events that have transpired since).

When it comes to escaping Katrina's waters (Motif A1020, Escape from deluge; and/or Motif A1021, Deluge: escape in boat [ark]), the motifs incorporated are apparently hopeful, but survivors still use them to offer critical commentary on the local events and circumstances in which this flood transpired. Some survivors, such as Porché, focus on their trust in God as an explanation for their rescue, echoing those myths that reward the faithful: "all I did was kept praying, asked the Lord if he'd ever get me out of this situation, and he did. And I was patient, and I just keep his will and just pray" (SKRH2006a). Others offer divine intervention as an explanation for their rescue by other people. One woman, for example, describes how a stranger in a boat approached her flooding home, offering to bring her to safety, and she told him to go on ahead because she was awaiting rescue by helicopter. It dawned on her soon after that this man in the boat had been sent by God; then, thankfully, he returned for her. She also realizes that had she waited for the helicopters as she intended, she likely would not have survived. Still others attribute their own remarkable strength to God's will and assistance. Joan Britton, for instance, realizes—during a chaotic struggle to get her family on a departing bus—that her daughter and son-in-law have been left behind:

> So I push my way back off the bus, through all those people—I know it was nobody but God—how did I get back off, it was hard enough getting on. I pushed my way back off the bus. I ran to the middle of I-10 and I grabbed their hands, and I told them, "Get on, come on, get on." And we got back on the bus. (SKRH 2006b)

In this instance, Britton credits God with her ability to navigate a seemingly impossible situation to help her loved ones escape the flood. In both cases of the boat and the bus, the narratives incorporate the motif of being spared from the flood by God. They also, however, signal a noticeable absence as these women were not saved from the flood by official rescuers. If the civilian boat had not returned, the first narrator believes she would have drowned waiting for a helicopter. In Britton's story, if God had not helped her get her family on the bus, nobody else would have. The hopeful allegory of escape from the flood has a dark counterpart in the implied and alternative narratives—of what *could* have happened and what did happen to those left behind.

The story of Katrina told by Henry Armstrong, about himself and his elderly mother, approaches the escape motif at first but ultimately departs from it in a radical, telling way. Armstrong says from the start of his story that he had bargained with God to ensure his mother's safety: "I talked to God and I tried to make a deal with him, and it seemed that the deal worked. Because I told him if he let me get my mother out of this I'd be forever committed to her, which I was already" (Ancelet, Gaudet, and Lindahl 2013, 136). His comment that he "was already" devoted to his mother explains to listeners why he is worthy of this divine favor and echoes many other survivors who place ultimate importance on their loyalty to and care for their family members.[8] In referring to his deal with God, Armstrong alludes to the mythological motif of a deserving human (and his family) spared the wrath of the flood. However, his story takes an extensive detour between the deal and the deliverance as he describes awful suffering on the highway overpass where he and his mother awaited rescue for three days along with other members of his family and thousands of other people. Nonetheless, he finds a kernel of hope in this: "a lot of it had to be God's work because it taught us to be humble, it taught us to be kind to one another, it taught us to help one another *any* way we could, because we were all in the same boat, so to speak" (Ancelet, Gaudet, and Lindahl 2013, 140). Here, instead of climbing aboard his own promised ark, he remains in "the same boat" with his suffering neighbors in which they engage in the creative and selfless task of rescuing one another.[9] Lindahl (2013, 251), writing about Armstrong's story and others like it, observes, "essential to the transforming endurance of all the survivors, old and young, is a recognition that their strength comes to them through others. . . . Thus, the rescuers repeatedly give the rescued the credit for the rescue." Indeed, even when these narrators are giving God the credit for escape, they are emphasizing human actors and pointing out the admirable qualities that makes rescue warranted and worthwhile. Just as the motif of punishment is tailored to fit local concerns about crime, corruption, and the failure to protect vulnerable people from Katrina's flood, the motif of escape is made to demonstrate how certain qualities such as faith, devotion to family, resourcefulness, and care for others ought to be the redeemable qualities in the post-deluge city. Flood survivors' uses of mythological punishment and escape motifs illustrate both the enormous scale of their experiences and the specificity of their struggles and hopes. They engage with the allegorical in a way that allows them to make sense of the personal, but they do not offer any universal meaning in the process.

Conclusion

In his study of Norwegian media coverage of Hurricane Katrina, folklorist Kyrre Kverndokk identifies major discursive tropes in how news outlets made sense of the disaster. He begins with a much older calamity, the 1755 Lisbon earthquake. Kverndokk (2014, 79) writes:

The Lisbon earthquake and Voltaire's polemic writings on the disaster are often referred to as a turning point in the [sic] western thought. It changed the way of thinking about disasters, nature, evil and morals (e.g., Löffler, 1999). In her now classic book *Evil in Western Thought*, the philosopher Susan Neiman writes: "Since Lisbon, natural evils no longer have any seemingly [sic] relation to moral evils; hence they no longer have meaning at all. Natural disaster is the object of attempts at prediction and control, not of interpretation" (Neiman, 2002: 250). If this is the case, it is tempting to ask a rather naive and simple question; if natural disasters really are fundamentally meaningless, how do we then make them understandable?

Kverndokk turns to media narratives to find answers about how meaning is assigned to modern disasters. He concludes that Katrina, like other contemporary "natural" catastrophes, is framed in terms of "the collapse of civil society," "social vulnerability," or "extreme weather and global warming" (Kverndokk 2014, 80). As far as his examples show, media narratives do not employ the motif of disaster as punishment from a deity, though Kverndokk (2014, 85) makes a compelling argument that there are parallels between the divine retribution narrative and the "nature strikes back" narrative of climate change: "Like God was considered as a rightful punisher in early modern Europe, nature, in this kind of late modern popular discourse is presented as an autonomous and rightful punisher. Nature and morals are again intertwined." Likely because he is focused on media narratives, Kverndokk (2014, 85) categorizes the flood as a punishment motif strictly in "early modern disaster interpretations."

Other scholars have also identified the Lisbon earthquake as a critical moment for European Enlightenment thought:

> The Lisbon disaster is widely regarded today as a key juncture in shifting understandings of the place of religion in relation to disasters and relief. Today, the dominant "Secularizing Interpretation" (Nichols 2014) . . . locates the disaster as a decisive moment in the emerging Enlightenment critique, and eventual displacement, of theological responses to disaster in favor of sociological explanations and an appeal to new technologies of bureaucratic management. (Fountain and McLaughlin 2016, 3)

As Philip Fountain and Levi McLaughlin go on to point out, however, religious thinking and—more to their point, the intersections of religion and disaster relief—remain significant in how people respond to catastrophe.

In the case of the Katrina survivors quoted above, whether they profess religious faith or not (and most do), they draw on motifs familiar from sacred mythology. Contrary to how a "modern" explanation of disaster might be expected to appear—and does appear in official accounts, according to these scholars—flood motifs, particularly that of the flood as punishment, are alive and well in contemporary vernacular interpretations of catastrophe. And these motifs are no static remnant of an earlier age, but rather a powerful rhetorical resource serving survivors in at least two ways: first, they offer a category of experience appropriate in scale for these narrators, faced with an otherwise uncategorizable (and therefore potentially untellable) experience (Shuman 2005); second, they enable reflection on what is valuable and what is not valuable before and after the storm—what ought to remain in ruins and what ought to be salvaged.

Returning to the words of Darenbourg, who remarks on the "biblical proportions" of Katrina's flood, we find a final, instructive juxtaposition of promise and failure. Though he has explicitly called to mind the flood of the Old Testament, he offers no rainbow sign of hope to follow it. Rather, he comments later, "I think it was avoidable. The destruction itself was avoidable, because if the levee systems were reinforced they would not have broken" (Ancelet, Gaudet, and Lindahl 2013, 243). This echoes Hansen's (2009, 248) claim that the "principle of retribution" often carries the subtext "it could have been otherwise." In Darenbourg's view, the "otherwise" was not dependent on the sins or salvation of those most severely punished, but rather the inaction of those who should have protected people from the effects of the punishment. Invoking the allegorical calls attention to the ruins not of the city and its people, but of the officials and infrastructure that failed them. Implicitly, we might hear the hope that in our warming, disaster-prone world, we will get it right next time.

Notes

1. SKRH was a large-scale documentation project orchestrated by Texas-based folklorists Carl Lindahl and Pat Jasper and led by survivors. Survivors participated in training workshops and received compensation for their work interviewing their fellow survivors. SKRH hired me to help transcribe interviews between 2009 and 2011. For more on SKRH, see Ancelet, Gaudet, and Lindahl 2013; Lindahl 2012; and Rowell 2006.

2. For an excellent discussion of the invisibility of other parts of the Gulf Coast affected by Katrina, especially Mississippi, see Ingram 2019.

3. Although all the interviews quoted in this chapter are part of the same SKRH database, the sources cited and naming conventions vary: in cases where I have permission from interviewees to quote their words, I do so directly from interview transcripts, using their name or a pseudonym depending on their stated preference and including the SKRH accession number. In cases where I was unable to obtain explicit permission, I cite, whenever possible, from previously published interviews, providing the bibliographic information

for the publication rather than the interview itself. In those few cases where I was unable to obtain explicit permission regarding an unpublished portion of an interview, I conform with the standards of previously published texts in using either the real name or a pseudonym. I do not in any cases include quotations from or names of interviewees who never consented to be published, though I do describe patterns from their interviews in quantitative and generalized terms. Lindahl (2012, 174) explains the permissions associated with the SKRH database as follows: "All narrators whose interviews are included in the database gave their signed permission for researchers to study their interviews. The great majority were also willing to share their interviews in publications, radio broadcasts, and other forms. But because it is one of the project's guiding principles that the narrators always own their words, researchers wishing to publish a narrator's words are required to obtain permission every time that they intend to identify the interviewee by name." Due to the time passed since the creation of the database and the changed contact information of most of its contributors (who were living in temporary housing at the time of their interview), this permission is sometimes not possible to obtain.

4. Dundes's (1988, 178) own contribution to the volume is a psychoanalytic study of the flood myth in symbolic terms, whereby "the pain of parturition becomes transformed into a male fantasy of inflicting wholesale destruction on the world." For a recent take on Dundes's approach, see Silverman 2016.

5. I also lived in New Orleans during the time period of Hurricane Katrina. I have written elsewhere about occupying the role of both survivor and ethnographer (Horigan 2017).

6. False claims about the levees's safety call to mind Motif Q221.4.1, Dam builder remarks that God Almighty could not sweep completed dam away. The whole structure gives away, disappears (Thompson 1955–58).

7. This interview (SKRH 2006e) is the sound recording afc2008006_sr103 in the Surviving Hurricanes Katrina and Rita in Houston Collection (AFC 2008/006) in the American Folklife Center at the Library of Congress. Many thanks to Judith Gray, coordinator of reference services at the AFC, for locating this information.

8. I have written elsewhere, with Sheila Bock, about rhetorical negotiations of family relationships in contexts of stigma (Bock and Horigan 2015). On the importance of family in Katrina survivors' narratives, see also Lindahl 2012 and Ancelet, Gaudet, and Lindahl 2013.

9. On creativity and improvisation in survivors' responses to hurricanes, see Ancelet, Gaudet, and Lindahl 2013, especially the editors' introduction and Barry Ancelet's essay, "Storm Stories: The Social and Cultural Implications of Katrina and Rita."

References

Ancelet, Barry Jean, Marcia Gaudet, and Carl Lindahl, eds. 2013. *Second Line Rescue: Improvised Responses to Katrina and Rita*. Jackson: University Press of Mississippi.

Bock, Sheila, and Kate Parker Horigan. 2015. "Invoking the Relative: A New Perspective on Family Lore in Stigmatized Communities." In *Diagnosing Folklore: Perspectives on Health, Trauma, and Disability*, edited by Trevor J. Blank and Andrea Kitta, 65–84. Jackson: University Press of Mississippi.

Cooley, Peter. 2006. "Adam after the Hurricanes." In "American Tragedy: New Orleans under Water," edited by Charles Henry Rowell. Special issue, *Callaloo* 29, no. 4: 1280.

Dundes, Alan, ed. 1988. *The Flood Myth*. Berkeley: University of California Press.

Fountain, Philip, and Levi McLaughlin. 2016. "Salvage and Salvation: Guest Editors' Introduction." In "Salvage and Salvation: Religion and Disaster in Asia," edited by Philip Fountain and Levi McLaughlin. Special issue, *Asian Ethnology* 75, no. 1: 1–28.

Frymer-Kensky, Tikva. [1978] 1988. "The Atrahasis Epic and Its Significance for Our Understanding of Genesis 1–9." In Dundes 1988, 61–87.

Gross, Terry. 2006. "Pastor John Hagee on Christian Zionism, Katrina." *Fresh Air*. National Public Radio. Philadelphia, PA: WHYY-FM, September 18, 2006.

Hansen, William. 2009. "Poverty of Cause in Mythological Narrative." *Folklore* 120, no. 3: 241–52.

Horigan, Kate Parker. 2017. "Critical Empathy: A Survivor's Study of Disaster." *Fabula* 58, no. 1–2: 76–89.

Horigan, Kate Parker. 2018. *Consuming Katrina: Public Disaster and Personal Narrative*. Jackson: University Press of Mississippi.

Ingram, Shelley. 2019. "The #Landmass between New Orleans and Mobile: Neglect, Race, and the Cost of Invisibility." In *Implied Nowhere: Absence in Folklore Studies*, by Shelley Ingram, Willow G. Mullins, and Todd Richardson, 125–33. Jackson: University Press of Mississippi.

Kelsen, Hans. [1943] 1988. "The Principle of Retribution in the Flood and Catastrophe Myths." In Dundes 1988, 125–49.

Kluckhohn, Clyde. 1959. "Recurrent Themes in Myths and Mythmaking." *Dædalus* 88, no. 2: 268–79.

Kverndokk, Kyrre. 2014. "Mediating the Morals of Disasters." *Nordic Journal of Science and Technology Studies* 2, no. 1: 78–87.

Lindahl, Carl. 2012. "Legends of Hurricane Katrina: The Right to Be Wrong, Survivor-to-Survivor Storytelling, and Healing." *Journal of American Folklore* 125, no. 496: 139–76.

Lindahl, Carl. 2013. "Epilogue: A Street Named Desire." In *Second Line Rescue: Improvised Responses to Katrina and Rita*, edited by Barry Jean Ancelet, Marcia Gaudet, and Carl Lindahl, 248-60. Jackson: University Press of Mississippi.

Noyes, Dorothy. 2021. "Talking about the Weather: Common Sense, Common Sensing, Commonplaces." *Journal of American Folklore* 134, no. 533: 272–91.

Rowell, Charles Henry, ed. 2006. "American Tragedy: New Orleans under Water." Special issue, *Callaloo* 29, no. 4.

Shuman, Amy. 1993. "Dismantling Local Culture." In "Theorizing Folklore: Toward New Perspectives on the Politics of Culture," edited by Amy Shuman and Charles Briggs. Special issue, *Western Folklore* 52, no. 2–4: 345–64.

Shuman, Amy. 2005. *Other People's Stories: Entitlement Claims and the Critique of Empathy*. Champaign: University of Illinois Press.

Silverman, Eric K. 2016. "The Waters of Mendangumeli: A Masculine Psychoanalytic Interpretation of a New Guinea Flood Myth—and Women's Laughter." *Journal of American Folklore* 129, no. 512: 171–202.

[SKRH] Surviving Katrina and Rita in Houston Database. n.d. Houston Folklore Archive, University of Houston, Houston, TX.

[SKRH] Surviving Katrina and Rita in Houston Database. 2006a. Bernie Porché (USKR-LTU-SR03).

[SKRH] Surviving Katrina and Rita in Houston Database. 2006b. Joan Britton (SKR-NE-SR10).

[SKRH] Surviving Katrina and Rita in Houston Database. 2006c. Larry X (USKR-LTU-SR08).
[SKRH] Surviving Katrina and Rita in Houston Database. 2006d. Patrice M (SKR-AD-SR02).
[SKRH] Surviving Katrina and Rita in Houston Database. 2006e. Patrice M (SKR-SS-SR16).
[SKRH] Surviving Katrina and Rita in Houston Database. 2007. Lester Harris (SKR-CJ-SR10).
Smith, Ryan. 2010. "Pat Robertson: Haiti 'Cursed' after 'Pact to the Devil.'" *CBS News*, January 13, 2010. https://www.cbsnews.com/news/pat-robertson-haiti-cursed-after-pact-to-the-devil/.
Solnit, Rebecca. 2009. *A Paradise Built in Hell: The Extraordinary Communities that Arise in Disaster*. New York: Penguin Books.
Thompson, Stith. 1955–58. *Motif-Index of Folk-Literature: A Classification of Narrative Elements in Folktales, Ballads, Myths, Fables, Mediaeval Romances, Exempla, Fabliaux, Jest-Books, and Local Legends*. Rev. and enl. ed. 6 vols. Bloomington: Indiana University Press.

Chapter 3

IN THE BONES

Prognostication and Weather in the Twenty-First Century

Willow G. Mullins

Popular jokes about the weather reveal a complicated relationship between meteorological forecasting and belief. The internet, like daily conversation, is full of jokes at the expense of meteorologists: "Meteorologist—The Only Job Where You Can Be Wrong Every Day," appears across a range of memes and products (e.g., Meteorology Gifts and Shirts n.d.); "The Weather or Not Channel" labels a cartoon depicting dubious forecasting (Piraro 2013); and a faked child's book cover reads "The Meteorologist Who Cried 'Blizzard': A Story of Exaggeration and Deceit" (Meme n.d.). Even comedian Bill Murray has had a go: "Fool me once, shame on you. Fool me twice, shame on me. Fool me 350,000 times, you are a weatherman" (quoted in Satterfield 2014, n.p.). Another comic strip by Wiley shows a spinner with different weather possibilities behind a door labeled "Weather Center" and carries the tagline "Your suspicions confirmed" (quoted in Panovich 2014, n.p.). This is not a chapter about jokes though; it's a chapter about prediction.

These jokes reflect weather-forecast consumers' frustrations with the difference between predictions and what happens. You planned a hike because the local weather report predicted fair skies but found yourself drenched and stuck in the mud. You wore jeans only to swelter. Yet, according to the United States' National Oceanic and Atmospheric Administration (NOAA), five-day forecasts are 90 percent reliable and are getting more reliable all the time. Charlotte, North Carolina, meteorologist Brad Panovich (2014) points out that statistics are much worse in a lot of other fields—doctors misdiagnose about 26 percent of illnesses, and professional athletes miss their shot far more often than they make it.

As with much folklore, the jokes reveal more about the people telling them than the object of their derision. Most members of the weather-forecast-consuming public, a group that includes me, know comparatively little about how weather is scientifically forecast. Those jokes reveal a gap between the public understanding of the weather as a lived and embodied experience and the complexities of weather forecasting, between the weatherman[1] we see and the vast network of meteorologists and climatologists and mathematicians that we don't. Andrew Blum (2019, passim) calls this gap "the weather machine," the space in which weather is interpreted by experts, and weather predictions—those tiny reductions of temperature and precipitation most of us rely on—are constructed. Despite all the science, those weather reports face a steep challenge to their credibility. Many of us feel at a loss when presented with isobars and wind charts, but we inherently know what the weather *feels* like.

Humans experience weather both physiologically and psychologically. Not only do we know the sensation of the wind on our cheeks or the sun warming our backs, but we also know the sense of happiness on a warm spring day after the gray cold of winter weeks and the coziness and low energy of rainy autumnal afternoons that send us seeking warm drinks and quiet corners in which to drink them. The embodiment of the weather goes further though. Weather changes the way we behave—impacting crime and suicide statistics, for example—in ways that can best be explained through embodied cognition, the idea that "the mind must be understood in the context of its relationship to a physical body that interacts with the world" (Margaret Wilson quoted in Spackman and Yanchar 2013, 47).

However, it is not just that weather causes sensation but also that sensation can predict the weather. In another comic, an old *Far Side* cartoon by Gary Larson, three old men sit in rocking chairs on a porch. A dark cloud shadows the background. One man claims there is a storm coming, he can feel it in his knee. His knee appears abnormally large, bulging cartoonishly. The next declares a blizzard is more likely; he can feel it in his swollen hand, which looks more like an inflated rubber glove. The last, with a bulbous head, remarks, "Well, *somethin's happenin'* . . . there goes my head" (G. Larson 2019). Titled "Front Porch Forecasters," the cartoon invokes the commonality of embodied weather prediction, a sensation that those who get headaches before storms or "feel it in their bones" know only too well. One final joke pulls together the embodied prediction of weather and the trust in meteorology:

> Why did the skeleton know it was going to rain?
> Because he read the fucking forecast. What do you think?

This last joke casts its weight fully on the side of science, rejecting vernacular, intuitive forecasting and the sayings that accompany it. The swear word makes it all seem so obvious: of course, the skeleton read the forecast because that is what rational skeletons do. After all, if a skeleton cannot feel it is his bones, then who really can?

The human relationship to the prediction of weather, represented by both sets of jokes, suggests a negotiation between belief and experience. While forecasting the weather may be defined as a scientific paradigm and methodology, "usually a result of study and analysis of available pertinent data" (*Merriam-Webster* n.d.), its listed synonyms suggest something else: augury, prediction, prognosis, soothsaying. Weather forecasting is not foretelling or fortune telling; it is a matter of trust and belief. We, the forecast consumers, trust that meteorologists have the science handled, and, like the skeleton, we believe, and act on, the weather forecast. For many, the weatherman acts like a modern-day diviner, reading the signs of the skies.

However, there is another side of prediction too, one that recognizes, perhaps subconsciously, the embodied experience of weather and its effects within our systems. This experiential understanding of weather also relies on trust and belief in our bodies and in our minds' ability to recognize the patterns of our bodies in relation to the world they inhabit, patterns most often encoded in folklore. Despite the advances of modern science—meteorological, medical, and cognitive—that offer explanations for the ways we experience the weather and cogent and trustworthy predictions of the weather to come, many of us believe weather forecasting in much the same way that most of us believe germ theory: we have not seen germs under a microscope but trust that bacteriologists have, and we combine that belief with a sense that cleanliness will keep us from getting sick and with cultural pressure, so we wash our hands (Pimple 1990, 52). We may not understand the mathematics of wind currents, but we believe the weatherman does, and we dress accordingly.

Meteorology and the Mantic Arts

Though the weather forecast is a comparatively recent phenomenon, weather prediction has a long history. To understand how belief colors current weather forecasting and its reception by the lay population, it is helpful to understand that weather science was once much closer to theology. The history of weather forecasting is a story of observation and theory made manifest through technological advances. Throughout this history, people have understood what the weather *could* do, thanks to vernacular and embodied conceptions of the weather, before they could precisely describe what it *would* do.

Since weather is an integral part of life, weather knowledge is as old as the experience of weather itself. There is documented evidence that humans have been trying to predict the weather at least since the first millennium BCE. Aristotle might have written *Meteorologia* outlining his methods of weather prediction around 340 BCE, but the Babylonians and Chinese had already been using clouds and haloes around the moon and the sun to make short-term forecasts. These techniques offered a highly localized understanding of weather that was largely encoded in folk meteorology and passed down along cultural and occupational lines. To be clear, though, to draw the conclusion that weather knowledge was a result of farmers and sailors needing to know the weather, as many early weatherlore collectors have argued, suggests a modern understanding of the primacy of professions over the realities of lived experience (see Garriot 1903; Inwards [1869] 2013). It is difficult to escape experiencing the weather, regardless of one's profession. Weather prediction, before it became a matter of math and modeling, combined that experience with watching the weather and trying to predict it with whatever tools were available within the technological and theological understandings of the day.

As a result, since before forecasting has been a science, in modern terminology, it has been a form of divination—a form of objective knowledge based on the observation of natural phenomena that requires a specialist to interpret them (Boyer 2020, 100; Silva 2016, 507). While sciences like meteorology and divination may seem leagues apart to contemporary thinkers, they have not always been understood as separate in the West. For ancient Greeks, weather and the "inclinations of the divine" seemed ubiquitous, a constant topic of human inquiry (Struck 2013, 1).[2] Skipping ahead some 1200 years or so, the connection between divination and the weather remained intact. In the early medieval period, meteorology and medicine were categorized along with astrology and the other mantic arts, that is, those disciplines concerned with prophecy and divination (Fidora 2013, 517–18). Each of these disciplines had bearing on the others and needed to be read together. Meteorology, in particular, was connected to other studies of the sky through a paradigm of correspondences. Augury—the study of birds in flight, aeromancy—divination based on weather, and astrology all influenced one another. For example, not only was the weather of autumn considered dependent on the weather experienced in the summer, but it was also dependent on the nature and alignment of the planets and an observation of birds (Fidora 2013, 530–31).

These allied disciplines of medicine, meteorology, and astrology used common methodologies for common ends. Practitioners objectively observed naturally appearing signs in order to determine what likely course events would take. The sources of signs were perceived in the Middle Ages as "[partaking] the knowledge of God," using the natural world, God's creation, to determine natural events on earth (Fidora 2013, 519). The objective was prognostic. As anthropologist Sónia Silva (2016, 508) explains, "divination and positivist science share a similar emphasis on objectivity, detachment, observation." It would not be until the eighteenth century that these disciplines started to pull apart. And if we look back at that definition of "divination," objective knowledge based on observations of nature requiring an expert to interpret, it sounds a lot like any scientific field. What has changed is not *what* meteorologists do but rather the technology they use to do it. There is a case to be made that meteorology has never really left its associations with divination behind.

Even with the scientific technologies of the European Renaissance, like the barometer, hygrometer, and thermometer, that helped make recording the weather more precise, much weather prediction remained based on personal observations and folk knowledge, at least until the mid-nineteenth century. The first weather forecasts in the West to be called such were pioneered by British Admiral Robert FitzRoy, erstwhile captain of the *HMS Beagle*. FitzRoy established what would become the British Meteorological Office in the 1850s, and the first forecast appeared in *The Times* in London on August 1, 1861. It listed temperatures, cloud cover, and wind direction for cities around England (Moore 2015). These daily forecasts were an extension of storm warnings FitzRoy had been publishing for a few years, but neither warnings nor forecasts would have been possible without the invention of the telegraph.

Without the telegraph, weather prediction had been a local affair; with the telegraph, scientists could send warnings to places they feared might be set for storms before the storm reached them. Not that such telegraphed predictions were always correct, setting the stage for early versions of the jokes about meteorological inaccuracy that began this chapter. As the *Cork Examiner* (Ireland) commented in response to one such report: "Yesterday, at two o'clock, we received by telegraph Admiral FitzRoy's signal of a southerly gale. The gallant meteorologist might have sent it by post, as the gale had commenced the day before and concluded fully twelve hours before the receipt of the warning" (Moore 2015). Meteorological snark appears to be as old as the forecast itself.

In the United States, weather forecasting was a science of observation rather than prediction, but it was also a matter of national pride. When the Smithsonian opened its main building on the National Mall in 1856, the main hall featured a weather map. Observers around the country would telegraph their weather observations into the Smithsonian each morning, and the map would be updated accordingly. The map was symbolic of technological advancement, of national unity, and of expansionism (Blum 2019, 15–16). Starting in 1870, the US Army Signal Corps tasked a group of twenty-four observers with telegraphing temperature, barometric pressure, cloud cover, wind speed, and direction to Washington, DC, each morning (Williams 1986, 71). This practice would form the basis of the US Weather Bureau, which later split from the Army Signal Corps (NOAA n.d.). Despite part of the bureau's purpose being to warn of coming hurricanes, the bureau tended to dissuade the use of the words "hurricane" and "cyclone" lest they create panic or later be proven wrong (E. Larson 1999, 106–8). Indeed, the Weather Bureau hesitated to predict weather at all as a bad forecast was believed to impugn their reputation (E. Larson 1999, 69).

Not until after the turn of the twentieth century do the first attempts to *predict* weather, using mathematical models, appear. These models were notoriously inaccurate, however, due to a lack of computing power. As the story is told, while the mathematical models to predict weather in the early twentieth century existed, it would have required a stadium filled with mathematicians under the conductor-like guidance of a head climatologist to do the necessary calculations (Graham, Parkinson, and Chahine 2002). It would take the advent of modern computers to start to forecast weather. To put this history into the perspective of the humanities, when the Weather Bureau began getting down to the business of predicting the weather, formalism and functionalism held primacy in American folklore theory (see Bronner 1986, 74–106).

Yet until the middle of the twentieth century, many believed "men should not try to predict the weather, because it was God's province" (E. Larson 1999, 31). The advances in forecasting since then have been exponential and closely linked to the advances in technology: for every decade since the 1960s, the forecast was able to be predicted one day further out, so "a six-day forecast today is as good as a five-day forecast a decade ago; . . . today's six-day forecast is a s good as a two-day forecast in the 1970s" (Blum 2019, 3). However, most of us accept this science-based weather prediction on faith. We do not see that network of scientists around the world building models and running equations to tell us whether it will rain. Instead, we see wispy clouds gather and grow, we feel the wind rise, and we feel the chill. We look up the weather forecast built by climatologists, and then, we add a layer based on our experience of the weather itself.

Belief and the Weatherman

It is not that meteorology and divination remain linked in a scientific paradigm or that the general public treats their weather report with the same regard as their horoscope. Rather, the connection appears in the ways that many people believe in and interact with meteorologists in their communities, meteorologists whom most people do not personally know but whom they nonetheless basically trust. Belief forges a crucial link in this relationship. Folklorists tend to place belief in opposition to scientific knowledge: "as a process of knowing that is not subject to verification or measurement by experimental means within the framework of a modern Western scientific paradigm" (Marilyn Motz quoted in Mullen 2000, 120). Since the technological advances of the last few centuries, weather can be verified and measured. Even on the vernacular level, home "weather stations" quantifying the weather feature in many American kitchens. At the very least, a thermometer can be found suctioned onto the window. This quantification would place the weather forecast outside the genre of belief, but it seems somewhat arbitrary. At the very least, it privileges Western scientific rationalism over folk knowledge to consider quantitative knowledge as inherently different from other forms of knowledge.

Two considerations help make sense of the relationship between meteorology and belief. First, our understanding of the weather takes place within a spectrum that also includes meteorology in addition to vernacular and experiential weather. David Hufford (1994, 119) argues that the majority of people practice elements of both vernacular medicine and institutionalized medicine, moving between them depending on their needs of the moment. The same might be said of weather forecasts, as we combine information from multiple sources to decide accordingly. Second, the key component of Motz's definition above might lie in the word "process," which describes belief as an active practice of *knowing*. As Motz (1998, 349) argues, it can be more useful to think of belief as a verb rather than a noun—the process of "how people believe." That process of belief in meteorology occurs through relational experiences of benevolence and trust between the forecast-consuming public and the meteorologist. Meteorologists must build trust through appeals to vernacular weather language and experiences of weather—"good picnic weather" or "sweater weather" or, to quote both my grandfather and my local meteorologist, "good weather for ducks."

There is a knowledge that there is more to meteorology than these platitudes though. Like priests, meteorologists are expected to be competent readers of signs, translators of the weather into lay language. We trust that they have the science, the mathematical models, and the quantitative reasoning handled, but that trust can be fragile. In 1999, Ireland's television station briefly replaced their meteorologists with presenters who lacked the requisite scientific pedigree only to be met with public outcry. The meteorologists returned to their posts of interpreting the signs within a month (Ingle 1999). How much the viewership knew of the specific rigors of earning a meteorology degree likely mattered less than knowing that meteorologists had been tested by their profession. Presenters had not. Meteorologists hold arcane knowledge that no mere television presenter can replicate, and their viewers rely on them to interpret that knowledge.

And their viewers watch them for signs. When the weather gets serious, American weathermen loosen their ties. The dress code of the American television meteorologist is business formal. Until a storm approaches. Then jackets are shed, ties loosened, and shirt sleeves rolled. Forecast viewers measure the extremity of the weather by the lowering formality of the weather oracle, like the Catholic faithful watch the various colors of smoke emitted during the papal conclave for a message. Tom Terry, a weatherman in Orlando, Florida, reposted a categorization of dress level sent to him by a viewer that plays on the categories of hurricanes that Orlando faces:

Tom Terry Hurricane threat level:

1. Jacket on, normal conditions
2. Jacket off, sleeves down, getting serious
3. Jacket off, sleeves rolled up, get ready, power's going out soon
4. Jacket off, sleeves up, sweat stains, and a red bull in hand: I haven't slept in 48hrs. Fire up the generator.

Let's hope I keep my jacket on this season. (Terry 2014, n.p.)

"Bad-weather-casual dress" puts the meteorologists in contrast to the buttoned-up news anchors, but it also makes them more trustworthy and reveals a believed benevolence on the part of the weathermen, that the safety of viewers matters more than sartorial compliance.

Also like priests, meteorologists are expected to have our best interests at heart (Boyer 2020, 101). Even the jokes that opened this chapter document a kind of bumbling ignorance rather than a willful desire to lead the public unwittingly into blizzards. At worst, the meteorologist makes a convenient scapegoat.[3] In other words, if you plan a hike for your friends on Saturday because Saturday's prediction is sunny and warm, you can at least save some face when you all arrive at the end of the hike cold and wet.

The Seers

The result of this belief in the weathermen, even as jokes are told about their errors, is that they can achieve folk-hero status as seers of the weather. Good meteorologists become local icons. Last tornado season, I was chatting with a friend about the weather. She had recently moved to Tulsa, Oklahoma, the heart of Tornado Alley. Around the time we talked, a number of tornadoes had been spotted in the area during the previous few days. She told me that a local meteorologist, Travis Meyer of KOTV Channel 6 in Tulsa, had become a meme in the wake of the storms. According to legend, he had stayed at the station's weather center for more than twenty-four hours, through the worst of the storms, to provide constant updates. Such behavior is not unknown among Tornado Alley meteorologists but is legend making all the same. Afterward, Tulsa's Philbrook Museum offered Meyer a day at the museum by himself to relax from the stress of emergency forecasting. Meyer has stated that staying through the course of a storm might be difficult, but he "tries to help as many people as he can" (TTV staff 2019, n.p.).

The viewers have rewarded him with their belief in him—both through their viewership and in semi-tongue-in-cheek iconography, offerings to a vernacular, secular prophet. In memes and local Tulsa products, Meyer began to pop up all over in ways that overtly drew on both religious and pop-cultural cult images. A local brewery put his likeness on prayer candles (TTV staff 2019, n.p.). Meyer appeared as both a Jesus figure and an Eastern Orthodox saint; as Poseidon, reporting on flooding; as cult-status painter Bob Ross, painting a tornado, of course; on the famous throne of swords from the HBO show *Game of Thrones*, retitled "Game of Meteorologists"; and on cupcakes, with a chocolate twist (Owen 2019, n.p.). A crowdfunding page raised over one thousand dollars to "buy Travis Meyer and his Storm Team a beer" (Owen 2019, n.p.). So iconic of Tulsa's weather has Meyer become that he even made a cameo in the young-adult, fantasy novel *Lost* by P. C. Cast and Kristin Cast (2018). As for Meyer, he was touched: "I don't know if I was just really tired or if it was just the moment, but that was probably one of the best moments in my life" (TTV staff 2019, n.p.). Most meteorologists are not feted with such fandom, but many do have similar cult followings: Jim Cantore, of the Weather Channel, has Facebook pages dedicated to his fans.

The lionization of meteorologists is not restricted to Oklahoma, but it is notable there. Before Travis Meyer, there was Gary England. England worked in the Oklahoma City market for over forty years, retiring not long after the series of tornadoes in 2013 that would reshape the city. He may not have as much kitsch produced in his name as Meyer, but his reputation was similar. He was known for his "folksy language and homespun colloquialisms" while delivering the forecast (Bailey 2015, 89), so much so that he spawned his own mythology, including both a story about a "thunder lizard" who created chaos on the plains and a devotional belief in his forecasts. Journalist Holly Bailey (2015, 74), writing in *The Mercy of the Sky* about the 2013 tornadoes, describes England and the other Oklahoma City meteorologists in magico-religious terms:

> wizards of the weather, shamans of the storm who could conjure an alchemical combination of wisdom and magic. In them one could detect varying degrees of the fire and brimstone of a Holy Roller preacher desperate to save souls from the blazing pits of hell. . . . Oklahoma's

weather gods also had to possess one essential skill that was even harder to attain: the art of knowing the weather so well that they could somehow collect the clues to anticipate its next move. To some of us it seems almost like magic.

Bailey's words evoke the deep connection between weather and magic, but more than that, they manifest how much of that connection lies in prognostication. Weathermen are not craftsmen in these metaphors, they are not building beauty or usefulness out of raw material; nor are they doctors, reading the signs of sickness and deploying their knowledge in battle.[4] Weathermen predict, and thus they are wizards or shamans or preachers or gods.

As in the medieval Christian church, weather prediction requires a network of such holy people employed in differing ways. Media meteorologists function like priests in that they translate the complexities of the weather to the general public of a community they typically live among. Weather prediction as a scientific enterprise and climatology, however, relies on the local observations of thousands of other people. The US National Weather Service employs close to twelve thousand volunteer weather observers. Other countries have similar cadres of weather observers, some employed, some volunteer. Most of these observers are roughly what one might expect—farmers, science teachers, corporations. Some, though, operating in the world's more remote locations, seem closer to holy hermits.

The television meteorologist may in Bailey's words be a "weather shaman," but there is something mystical about those who measure the weather alone. They capture the imagination in their dedication to the weather. Iceland runs seventy-one weather-observation stations around the country. While for many American weather observers, the job requires twice-a-day readings, in Iceland, the weather must be recorded every three hours. And so, every three hours around the clock, someone, like Marsibil Erlendsdottir, goes out in the snow and wind on the northeastern edge of Iceland to a no-longer-manned lighthouse to measure the weather (Skubatz 2021). There is something lonely and compelling in this devotion. In the context of the belief and prediction, the remote weather observer evokes the monastic hermit of Christian medieval literature. There seems to be an inheritance from the ninth and tenth-century hermits of Skellig Michael off the coast of Ireland, offering their daily prayers, to the modern staff at Giles Weather Station, 750 kilometers from the nearest town in central Australia, offering their daily weather forecasts. There is a devotional aspect to the daily marking of the weather itself. The weather changes, but the weather doesn't go anywhere.

The Visceral Weather

Perhaps part of the reason that meteorology is so often perceived in terms of belief lies precisely in the ways that we feel the weather on and within our bodies. Like belief, weather operates in a space that can be difficult to put into words. My grandmother played tennis and sometimes wore a sock with the toe cut off pulled up around her elbow. As a child, I thought the two were connected until, one rainy day, I asked about the sock since she couldn't have been playing tennis. "Old Arthur's come to visit," she replied. I suspected this was an adult joke I was somehow missing or maybe that she was referring to her tennis partner, though I could make no sense of Arthur's connection to the sock or where in her house he might be hiding. A year or so later, though still mystified, I commented that she had her elbow sock on: "Arthur's back?" It was the rain, she explained, storm's coming, and she could feel it in her elbow. Talking to my grandma, I recall thinking it would be some trick if I could predict the weather and feel it in some part of my body like a train rumbling in the distance. Pain in the body, instability in the physiological system, signaling bad weather or instability in the atmosphere demonstrates the correspondence of sympathetic magic and serves as a reminder of the connection between humans and their environment (Magliocco 2004, 118). To know in advance felt mystical; it felt powerful. Only later would I learn about arthritis.

Medically, my grandmother wasn't wrong, nor was she alone in her predictive abilities. I have had friends, bosses, and even a couple of coffee-shop baristas comment on their (or their customers') weather aches and pains. Arthritis pain and migraines connected to barometric pressure changes, often just before storms, are well documented phenomena, caused by the pressure imbalances between fluid and air inside the body and the surrounding atmosphere (see Beck 2013; Huizen 2020). The connection between aches and storms has even led to the development of a hyper-specific mobile weather app that alerts users to pressure changes before they happen (Cirrus HealthCare 2021, n.p.). Pressure isn't the only problem though, as other maladies also correlate to the weather. Rapid decreases in temperature coincide with increased risk of cardiac arrest, a 7 percent increase for every 10-degree drop in Celsius (Beck 2013, n.p.). And storms change how we relate to others too. Charting behavior of preschoolers against National Weather Bureau reports, researchers found that young children spend more time with peers and adults during unstable weather and that girls were more impacted than boys (Essa, Hilton, and Murray 1990). Hot weather increases the number of assaults, rapes, and domestic violence, though tempers are quicker in cold weather (Cohn 1993).

However, it is the wind that seems to have the most profound effects on both mind and body. In several parts of the world, warm winds preface headaches, malaise, depression, and an increased number of suicides. The phenomenon seems notable in places where winds are drawn down mountain slopes into closed valleys, as in the case of foehn winds, like the Santa Ana in California, those in Leukerbad, Switzerland, or those in the Tatra mountains in Poland (Koszewska et al. 2019; Strauss 2007). Raymond Chandler remarks about the Santa Anas, "On nights like that, every booze party ends in a fight. Meek little wives feel the edge of the carving knife and study their husbands' necks. Anything can happen" (quoted in Ulin 2014, n.p.). Joan Didion, who quotes Chandler, went further still: discussing a Santa Ana in 1957 that lasted nearly two weeks and reached hurricane force, Didion ([1969] 2017, 217) records wildfires, unusually high numbers of traffic deaths, a murder-suicide, and a woman thrown out of a car. The embodiment of weather prediction moves into the eerie. It is not simply my grandmother's sense that her elbow aches and so the weather will change, but that something wicked this way comes: the unstoppable and unavoidable weather.

Weather combines knowledge and belief with the experiential. We believe the local meteorologist in part because they confirm our own experiences of weather. Hufford (1995, 11) recasts folklorists' understanding of belief as "rationally developed from experience." This shift helps to move academic studies of belief away from earlier descriptions that tend to either romanticize belief within a naturalistic paradigm or pathologize it within a scientific one (see Mullen 2000). We have a lifetime of experience with the weather, shaped by where we live and our individual physiognomies. We interpret that technological observation in terms of what "feels real" to each of us (Mills 1993, 179)—as anyone who has argued with roommates, office mates, or family over the thermostat knows well. Drawing on the work of feminist and phenomenological belief scholars like Elaine Lawless and Hufford, Margaret Mills's folk knowledge of what "feels real" describes experiences as incontrovertible to the person experiencing them. At the extreme end of weather—the hurricane, the blizzard, the tornado, the Santa Ana, researchers in atmospheric science trying to figure out the best way to warn the public of impending weather risks document a clear link between individuals' past experiences with extreme weather and their likelihood to heed and act on warnings of future weather events (Demuth et al. 2016).

More than simply experience, however, the feeling of weather is visceral. What "feels real" in the weather is kinesthetic and proprioceptive, rooted in the body of the subject themselves in a nonverbal form. Tim Ingold's (2007, S29) "weather world" describes this sense of what is felt but not necessarily touched or looked at: the weather "infuses our entire being." To feel the weather embodied in the self, Ingold (2007, S29–S30) argues, suggests a linking of the body to the larger rhythms of the earth, a "commingling," even "an immersion in the fluxes of the medium: in sunshine, rain, and wind." Ingold's aim is to question the limitations of separating material culture from people, but the weather, in particular embodied weather, suggests that it is more than a medium, more than immersive. To feel the weather—not as rain on the face or wind pulling at clothes, but deeply inside the bone—implies a cosmological unification of person and atmosphere that is not happening in the present moment or lingering from the past after one comes inside. To feel the weather as my grandmother felt it is prognostication, to feel the weather yet to come.

The embodied prediction of weather temporalizes belief in an immediate present and future. In religious structures, belief often invokes cosmologic time, almost in opposition to the highly immediate and localized experience of the weather. Embodied weather prediction, however, kaleidoscopes time—it is eternal, immediate, and still in the future. Embodied weather prediction like my grandmother practiced—half joke, half serious—requires a belief in signs and portents that the future can be predicted. It is built from a faith in vernacular phenomenological, immersive, and integrated knowledge of weather and, as she switched on the weather radio that sat by her kitchen sink, trust in the words of the weathermen.

We trust in the benignity of meteorologists, and that trust lets us believe that if they predict rain, we should take our umbrellas. We believe that the abstract of the weather forecast we hear is supported by reams of math and science most of us shall never see but that it is all we really need to make informed decisions about our immediate futures. Most of us respond to meteorologists as though they were seers of old, someone to consult to know what the future holds, but whose methods might be best left mysterious.

Notes

1. I use the term "weatherman" here to designate the role of meteorologist filled predominantly by men on American television. While the number of weather presenters is more equally split between men and women in some markets, a profound gender bias exists in the field such that 92 percent of chief meteorologists and 89 percent of evening weather presenters are male (Cranford 2018). Thus, there remains in the cultural imagination a view of weather forecasting as a specifically male occupation. More important to this chapter, the idea of the "weatherman" as someone who reads the atmosphere and makes predictions remains culturally relevant.

2. However, divination was not perceived as occult or magical as it generally is today. Greek texts used starkly different words to describe magic and divination, understanding the

latter as "one of the useful arts" akin to mathematics or observational sciences, like agriculture and meteorology (Struck 2013, 3–4).

3. There is a notable exception to the good-meteorologist stereotype in the legends and conspiracy theories that detail purported attempts to control the weather. Beyond seeding clouds to induce rain, however, most weather-modification techniques remain the purview of fiction.

4. The battle metaphor of disease is largely Western in origin and use, with sometimes problematic consequences for patients and epidemiology (Martin 1990).

References

Bailey, Holly. 2015. *The Mercy of the Sky*. London and New York: Penguin.

Beck, Melinda. 2013. "How Your Knees Can Predict the Weather: Granny Was Right. Scientists Find Link between Achy Joints and the Forecast." *Wall Street Journal*, October 14, 2013. https://www.wsj.com/articles/how-your-knees-can-predict-the-weather-1381792289.

Blum, Andrew. 2019. *The Weather Machine: How We See into the Future*. London: Vintage.

Boyer, Pascal. 2020. "Why Divination? Evolved Psychology and Strategic Interaction in the Production of Truth." *Current Anthropology* 61, no. 1 (February): 100–112.

Bronner, Simon J. 1986. *American Folklore Studies: An Intellectual History*. Lawrence: University Press of Kansas.

Cast, P. C., and Kristin Cast. 2018. *Lost*. Ashland, OR: Blackstone.

Cirrus HealthCare. 2021. "WeatherX Forecast." Apple App Store, Vers. 4.8.1. Mobile Application. Accessed September 5, 2022. https://apps.apple.com/gb/app/weatherx-forecast/id1125533216.

Cohn, Ellen G. 1993. "The Prediction of Police Calls for Service: The Influence of Weather and Temporal Variables on Rape and Domestic Violence." *Journal of Environmental Psychology* 13, no. 1: 71–83.

Cranford, Alexandra. 2018. "Women Weathercasters: Their Position, Education, and Presence in Local TV." *Bulletin of the American Meteorological Society* 99, no. 2: 281–88.

Demuth, Julie L., Rebecca E. Morss, Jeffrey K. Lazo, and Craig Trumbo. 2016. "The Effects of Past Hurricane Experiences on Evacuation Intentions through Risk Perception and Efficacy Beliefs: A Mediation Analysis." *Weather, Climate, and Society* 8, no. 4 (October): 327–44.

Didion, Joan. [1969] 2017. *Slouching Towards Bethlehem*. London: 4th Estate.

Essa, Eva L., Jeanne M. Hilton, and Colleen I. Murray. 1990. "The Relationship between Weather and Preschoolers' Behavior." *Children's Environments Quarterly* 7, no. 3: 32–36.

Fidora, Alexander. 2013. "Divination and Scientific Prediction: The Epistemology of Prognostic Sciences in Medieval Europe." *Early Science and Medicine* 18, no. 6: 517–35.

Garriott, Edward B. 1903. *Weather Folk-Lore and Local Weather Signs*. Honolulu: University of the Pacific Press.

Graham, Steve, Claire Parkinson, and Mous Chahine. 2002. "Weather Forecasting through the Ages." National Aeronautics and Space Administration (NASA), February 25, 2002. https://earthobservatory.nasa.gov/features/WxForecasting/wx3.php.

Hufford, David. 1994. "Folklore and Medicine." In *Putting Folklore to Use*, edited by Michael Owen Jones, 117–35. Lexington: University of Kentucky Press.

Hufford, David. 1995. "Beings without Bodies: An Experience-Centered Theory of the Belief in Spirits." In *Out of the Ordinary: Folklore and the Supernatural*, edited by Walker Barbara, 11–45. Boulder: University Press of Colorado. https://doi.org/10.2307/j.ctt46nwn8.6.

Huizen, Jennifer. 2020. "What You Should Know about Barometric Pressure and Headaches." Medical News Today, May 21, 2020. https://www.medicalnewstoday.com/articles/320038.

Ingle, Roisin. 1999. "TV Weather Set for Return of Professional Presenters." *Irish Times*, October 28, 1999. https://www.irishtimes.com/news/tv-weather-set-for-return-of-professional-presenters-1.243733.

Ingold, Tim. 2007. "Earth, Sky, Wind, and Weather." *Journal of the Royal Anthropological Institute* 13: S19–S38. http://www.jstor.org/stable/4623118.

Inwards, Richard. [1869] 2013. *Weather Lore: A Collection of Proverbs, Sayings, and Rules Concerning the Weather*. Cambridge, UK: Cambridge University Press.

Koszewska, Iwona, Ewelina Walawender, Anna Baran, Jakub Zieliński, and Zbigniew Ustrnul. 2019. "Foehn Wind as a Seasonal Suicide Risk Factor in a Mountain Region." *Psychiatria i Psychologia Kliniczna* 19, no. 1: 48–53.

Larson, Erik. 1999. *Isaac's Storm*. New York: Vintage Books.

Larson, Gary. 1991. "Front Porch Forecasters." *The Far Side*, April 24, 1991. Comic strip. https://www.newspapers.com/clip/958000/the-far-side-front-porch-forecasters/.

Magliocco, Sabina. 2004. *Witching Culture: Folklore and Neo-Paganism in America*. Philadelphia: University of Pennsylvania Press.

Martin, Emily. 1990. "Toward an Anthropology of Immunology: The Body as Nation State." *Medical Anthropology Quarterly*, n.s., 4, no. 4: 410–26. http://www.jstor.org/stable/649224.

Meme. "The Meteorologist Who Cried 'Blizzard.'" n.d. Accessed September 5, 2022. https://me.me/i/3-the-meteorologist-ho-cried-blizzard-a-story-of-exaggeration-11258729.

Merriam-Webster. n.d. s.v. "forecast." Accessed September 5, 2022. https://www.merriam-webster.com/dictionary/forecast#:~:text=Definition%20of%20forecast&text=1a%20%3A%20to%20calculate%20or,company%20is%20forecasting%20reduced%20profits.

Meteorology Gifts and Shirts. n.d. "Meteorologist—The Only Job Where You Can Be Wrong Every Day." Accessed September 5, 2022. https://www.redbubble.com/i/poster/Meteorologist-The-Only-Job-Where-You-Can-Be-Wrong-Every-Day-by-jaygo/34638677.LVTDI.

Mills, Margaret. 1993. "Feminist Theory and the Study of Folklore: A Twenty-Year Trajectory toward Theory." *Western Folklore* 52, no. 2–4: 173–92. https://doi.org/10.2307/1500085.

Moore, Peter. 2015. "The Birth of the Weather Forecast." *BBC Magazine*, April 30, 2015. https://www.bbc.com/news/magazine-32483678.

Motz, Marilyn. 1998. "The Practice of Belief." *Journal of American Folklore* 111: 339–55.

Mullen, Patrick B. 2000. "Belief and the American Folk." *Journal of American Folklore* 113, no. 448: 119–43.

[NOAA] National Oceanic and Atmospheric Administration. n.d. "History of the National Weather Service." https://www.weather.gov/timeline. Accessed September 5, 2022.

Owen, Kirsti Roe. 2019. "King of the Soakpocalypse: How the Cult of Travis Meyer Gives Us Life." Tulsa Kids, May 30, 2019. https://www.tulsakids.com/king-of-the-soakpocalypse/.

Panovich, Brad. 2014. "Perspective on the Accuracy of Meteorologists." WXBrad, May 19, 2014. https://wxbrad.com/perspective-on-the-accuracy-of-meteorologists/.

Pimple, Kenneth. 1990. "Folk Belief." In *The Emergence of Folklore in Everyday Life*, edited by George H. Schoemaker, 51–56. Bloomington, IN: Trickster Press.

Piraro, Dan. 2013. "The Weather or Not Channel." *Bizarro*, July 17, 2013. Comic strip. http://bizarrocomics.com/files/2013/07/bz-panel-07-17-13.jpg.

Satterfield, Dan. 2014. "A Perspective on the Accuracy of Meteorologists." Dan's Wild Science Journal, September 3, 2014. https://blogs.agu.org/wildwildscience/2014/09/03/perspective-accuracy-meteorologists/.

Silva, Sónia. 2016. "Object and Objectivity in Divination." *Material Religion* 12, no. 4, 507–9, https://doi.org/10.1080/17432200.2016.1227638.

Skubatz, Marzena. 2021. "Monitoring the Weather at the End of the World." *New York Times*, January 25, 2021. https://www.nytimes.com/2021/01/25/travel/remote-weather-station-iceland.html.

Spackman, Jonathan, and Stephen Yanchar. 2013. "Embodied Cognition, Representationalism, and Mechanism: Review and Analysis." *Journal for the Theory of Social Behavior* 44, no. 1: 46–79.

Strauss, Sarah. 2007. "An Ill Wind: The Foehn in Leukerbad and Beyond." *Journal of the Royal Anthropological Institute* 13, S1: S165–S181.

Struck, Peter. 2013. "Arthur O. Lovejoy Lecture: A Cognitive History of Divination in Ancient Greece." *Journal of the History of Ideas* 77, no. 1 (January): 1–25.

Terry, Tom. 2014. "Tom Terry Hurricane Threat Level." Facebook, June 18, 2014. https://www.facebook.com/profile/100044383991161/search/?q=jacket%20off.

TTV staff. 2019. "The Weather Wizard of Tulsa: Travis Meyer on Saving Lives and Becoming a Meme." *Tulsa People*, June 5, 2019. https://www.tulsapeople.com/the-weather-wizard-of-tulsa/article_3e2e9630-863c-58f3-bb1d-4f98f0cfdb58.html.

Ulin, David L. 2014. "The Santa Ana Winds and the Literature of Los Angeles." *LA (CA) Times*, May 14, 2014. https://www.latimes.com/books/jacketcopy/la-et-jc-the-santa-ana-and-the-literature-of-los-angeles-20140514-story.html.

Williams, Gordon. 1986. "The Weather Watchers." *The Atlantic* 257, no. 3 (March): 69–71.

Chapter 4

CONTRAILS TO CHEMTRAILS

Atmospheric Scientists Respond to Challenging Belief Narratives

Anne Pryor

Becoming regular monthly guests on the *Larry Meiller Show* was a happy achievement for Steve A. Ackerman and Jonathan E. Martin.[1] Both professors of atmospheric science at the University of Wisconsin–Madison, they first wrangled a one-time spot in 1997 by repeatedly asking the producer of this popular call-in radio show for an invitation. The show's producer at the time, Jim Packard, preferred to book single guests, but Ackerman and Martin knew they would be good complements to each other. "I could answer questions about atmospheric physics and Jon could talk about dynamics; that way we'd be able to cover everything," Ackerman recalls (2020b, n.p.). When another guest canceled unexpectedly and Packard needed to fill the last-minute opening, the two scientists were luckily both available despite short notice. They knew that first show had gone well when, at the end of the forty-five minutes, host Larry Meiller said, "We'll have to have you guys back," a coveted endorsement that Ackerman (2020b, n.p.) remembers with delight.

The *Larry Meiller Show* is a ninety-minute program aired weekdays from 11 a.m.–12:30 p.m. over eighteen stations in the Wisconsin Public Radio (WPR) network. The show is typically divided into two segments, each forty-five minutes in length with distinct topics; Ackerman and Martin now occupy the second half on the last Monday of each month. The show's goal is to promote lifelong learning on useful topics through casual interactive conversations between invited guest, listeners, and the host. The host, Meiller, is professor emeritus in Life Sciences Communication at UW–Madison who has built a broad and loyal listening audience over his long radio career. His was the first show on WPR to introduce the call-in format (1978) and is the longest running call-in show of any type in Wisconsin (Wisconsin Broadcasters Association Foundation 2017, n.p.).

Since that first appearance, Ackerman and Martin have become regular monthly guests. Thanks to creative marketing by the Why Files at UW–Madison, the pair are known as the "Weather Guys," a moniker they use on the radio, for a weekly newspaper column they write for the *Wisconsin State Journal* newspaper, and when they give joint public appearances to community groups.[2]

The high level of public outreach by the Weather Guys is a commitment the two share philosophically and pursue in addition to the required dimensions of their jobs. Both are full professors with research programs, administrative responsibilities, and courses to teach, yet both fully embrace the guiding principle of their university, known as the "Wisconsin Idea" (Drury 2017, passim). The Wisconsin Idea is the iconic philosophical foundation of the University of Wisconsin system credited to University President Charles Van Hise, put into action in the Wisconsin legislature under Governor Robert M. La Follette, and named the "Wisconsin Idea" by author Charles McCarthy (1912). This philosophy is one of service to the state, described by University President John Bascom (president from 1874–1887 and mentor to both Van Hise and La Follette) as "seeking knowledge not in order to seek personal privilege but in order to seek a better world for the common good" (quoted in Drury 2017, 22). Popularly summarized as "the boundaries of the university are the boundaries of the state," the Wisconsin Idea is embraced as core to the university (see, respectively, Stark 1995, 101–2; University of Wisconsin n.d.). It is invoked and enacted through diverse efforts, including public outreach, which is the foundation for the *Larry Meiller Show* and an inspiration for Ackerman and Martin's public work.

Ackerman and Martin are eager to communicate directly with the public for multiple reasons. Martin (2020, n.p.) explains: "The main reason I'm thrilled to have this opportunity, and I would never give it up, is we get to communicate how it is we do our jobs without bragging about it. And that removes this notion of the exclusivity [of science]." Ackerman (2020a, n.p.) further reflects: "Building on the public's near-universal fascination with weather is an opportunity to show how science works. Plus, it's fun, and we learn. Callers sometimes have local knowledge of weather that contributes to our understanding of Wisconsin's weather."

Communicating science effectively to the public is the topic of a growing field within academia and science professions (National Academies of Sciences, Engineering, and Medicine 2017); Ackerman even developed and cotaught a course on the topic with fellow faculty from the journalism and engineering departments. Articulated principles include careful attention to language and avoidance of jargon, the importance of finding a storyline, and the value of listening (Yuan et al. 2017). Ackerman (2020a) reflects that Meiller helps to keep jargon to a minimum through his clarifying questions. The call-in format of the show provides a conversation with the individual and helps to determine the caller's level of knowledge, which in turn guides how the Weather Guys frame their answer. He explains: "It is the caller's interest that's at the heart of how the science is shared and applied. We try to find out as much from them as we can so we can engage at their level of understanding and language" (Ackerman 2020a, n.p.).

Over the more than two decades of the Weather Guys radio program, the show has established a friendly, welcoming, and fun tone. The three typically begin with a jovial conversation of light banter with perhaps a weather-related, personal-experience story. Regular listeners know that Meiller is an avid fisherman and golfer; some shows commence with stories from his outdoor experiences. Ackerman and Martin welcome personal narratives as opportunities to bring science to bear on everyday experiences. Their own modeling of narrative-based weather discussion creates space for callers to frame their questions within their own experiences. Here is an excerpted example of the breezy yet informative camaraderie of the show.

> **Meiller:** Let's see. Last week, I was out golfing, late afternoon. I heard some thunder. Didn't see any lightning, but we continued on, and it actually kind of cleared up after the fact. But if you hear thunder, can you be struck by lightning?
>
> **Martin:** I think the prudent answer to that is "yes." There might be somebody who would say there might be circumstances under which you might hear thunder from a very distant place, and you've got no risk at all of a lightning strike, but I just think that's not safe. It's more prudent to be careful. If you hear some thunder, then you're probably somewhat under the gun, some degree of being under the gun. What do you think about that, Steve?
>
> **Ackerman:** Yeah, definitely. In fact, Larry, I can't believe you're on the radio saying, "Yeah, we heard it and kept playing!"
>
> **[All of them laugh.]**
>
> **Meiller:** [Said defensively] Well, it looked like it was gonna clear up!
>
> **Ackerman:** You're supposed to be setting a good example!
>
> **[More laughter.]**
>
> **Meiller:** I had a good round going.
>
> **Ackerman:** Well, that's it. We all. . . . Well, if you can hear thunder then generally you're at risk of getting hit by it. Now, what is the probability? There you have to balance it with your own decisions about whether or not you want to take that risk—but if you're having a good golf game, it's probably really hard to walk away.

Meiller: [Laughs.]

Martin: I think that MSCR[3] softball and West Madison Little League baseball have the rule that if you hear thunder on the field, you're off the field.

Ackerman: Right.

Martin: And, certainly, if you see the lightning first before you hear the thunder, you're off the field as well.

Ackerman: They leave that decision now, I think, to the umpires. I can remember a couple of years ago, before they had that rule, we were in the bottom of the sixth, and we were winning. We heard lightning and were like, "Nooo, let's just play one more inning."

Martin: [Laughs.] That was the other team.

Ackerman: We wanted to walk away with the win.

[Laughter.] (Meiller 2007b, n.p., sic throughout all Meiller quotations)

This casual interchange is based on sharing abbreviated or referential personal-experience narratives; the constrained performance space of the radio program precludes fully developed versions of the stories about Meiller's golf game, Ackerman's softball game, or Martin's role as a youth baseball coach. These stories are platforms for linking weather science with lived experience and to create an intimate welcoming space for callers. Since stories beget more stories, this exchange was followed later in the show by multiple lightning stories set on Wisconsin's various (over fourteen thousand) lakes: Ron in West Bend told how lightning struck an aluminum fishing boat in Governor Dodge State Park, Marv in Menominee spoke of "lightning out of the blue" hitting a fishing boat from miles away, Steve in St. Germaine told of a lightning strike while fishing on the Eagle River chain of lakes, and Meiller recalled an experience about a sudden storm while fishing on a large northern Wisconsin lake (Meiller 2007b, n.p.). Listening to the show, one not only learns about weather and climate but about occupational and recreational practices in Wisconsin and bordering states.

Along with lightning, other questions frequently raised by callers relate to the topics of temperature change at sunrise, relative humidity and dew point, and "storm splitting" (Ackerman and Martin 2013, n.p.). Other subjects are frequently folkloric. Some people question common assumptions: "Can it be too cold to snow?" The answer was "no" (Meiller 2007e, n.p.). Others ask for advice: Can they reschedule an annual community festival from the first weekend in June to another, less rainy weekend? The answer was "yes" (Meiller 2007e, n.p.). Many questions are based on personal observations of phenomena: Ackerman once identified a caller's upside-down rainbow over Lake Michigan as a circumzenithal arc in a rare use of fancy vocabulary (Meiller 2007c, n.p.), and Martin and Ackerman brainstormed possible climatological reasons for why deer don't "yard up" in the winter like they did twenty years ago, an observation made by Pat in Three Lakes (Meiller 2007d, n.p.). Some calls refer to seasonal practices, such as how Wisconsin tobacco farmers know about the typical January dry period and use it to finish their harvest (Meiller 2007a, n.p.).

Ackerman and Martin are passionate about their chosen profession, its continued advancement, and the benefits it brings to society. Both regularly contextualize the positive social impact of weather-satellite technology and forecasting abilities that have become accepted contemporary norms, save lives, and benefit multiple dimensions of society. They use the opportunity of their radio platform to remind listeners of the extraordinary nature of these advances. This passion for the science of weather and climate is one of the elements that makes them compelling guests, according to Meiller: "They have a lot of enthusiasm for their topics . . . I can see it in their eyes; listeners can hear it in their voices. It's an important thing about radio—you have to have enthusiastic guests, people who really like what they do and are passionate about it because they can convey that to the audience through their voices. And they do" (Ackerman and Martin 2013, n.p.).

The appreciation is mutual between guests and host. Both atmospheric scientists give Meiller much credit for the show's success. Ackerman (2020b, n.p.) reflects, "Larry creates an intimate space where we can all engage in a conversation about a question or experience." Martin (2020, n.p.) expands on this:

> He's got a goal in mind, it seems to me. That goal is to encourage absolute freedom in talking about whatever the ideas are revolving around the show that he's working on. And he's really great at that. He's kind to the people. He's considerate to them. He invites us to be considerate in inviting ways to talk back. He's excellent. I think that the show would be nowhere near as fun nor anywhere near as effective if it were anybody else running it.

In all, these multiple elements combine for a popular show, as evidenced by twenty-three years of monthly programs, full phone lines every month, and comments like the following one from callers, prefacing their questions:

> **Ellie:** I have many, many favorites on your programs, Larry, but this is absolutely one of them.
> **Meiller:** Thank you.
> **Martin:** That's so nice of you to say that.
> **Ackerman:** Just for that we're gonna answer your question correctly! [Laughs.] (Meiller 2008, n.p.)

Chemtrails

That absolute freedom for callers to raise any question and have it met with respectful consideration is rarely but occasionally strained, thanks to one topic—chemtrails. The term "chemtrail" is derived from "contrail," the name for the condensation trail of jet engine exhaust. "Chemtrail" substitutes "chemical" for "condensation" in the portmanteau. The following is a sample call that lays out some key elements of chemtrail belief, with a particularly gentle response from Martin.[4]

> **Meiller:** Lee in Green Bay up next. Hi, Lee. Thank you for calling.
>
> **Lee:** Yeah, hi. Umm, I'll just be real [sic] brief about it. I know nobody wants to talk about it if it's connected with the media or especially the government but... chemtrails. The spraying to control the weather by jets. And I know it's not exhaust. Because they were spraying here where I live Friday.
>
> **Martin:** How did you know that?
>
> **Lee:** They were crisscrossing. And a regular exhaust from a jet—certainly, whether it's ice particles or whatever, it's not going to spread. It's not going to run from almost one horizon to one horizon. It's not quite that long, I'm sure.
>
> **Martin:** I know what you mean. They sometimes get that long and sort of, they look like a Tootsie Roll that's been squashed, right?
>
> **Lee:** [Said hesitantly] Yeah, some of them do. It depends on the way the winds are high up above. But you can just see how that stuff is spreading. And the normal exhaust from a jet doesn't do that. And I know they don't want to admit it. People that are connected to the news media, from what I've heard, are forbidden to talk about it. But, I mean, it's happening. They're trying to control the weather. And they've been doing this for... I don't know.
>
> **Martin:** Well, let me suggest a couple of things here, Lee. First of all, if anyone's been trying to mount a campaign to control the weather, it's been the most miserable failure of all campaigns ever launched because we can't control the weather, and if we could, we would never have concerns about droughts or we would never have concerns about floods or too much snow or whatever it is. So that leads me to conclude that, first of all, that's not likely to be what's going on.
>
> I don't have a horse in this race. I don't have a dog in this fight. I'm just trying to be as logical about what I see as I can be. And I find that, on certain days—and we were just talking about this with a caller; you may have heard—we were talking about how jets can leave a trail

on some days and not on another one. And that's just because of the conditions in the atmosphere through which they fly.

And then, in addition to leaving a trail, the trail can be massaged by the winds that are part of the environment. They can definitely spread it out and make it deeper and all kinds of different other things. So, at a minimum, there are alternative explanations that need to be as thoroughly investigated as any of the investigations that lead a person to believe that there's a conspiracy going on because I suspect that there is no conspiracy.

Meiller: It would be pretty difficult, pretty difficult to do, but you know...? That's not to say it hasn't been tried.

Martin: Well, there was a long period of time in the fifties, right after World War II, where weather control was actually the goal of numerical weather prediction. It was actually a sense that learning how to predict the weather would give us a sense of how to control it. There were experiments along the lines of cloud seeding and various others that had a long, long history that were not particularly, that were not successful at all. In fact, they were places for various people to run sort of financial scams on farmers and so on. And it was really a bad deal. And we've exposed that through the scientific method, that this stuff is not reasonable, and you can't really conclude anything from it. And so, the scientific method is the friend of the farmer in that case, and some people will say, you know, there's something going on behind all our backs. It's not really happening.

Meiller: Thanks again for calling. We appreciate it. (Meiller 2015, n.p.).

This conversation raises several main areas of interest regarding chemtrail calls on the Weather Guys program: elements of the chemtrail theory, including geoengineering concerns, conspiracy beliefs, and how scientists respond.

Chemtrail Theory

Lee defines "chemtrails" succinctly as "spraying to control the weather by jets." His definition leaves out details included in more articulated variants of this theory, found on internet sites, chat groups, radio programs, and videos. Those extra details might include naming who is doing the spraying, what substance is being sprayed, why this is occurring, and its consequences. Such details shift between tellers and over time.

Fully articulated chemtrail-narrative variants might identify the substance being sprayed as barium, aluminum, lithium, nano-aluminum coated fiberglass particles, pathogens, or desiccated blood (Cairns 2016; McGrath and Perritano 2011). The actors behind the spraying might be specifically identified, such as the US Air Force, NASA, or the CIA; more generally identified, such as the US military or the New World Order; or vaguely and ambiguously referred to as "global elites" or even "they," as we find with Lee. The results of the spraying might be the cause of "drought in Africa, forest fires, bee decline, fisheries collapse, increases in Alzheimer's and autism, extreme weather events, reduction of arctic sea ice, and species extinctions, among other ills" (Cairns 2016, 76).

The purpose of the spraying is relayed variously as forcible vaccinations, the testing of pesticides, determining the effects of particular chemicals on humans, weeding out the sick and elderly, conducting mind-control experiments, sterilizing the population, manipulating the weather for national defense, mitigating climate change, or restoring the ozone layer (McGrath and Perritano 2011). The earliest concerns seem to be focused on simply the loss of blue sky, according to videos about chemtrails from the late 1990s. Patrick Minnis (2020, n.p.), a senior research specialist at NASA Langley Research Center, now retired, who led multiple studies of contrails and became a media go-to spokesperson on the topic, recalls some purposes: "killing the population by spraying . . . to block the sun . . . to turn us into mega-zombies. Most of the ones I heard were evil, except the one to combat global warming. That was kind of good."

Martin (2020, n.p.) recalls a purpose for the spraying related to agricultural warfare:

> [The first time I heard of chemtrails] was in the context of the radio show. I had never heard of it before that. So, the first time we ever heard it, we had to have the person explain what they meant by this and the idea that the government is spreading various types of unknown chemicals that will do damage to agriculture, or something like that, and there's a secret campaign to spread them in the entire atmosphere of the earth, as though you could somehow keep that secret [chuckles]. . . . I remember them talking about it in respect to agricultural warfare; us against Eastern Bloc countries, things like that. And there was never any explanation of how.

This particular secret goal is a purpose not found in more current chemtrail calls. For callers to the Weather Guys, the primary reason articulated is weather modification, probably due to the weather focus of this platform. In demonstration of the continuing evolution of the narrative, a summer 2020 variant found in a Facebook post links chemtrails to COVID-19 (Reiser 2020). All these differences in the details of the narratives speak to what Michael Vine and Matthew Carey (2017, 52) describe as a "remarkably plastic capacity to channel whichever fear or suspicion lies most readily to hand," a general quality of contemporary legends.

When relayed by a caller to the show, chemtrail narratives tend to be condensed and referential, much like a "dite" (see Library of Congress 2009, n.p.), in which the single word "chemtrail" refers to a deep set of beliefs and assumptions without articulating the actual narrative. Andrea Kitta (2019) notes this type of shift from specific to general in contamination and contagion narratives, a category of legends with many parallels to chemtrail narratives. This condensed articulation is occurring orally, while the deeper and broader webs of evidence to support the theory swell on the internet, a perfect medium for this type of exponential growth (Barkun 2003; Vine and Carey 2017).

Observations

In their examination of chemtrail belief, Jordi Mazon, Marcel Costa, and David Pino (2018) identify six main arguments found in chemtrail narratives, used as proof of their existence by believers. All are based on sensory observations, with five based on visual observations. These include: aircraft did not previously leave trails, which means the phenomenon started only a few years ago; chemtrails cannot be contrails because their occurrence is not consistent; sometimes two aircraft can be seen, only one of which is leaving a trail behind; interrupted chemtrails often appear—they are not a continuous structure; chemtrails form strange shapes in the sky; and toxic and anomalous smells and substances associated with chemtrails are detected at ground level. Lee's certainty about the existence of chemtrails is based on his interpretations of observations he makes, specifically of how jet exhaust clouds—contrails—behave in the sky: crisscrossing, spreading out, and running from horizon to horizon. Repeated observations suggest patterns and greater meaning to some callers. In an April 2020 call to the Weather Guys, Tony from Wausau describes attempts by himself and his wife to understand the larger patterns of contrails and weather observations they make regularly:

Meiller: Tony in Wausau. Your turn. Hi Tony.
Tony: Good morning!
Meiller: Hi. What can we do for you?
Tony: Yeah, my wife studied weather trends for some time, and she also noticed, when she went on walks during the day, there were a lot of chemtrails in the sky. And there's a lot of experts who agree that we're managing the weather by spraying certain things into the atmosphere. She noticed that there were many weekends in a row where

we'd have sun but then would rain on Monday. A good example is today; it's raining but sunny all weekend. And there would be three or four weekends in a row where it would be sunny, and then it would be raining during the week. This cannot be just a coincidence. We're almost convinced they're managing the weather. (Meiller 2020, n.p.)

Starting from the point of visual observation, the conclusion reached by Tony and his wife is that the observed weather patterns cannot be random but must be of external design. In their attempts to understand their personal experience and place it into larger patterns, the couple has turned to various other experts, unnamed in this instance, and is now turning to these two atmospheric scientists for advice in interpreting their experience. Martin takes the lead on the answer, addressing in turn unreliable sources, the fact that there are random natural events, and possible weather modification. He also acknowledges at the end of his answer that his tone is more strident than usual.

> **Meiller:** Well, are we able to manage the weather? Thanks, Tony.
> **Martin:** No. We are not. OK? There are no experts who think that this is the case. There are *so-called* experts who think this is the case.
>   And I'll call your attention to this: I was commenting on this to my class just before we had the COVID break in lecturing. We had four or five straight weekends where the only nice day of the week was Saturday or Sunday. And so, you're right. It isn't coincidence. It's just the natural progression of the way things go. It's lucky that we sometimes get a nice weekend. It's not always the case.
>   But there's no evidence at all and there's no way that it could even be done that we could be controlling the weather while still making short and medium range forecast errors. It's absolutely impossible. They do not sit together in the same universe. So that's my strident antianswer to that.
> **Meiller:** [Laughs.] OK.
> **Ackerman:** And I'll just agree with him because I *do* agree with him. (Meiller 2020, n.p., emphasis original)

These observations by callers attest to the embodiment of weather knowledge and the deep trust many weather observers place in what they see and feel (Dunne 2017; Weston 2017, 105–6). Atmospheric scientists operate from a different perspective, however, interrogating observations using physics, math, and the scientific method. They modify a theory based on additional verified information. Martin expresses his frustration with the observational interpretations aligned with chemtrail beliefs:

They cherry-pick their observations. They're not really scrutinizing the observations and being critical. . . . All the pictures on the chemtrail websites are pictures of interlocking or interspersing contrails. And they say, "We've never seen this before!" It's ludicrous to make that kind of statement because there's nothing to back up that we've never seen this before. . . . So, it's the lack of reflection, the lack of critical evaluation of the observations that might separate a scientific endeavor from one that is a more conspiracy-based endeavor. (Martin 2020, n.p.)

Chemtrails and Geoengineering

Both Lee and Tony identify the reason for chemtrails as weather manipulation. In both responses, Martin firmly counters that controlling the weather is not possible, but Meiller challenges that during the phone call from Lee when he comments, "It would be pretty difficult, pretty difficult to do, but you know. . . ? That's not to say it hasn't been tried." In his hosting role, Meiller raises an argument that is actually part of chemtrail belief: that attempts to modify the weather have been made before so to try again is not out of the realm of possibility. This is also one of the beliefs that make conspiracies believable: past conspiracies have occurred and been proven with enough time and research, which make new conspiracies just as possible (McArthur 1995).

Indeed, as chemtrail narratives have evolved over recent years, geoengineering has become an integrally linked concept, making it more difficult for scientists to easily and clearly refute the narrative. The notion of a chemtrail is one that they can unqualifiedly reject based on atmospheric physics; geoengineering requires a nuanced explanation. Martin discusses past failed attempts, framing them as scams that the scientific method helped to eradicate. In a different call two years later, again from Lee in Green Bay, Ackerman rejects chemtrails but can't do the same with geoengineering.[5]

> **Lee:** Yeah, good morning. I would like Steve to comment on geoengineering or what is known to some people as "chemtrails"—the spraying after some of the jets that takes place. I would just like to know more about this and why there isn't more discussion.
> **Ackerman:** I'm gonna separate that into his two questions: geoengineering, and then the other is chemtrails, which are two different topics. Chemtrails. . . it's not real. There are things called "contrails." And so, it's the jets that fly high. If the water vapor conditions up high are correct then—there's a lot of moisture that comes out from those jet

engines—if it's the right conditions, it will mix with the air around it and form tiny droplets. Very much like on a cold winter day when you breathe out, clouds form.

So, I know that there're conspiracies out there that it's the US government poisoning us, which is totally not true. I was involved in some experiments in the mid- and early nineties where we actually flew into those things to measure them because we were very concerned about the unintentional consequences on climate change: extra moisture-forming clouds and the clouds staying around.[6] So, we actually penetrated those and looked at those to see what they are, and they're water.

And you can actually see that their features are based on the type of gasoline that they are running: high sulfur or low sulfur. So, that was another interesting part of it.

The whole chemtrail thing is really just a conspiracy theory that has been disproven a number of times. They are contrails. They are the things that you see. But it's not the government poisoning us.

And then the geoengineering is a really interesting question because people are starting to look at ways that we might offset global warming that we're observing that's being exacerbated by human activities. And so, there was a whole National Academy study of looking at that, what to do [see National Research Council 2015a; National Research Council 2015b]. Which is great.

I think that it's something to look at to see if there are things that we might be able to do, but I personally tend to hedge on the cautionary side because we're changing the climate—not because we intended to; we unintentionally did that. And nature is very complex.

My analogy is, back in the forties and fifties, and to some degree in the sixties, there were attempts to do rain modification to try to get clouds to precipitate. There were a lot of experiments that were done on that. And we didn't really know the physics. There's no statistical evidence that it actually worked. But again, we were doing things that you don't really know what the unintentional consequences are.

With geoengineering, as the caller indicated, it would be a good thing to talk about. I know that one thing that people were talking about was to put a bunch of particles out in the atmosphere to reflect solar radiation out. That was one of the ones where I was, "I wouldn't do that!" because we don't know what the impact is. There are other things that could be done that maybe aren't as drastic. (Meiller 2017, n.p.)

Geoengineering, or deliberate manipulation of Earth's weather or climate, is an actual field of study and practice that, as Ackerman outlines, has existed for decades, has had mixed results, and raises grave concerns over unintended consequences. The chemtrail narratives that conflate chemtrails with geoengineering, as Lee does here—and which probably deserve the separate label of geoengineering narratives—typically identify chemtrails as the key tool in a secret program of large-scale weather and climate modification (Cairns 2016; Vine and Carey 2017).

Rose Cairns positions some geoengineering narratives as deliberate attempts by believers to establish more scientific legitimacy and create distance from chemtrails since that idea can be dismissed as a conspiracy theory. Geoengineering narratives reposition the original concerns about chemtrails under the more scientifically accepted term of "geoengineering" (Cairns 2016, 77–79). That difference in reaction to the two terms is present in Ackerman's response to Lee, with complete rejection of chemtrails and nuanced discussion of geoengineering.

Conspiracy Beliefs

In replies to callers' chemtrail queries noted above, both Ackerman and Martin use the term "conspiracy," either directly or indirectly, to label the chemtrail narrative, effectively dismissing it as not credible. In this additional example, Ackerman substitutes the term "paranoia" for conspiracy, acknowledging that to do so is less than ideal.

> **Meiller:** And Eric in Cashton, your turn now. Hi Eric.
> **Eric:** Hi there. This is kind of in addition to the caller two calls ago about the contrails. Um, I've seen a lot of stuff on the internet about the difference between contrails and chemtrails. I want to know if you are familiar with chemtrails. It just seems like government experiments at regulating weather and contrails, after the point in passage the contrails disappear after a while and the chemtrails tend to, they use chemicals to spray to create clouds, and if you watch chemtrails after a while, they tend to spread and create bigger clouds. They don't go away. Just wondering if you know anything about chemtrails.
> **Meiller:** What about these chemtrails?
> **Ackerman:** Yeah, chemtrails, they're kind of like—and I'm gonna get in trouble for saying this—but they're paranoia. There *aren't* such things as chemtrails. What he was describing in regard to the contrails spreading out, that actually is what we do observe.

Sometimes contrails don't, if you're following the jet. Sometimes contrails—you won't even see them. That means that the atmospheric conditions weren't right to form. Sometimes you'll see them, but they'll be really, really small, so they just dry up and evaporate. Other times—and usually if you're just on the border of a cirrus cloud system that's coming into your area, but you don't see them—you know, naturally they haven't generated a cloud—these contrails will stay around for a long time and then actually spread out. So, that spreading that we see is just actually normal contrails. (Meiller 2007c, n.p., emphasis original)

Lee from Green Bay introduced the topic of chemtrails by positioning it as a restricted, secret topic, later implying that the secrecy is empowered by unnamed forces powerful enough to muzzle the media. Secrecy is a core concept associated with chemtrails. Fully articulated discussions of chemtrails reflect a conspiracist worldview, one in which events do not happen randomly but by design (Barkun 2003, 3). Recall Tony from Wausau's comment: "This cannot just be a coincidence. We're almost convinced they're managing the weather."

Lee's concern about the media not being willing or able to discuss the topic reflects the power struggle over information's meaning and circulation, which Mark Fenster (2008, 116) identifies as being the heart of the struggle in conspiracy belief. Fenster posits that to prove the conspiracy, evidence is observed, collected, and interpreted but always in support of the preexisting belief. This was Martin's conclusion as well upon reflection of the calls the Weather Guys have received about chemtrails:

The conspiracy-theory-science view is very Aristotelian. I remember in a class I took in college, Aristotle's view of science was—you found evidence to support your theories. You didn't just collect the evidence and then sift through them to determine what were the underlying structural realities. . . . It does strike you when you listen to these people that they've got a grain of some understanding of science and that grain might be really the Aristotle model, but that is no longer the modern view. The Baconian view is different. It's the one we follow now.[7] (Martin 2020, n.p.)

Conspiracy theories look for evidence to support preexisting beliefs. They do not seek to solve a problem but to expose clandestine power and evil actions. In contrast, science accumulates evidence in order to draw conclusions. Scientists are compelled to abandon or alter an originating hypothesis should the evidence indicate so. They are compelled to examine all available evidence in their quest, a principal Martin refers to in his initial response to Lee: "I'm just trying to be as logical about what I see as I can be."

Scholars from multiple disciplines understand conspiracy narratives as expressing anxieties and a lack of control over matters greater than the narrator. Folklorists shy away from the term "conspiracy theory" probably because of the negative judgment embedded in the term. Instead, they approach these narratives as expressions of beliefs. What Kitta (2018, 409) describes for legends and fake news holds true for conspiracy narratives: "They express already established beliefs, and people share them *because* these stories already fit into their belief system either because they actually believe them or because they seem believable, demonstrating a form of confirmation bias." But that key concept, that the narrative is believable, is a major divergence for atmospheric scientists. For them, the idea of chemtrails and their associated idea of a secret conspiracy of weather manipulation is absolutely not believable.

Scientists' Response

Multiple efforts have been made by scientists to disprove the idea of chemtrails. In identifying the top six arguments in support of chemtrails, Mazon, Costa, and Pino (2018) refute each one from a physical perspective. Agencies implicated in the chemtrail theories have published fact sheets to help allay misinterpretations (Environmental Protection Agency 2000; United States Air Force 2014). While at NASA Langley, Minnis (2020) put in many personal hours arguing with chemtrail believers by becoming a troll on the chat site Chemtrail Central.

As for Ackerman and Martin, their response to chemtrail queries and narratives might vary in tone—from gentle to strident—but never in content. Their answers are consistently unequivocally against chemtrails. As Ackerman (2020b, n.p.) explains:

> Sometimes it's not about the caller; it's about the people listening. You're not necessarily trying to convince that person but to let other people listening in know that it's not a real thing. Same with climate-change skeptics.

Conclusion

Ackerman and Martin are among a growing cadre of scientists committed to communicating science to the public. That they work at an institution with a founding mission of public outreach supports this commitment; the medium of Wisconsin Public Radio, which shares that mission, is a valued platform for their efforts. Their pairing with the effective and trusted *Larry Meiller Show* enhances their outreach. The audio space created by Meiller, Ackerman, and Martin invites personal-experience narratives. Many are shared on a range of topics, although few are fully articulated performances. Even in their most abbreviated forms though, the stories provide insight into how people in Wisconsin physically interact with their environment.

The one narrative conveyed on the show that challenges the welcoming openness for which the host and scientists strive is chemtrails. Chemtrail narratives prove challenging because both scientists consider chemtrails a conspiracy theory and want to firmly squash it. They know that labeling it as such is problematic to the openness they aim to cultivate in the program, but they also know it is scientifically false; they value that truth more. Like other contemporary legends, chemtrail narratives are malleable holders for social anxieties; on the Weather Guys show, they express concerns connected to weather and climate. The shift to geoengineering narratives aligns those concerns more closely with climate-change anxieties. The chemtrail callers might be calling with genuine curiosity about what they're seeing in the sky and reading online, asking for expert help in making sense of the patterns they think they are observing. Or they might be using the radio show as a platform for inserting into this medium a forbidden topic that they entirely endorse and want to expose.

Atmospheric scientists Ackerman and Martin understand their oppositional role in relation to chemtrail narratives and practice that by debunking them when asked. For folklorists, this may seem harsh. As Tom Mould (2018, 376) explains:

> While the role of "debunker" remains a stigmatized one among folklorists—we are wary of criticisms about devaluing the importance of why people tell the stories they do and of dismissing the importance of very real anxieties, fears, and hopes—the era of "fake news" has forced many of us to reconsider our competing ethical obligations.

The Weather Guys provide a possible model for folklorists trying to forge a revised path related to legends that can be scientifically disproven and may cause social harm, that of respecting the narrator while actively disagreeing with the factual basis of the narrative.

Notes

1. For detailed information on Ackerman's work, see his online research description in Ackerman 2019. Ackerman is also the author's husband and a collaborator with her on the public folklore projects Wisconsin Weather Stories and Here at Home, cultural tours

provided by the educational network Wisconsin Teachers of Local Culture. For detailed information on Martin's work, see his online research description in Martin n.d.

2. *The Why Files: The Science Behind the News* (2017, n.p.) was "the world's first general science magazine written exclusively for the web," produced by a communications team at UW–Madison for twenty years, from 1997–2017 (Ackerman and Martin 2010). *The Why Files*'s artist, S. V. Medaris, created a compelling banner image for the Weather Guys, which Ackerman and Martin continue to use (Ackerman and Martin 2010).

3. Madison School Community Recreation.

4. Ackerman was absent from the show that day.

5. Martin was absent from the show that day.

6. Ackerman is referring to the field campaign Subsonic Aircraft: Contrail and Cloud Effects Special Study (SUCCESS), a multiaircraft effort led by NASA, conducted in April and May 1996 over the central and western United States. One goal was to "investigate the effects of subsonic aircraft on contrails, cirrus clouds and atmospheric chemistry" (Justice 2017, n.p.).

7. That chemtrail believers have "a grain of some understanding of science" is a conclusion also reached by Cairns (2016, 90–91), as well as Minnis. Like Martin, Minnis (2020, n.p.) concludes, "Their science is flawed. . . . I guess these are wannabe scientists who never got an education."

References

Ackerman, Steven A. 2019. "Steven A. Ackerman." University of Wisconsin–Madison. December 4, 2019. http://cimss.ssec.wisc.edu/wxwise/ack.html.

Ackerman, Steven A. 2020a. Interview with author. July 5, 2020.

Ackerman, Steven A. 2020b. Interview with author. October 24, 2020.

Ackerman, Steven A., and Jonathan E. Martin. 2010. "Why Are Clouds White?" *The Why Files: The Science Behind the News*, June 7, 2010. https://whyfiles.org/2010/why-are-clouds-white/index.html

Ackerman, Steven A., and Jonathan E. Martin. 2013. "The Weather Guys: Discussing Science via Regular Radio and Newspaper Appearances." Presentation at American Geophysical Union meeting, San Francisco, CA, December 9–12, 2013.

Barkun, Michael. 2003. *A Culture of Conspiracy: Apocalyptic Visions in Contemporary America*. Berkeley: University of California Press.

Cairns, Rose. 2016. "Climates of Suspicion: 'Chemtrail' Conspiracy Narratives and the International Politics of Geoengineering." *The Geographical Journal* 182, no. 1: 70–84.

Drury, Gwen. 2017. *The Wisconsin Idea: The Vision That Made Wisconsin Famous*. Madison and Belleville: University of Wisconsin Pressbooks and In Time Media.

Dunne, Carey. 2017. "My Month with Chemtrails Conspiracy Theorists." *The Guardian*, May 22, 2017. https://www.theguardian.com/environment/2017/may/22/california-conspiracy-theorist-farmers-chemtrails.

Ellis, Bill. 2018. "'Fake News' in the Contemporary Legend Dynamic." *Journal of American Folklore* 131, no. 522: 398–404.

Environmental Protection Agency. 2000. "Aircraft Contrails Factsheet." EPA430-F-00-005, September 2000. https://nepis.epa.gov/Exe/ZyPDF.cgi/00000LVU.PDF?Dockey=00000LVU.PDF.

Fenster, Mark. 2008. *Conspiracy Theories: Secrecy and Power in American Culture.* Minneapolis: University of Minnesota Press.

Justice, Erin, ed. 2017. "SUCCESS ESPO." NASA. Last modified November 29, 2017. https://espo.nasa.gov/success/content/SUCCESS.

Kitta, Andrea. 2018. "Alternative Health Websites and Fake News: Taking a Stab at Definition, Genre, and Belief." *Journal of American Folklore* 131, no. 522: 405–12.

Kitta, Andrea. 2019. *The Kiss of Death: Contagion, Contamination, and Folklore.* Logan: Utah State University Press.

Library of Congress. 2009. "Dites." January 5, 2009. https://id.loc.gov/vocabulary/ethnographicTerms/afset005327.html.

Martin, Jonathan E. n.d. "Jonathan E. Martin." Accessed August 29, 2022. http://marrella.aos.wisc.edu/Martin.html.

Martin, Jonathan E. 2020. Interview with author. June 27, 2020.

Mazon, Jordi, Marcel Costa, and David Pino. 2018. "Aircraft Clouds: From Chemtrail Pseudoscience to the Science of Contrails." *Mètode Science Studies Journal* 8: 181–87.

McArthur, Benjamin. 1995. "'They're Out to Get Us': Another Look at Our Paranoid Tradition." *History Teacher* 29, no. 1: 37–50.

McCarthy, Charles. 1912. *The Wisconsin Idea.* New York: Macmillan.

McGrath, Jane, and John Perritano. 2011. "What Are Chemtrails and Should You Be Scared of Them?" March 7, 2011. https://science.howstuffworks.com/transport/flight/modern/what-are-chemtrails.htm.

Meiller, Larry. 2007a. "The Weather Guys." *The Larry Meiller Show.* Wisconsin Public Radio. Madison: WPR, April 30, 2007.

Meiller, Larry. 2007b. "The Weather Guys." *The Larry Meiller Show.* Wisconsin Public Radio. Madison: WPR, June 25, 2007.

Meiller, Larry. 2007c. "The Weather Guys." *The Larry Meiller Show.* Wisconsin Public Radio. Madison: WPR, July 30, 2007.

Meiller, Larry. 2007d. "The Weather Guys." *The Larry Meiller Show.* Wisconsin Public Radio. Madison: WPR, August 27, 2007.

Meiller, Larry. 2007e. "The Weather Guys." *The Larry Meiller Show.* Wisconsin Public Radio. Madison: WPR, November 26, 2007.

Meiller, Larry. 2008. "The Weather Guys." *The Larry Meiller Show.* Wisconsin Public Radio. Madison: WPR, November 24, 2008.

Meiller, Larry. 2015. "The Weather Guys." *The Larry Meiller Show.* Wisconsin Public Radio. Madison: WPR, September 28, 2015.

Meiller, Larry. 2017. "The Weather Guys." *The Larry Meiller Show.* Wisconsin Public Radio. Madison: WPR, July 31, 2017.

Meiller, Larry. 2020. "The Weather Guys." *The Larry Meiller Show.* Wisconsin Public Radio. Madison: WPR, April 27, 2020.

Minnis, Patrick. 2020. Interview with author. July 7, 2020.

Mould, Tom. 2018. "Introduction to the Special Issue on Fake News: Definitions and Approaches." *Journal of American Folklore* 131, no. 522: 371–76.

National Academies of Sciences, Engineering, and Medicine. 2017. *Communicating Science Effectively: A Research Agenda.* Washington, DC: National Academies Press.

National Research Council. 2015a. *Climate Intervention: Carbon Dioxide Removal and Reliable Sequestration*. Washington, DC: National Academies Press.

National Research Council. 2015b. *Climate Intervention: Reflecting Sunlight to Cool Earth*. Washington, DC: National Academies Press.

Reiser, Albert. 2020. Text message to author. June 20, 2020.

Stark, Jack. 1995. "The Wisconsin Idea: The University's Service to the State." In *State of Wisconsin Blue Book: 1995/1996*, compiled by Wisconsin Legislative Reference Bureau, 100–194. Madison: Wisconsin Legislative Reference Bureau.

United States Air Force. 2014. "Contrails Facts." July 2014. https://www.epa.gov/sites/default/files/2016-10/documents/afd-051013-001.pdf.

University of Wisconsin. n.d. "The Wisconsin Idea." Accessed July 5, 2020. https://www.wisc.edu/wisconsin-idea.

Vine, Michael, and Matthew Carey. 2017. "Mimesis and Conspiracy: Bureaucracy, New Media and the Infrastructural Forms of Doubt." *Cambridge Journal of Anthropology* 35, no. 2: 47–64.

Weston, Kath. 2017. *Animate Planet: Making Visceral Sense of Living in a High-Tech Ecologically Damaged World*. Durham: Duke University Press.

The Why Files: The Science Behind the News. Last modified March 10, 2017. https://whyfiles.org/index.html.

Wisconsin Broadcasters Association Foundation. "Larry R. Meiller." June 16, 2017. https://www.wisconsinbroadcastingmuseum.org/hall-of-fame/larry-r-meiller.

Yuan, Shupei, Tsuyoshi Oshita, Niveen AbiGhannam, Anthony Dudo, John C. Besley, and Hyeseung E. Koh. 2017. "Two-Way Communication between Scientists and the Public: A View from Science Communication Trainers in North America." *International Journal of Science Education, Part B* 7, no. 4: 341–55.

Chapter 5

FROM CLOCKWORK WEATHERMAN TO ATOMIC ENVIRONMENTALIST

Máirt Hanley

This article is an attempt to look at the folk beliefs about weather that I encounter in my working life as a religious minister and how those beliefs connect with and compare with the religious beliefs of my congregation. I also explore how that belief is perhaps evolving and how its evolution connects with the church's theology and growing understanding of the weather, the environment, and climate responsibility. In doing so, I will outline my own lived experience and give some extracts from interviews with a bishop, a church environmentalist, and a successful farmer from my congregation.

I am a Church of Ireland priest (Anglican/Episcopalian) working in a rural parish where farming is prosperous. The church I am a member of and in which I work is a minority church. It is quite broad with both liberal and conservative elements. We are more "low church"—more to the Protestant end of the spectrum in the Anglican world—than our American cousins and, in an American context, would be more like the United Methodist Church than the Episcopal Church of the United States of America, our Anglican sister church.

Our climate is greatly influenced by the Atlantic, thus we don't have extremes of weather; however, day-to-day or even hour-to-hour weather can be quite variable. It is therefore a "small change" to everyday conversation, a safe subject on which everyone will have an opinion and a concern facing every social event. As such, the weather has been woven into the fabric of my working life. I have served in small communities whose livelihoods depend on the weather, and as I am from a fishing background, I have a particular understanding of this.

With small communities also come community events that, with limited resources and high importance, can be hugely affected by weather. For example, if a small parish has a fundraising barbeque, this might well be put together by five or six people giving time and resources with the hope of a hundred attendees, only to be canceled or "washed out" as you discover that morning that a weather front that was supposed to track north didn't and instead deposited a week's worth of rain on your event. In any small community group, parish or sporting or social, such events can constitute 10 percent or more of total year income. Thus, the stakes are high.

As a result, the weather is a very real and immanent part of my working life and community involvement. It is only natural, then, that parishioners of mine are in the habit of asking me to pray for the weather either because of their agricultural pursuits or because of some community event coming up that day or that week. Such prayers are often tagged on toward the end of the intercessions in my Sunday services, often adjoining a prayer for a local sports team. These prayers for sports teams and clear days for community events often take the form of a jovial postscript to God, a bit like the items one sometimes gets at the end of news broadcasts following the expression "on a lighter note...." These prayers for weather are part of the fabric of my liturgical round and parochial visits, though probably, it is fair to say, a very untheologically parsed part. My current bishop noted while visiting my parish that I am quite theologically thorough and reflective in the prayers I use for intercession; that is, the prayers addressed to God in the service usually ask for God's blessing in some way. It was quite a shock when the bishop, in a half-joking but quite serious way, told me that he did not approve of my praying for the weather.

Something I'd always done in a very unconscious way then came into sharp focus. It struck me that my prayers for rain or for the absence of it are a reaction not to my faith, the theology of my church, or holy scripture but more to the folk beliefs of the people in the pews. Yet, prayers for weather have also been part of a larger theological conversation within the church; the bishop was part of the committee who had removed them from more recent editions of the prayer book. What follows here is a little of my personal examination of my own belief, an explanation of the theological standpoint of my particular denomination, a brief look at some other churches' takes on the weather, and a final noting of what seems to be an evolving understanding of weather, climate, and the environment on behalf of those who sit in the pews of my rural Irish Protestant churches. The chapter is divided into two halves: the first details the theological context and background of the church's stance; the second is comprised primarily of two interviews and correspondence.

Axis Mundi

One of the books that I read as a theology student was Mircea Eliade's *The Sacred and the Profane* (1987), which seems to get underneath the theological process. It doesn't tell you about any particular ideology. What it does give you is a handle to understand any theology, and so I have found it invaluable in picking up the theological blind spots in my own faith journey. In it, Eliade examines the idea of the axis mundi, the point around which the world spins. Theologically, in Christianity, that point is not only a point in space but is, far more importantly, a point in time—the axis mundi being the crucified Christ on the Hill of Golgotha. If one wants to be a little more upbeat, one could say that the axis mundi is the death and resurrection of Christ. Indeed, as a pastor in a church that is both reformed and Catholic there is the sense that, every Easter, we remember that this is the point around which the universe spins.

However, each Christian community, while orbiting this point, also has its own spin. While the axis mundi for the universe might be the death and resurrection of Christ, the local axis mundi for most of the people, most of the time, is their church in which they worship. My experience of the congregations that I have served is that one of the reasons they want certain things to be prayed for is to bring those things to the axis mundi. In order for things to maintain their proper orbit, they must be reconnected to the central point, at least by being mentioned in the regular church services in the church building, the connection to which defines that congregation as a community. Put more simply, if it is part of who the congregation is, it must be part of what I say on a Sunday, even if it is only said in passing.

Part of what I do on a Sunday morning is to create a tableau of life, albeit an idealized one, containing all the important parts of that life. This is done so that important elements, or should I say elements that are important to my parishioners, may be represented at the act of worship on Sunday morning. Especially for the church-going Protestant Irish farmer, their week revolves around the one fixed point of church, the axis mundi, in temporal terms, of his or her working week. Just as the church is ideologically the axis mundi of their life, it represents the ultimate reality of God and is the intercessory and revelatory portal through which all of life is mediated. Thus, everything of value and everything which is an intrinsic part of my congregation's sense of being must be included in some way in the service. So, in the service, I bring before God everything from sports teams to politicians, from bulk milk prices to the activities of the fair-day committee.

The weather and its changeability are elements that seem to impact all these other elements, and so, the weather is very much woven into the fabric of Irish life—a key concern that touches on all others. How else am I to mention it unless to pray for specifics? Often, even when I am asked to pray for some other thing, they will tag on a climate request. In his memoir *The Rector Who Wouldn't Pray for Rain*, former Church of Ireland Rector Patrick Semple (2007,

140) muses on how people in town do not really understand the importance of weather to the farming community:

> If it is raining when they want fine weather . . . you'll hear all about it. When the weather is perfect you never hear a word . . . that situation never lasts for long. . . . I used to ask them did they really believe in a God up above the bright blue sky that would intervene in the weather if we prayed hard or often enough. The often indignant answer to this was almost invariably "Yes."

The ability for such a worldview to survive into the current age seems very much in line with the "watchmaker" view of creation. The famous argument from the Venerable William Paley[1] purports that if one found a watch, a complex and beautiful piece of machinery, dropped in a field, one would suppose that it had been created by somebody and dropped in its current location rather than suggest that the different elements randomly arrived at that location and put themselves together. This Victorian expression of the argument for the existence of God—that there must be a god because something is wonderful as the universe couldn't happen by chance—very much reflects the worldview of a society enthralled with mechanical devices. It also reflects a very dualistic worldview in which there are machines and there are operators. Especially in farming communities, this concept resonates where the communities are not only dependent on the machinery that they use but also so competent at the maintenance of the machinery they use. This is very much the case with older tractors and all the farmers in the place where I have my ministry: the Earth is seen as the ultimate piece of machinery created by God and worked by human hands. Like their tractors, farmers understand it and how it works and how to get the most out of it. However, the one element that is outside of that machinery and mechanism is the weather. This must therefore be a cosmic matter and only subject to the deity's will.

The argument that Semple uses in his book against such attitudes is the same one that former Archbishop of Armagh Richard Clarke put in his book *And Is It True?* (2000): namely, does a God who could intervene in the processes of nature and free will ever choose to do so in response to prayer? The example Bishop Clarke uses in his book is to consider the case of someone who always prays for a parking space while driving into town and claims that such prayers are always answered in comparison to the prayers of those being transported to Auschwitz: "For this could mean only one of two things. It might possibly mean that one has to have the line of communication to God exactly right, and that this involves membership of a particular religious institution, or the use of a particular choice of words" (Clarke 2000, 78–79).

While he was a speaker at a clergy conference that I was attending in 2006, in discussing the same question, he used the example of someone in the suburbs praying for a nice day for a garden party in comparison to those in the Horn of Africa praying for rain while there was an acute climate disaster unfolding in the area. The expectation of getting a good day for the garden party while being aware of a climate emergency in another part of the world relies on the thinking above, that God will act to bend the laws of nature in your favor if either you use the right form of words (which is surely magical thinking) or you belong to the right group (which is sectarian at best, racist at worst). But such thinking is totally debunked when one considers that even if we have the perfect words or are in the right club, we are also praying for the climate emergency, so why would God sort one out and not the other? The theological inconsistencies in such a viewpoint are evidently clear.

Bishop Clarke's solution to this theological conundrum is to follow the classic Roman Catholic theological line:

> If there is a purpose in prayer, even intercessory prayer, it can surely be only to bring the mind and the Heart of the supplicant into conformity with the will of God. There is good reason to be suspicious of the rigid distinctions between thanksgiving, intercession, and adoration. The purpose and motivation of all prayer should be the same. (Clarke 2000, 79)

The first step to aligning our will to the will of God must be to express our will. At the end of a conversation, you might hope to find agreement, by way of analogy, but before you can find agreement, you must first start the conversation. If we take the point that all prayer has the same aim and that distinctions between forms of prayer are suspicious, then surely, we can say the purpose of prayer is to make a connection with God: to affirm, express, and develop our relationship with God. It is to acknowledge the source of our, of all, being and to place ourselves in relation to that source, orientating ourselves in relation to the divine and making a "this-way-up" assertion for ourselves and for our world. As we seek to do this, we may seek to orientate not just ourselves in relation to God but also to orientate the different aspects of our lives, various activities and concerns, to place everything to which we ascribe value in relation to that which is the origin of all. In terms of asking for specifics, maybe the prayers don't need to be successful, but people feel that they need to try.

Clarke's example and the importance of prayers incorporating the values of the parishioners is reflected in an interview I did with a Roman Catholic colleague who worked in rural Africa in subsistence farming communities, where praying for the weather took on a far more vital role than whether or not a parish barbecue could go ahead. He stated that he would always pray for weather because it is so much part of the lives of his parishioners. He himself does not believe that praying for rain will deliver it, but he told me that he prays for rain on behalf of his parishioners for a number of reasons. First, he prays for the weather so that they know of his pastoral concern for them. Second, he prays for the weather so that they will feel that their yearnings have been poured out to God. Ultimately, though, his answer was that he would always pray for weather because it is so much a part of the lives of the people he was ministering to.

Praying for Rain

It occurs to me that essentially, in formulating prayers in Roman Catholic theology, the starting point doesn't so much matter as the direction of travel. The cleric with whom I was discussing this issue did not have a problem with a theologically inaccurate prayer so long as it would help lead the people toward a better belief. As a Protestant, I might say that he had no problem in praying 'wrong prayers' in order to move parishioners closer to a better theology. I'm sure that, as a Catholic, he would say there is no such thing as a wrong prayer—there are only more or less effective prayers, while in this sense the efficacy is only about whether it moves one toward the perfect faith as expressed by the mother church. This is perhaps reflective of the Roman Catholic standpoint that orthodoxy is kept safe in Rome rather than being a product of the church at large.

In Protestant churches this is not so. In many Protestant churches, orthodoxy is preserved in confessions, such as the Augsburg confession for Lutherans or the Westminster confession for Presbyterians. In my church, orthodoxy and in fact our theology are expressed through the prayer book. It is vital for us that our prayers printed in the prayer book can be theologically stood over. Anything that is said in prayers in the prayer book is something that the church is in fact saying to the world, to itself, and to its members. Our prayer book was last revised in the period through the 1990s with final publication coming in 2003. The revision is done by a committee of the general synod, which is the governing body of our church, known as the Liturgical Advisory Committee or LAC. The LAC, after it has recommended changes, brings it before the general synod that must then approve those changes. Michael Burrows, the person who was chair of the LAC during the time in which the prayers asking for particular weather were taken out of the prayer book, was a priest in the diocese of Cork and is now the bishop of the diocese of Cashel Ferns and Ossory, the diocese in which I now serve. So, the bishop who pulled me up for my weather-based prayers is also the person who might be thought of as being responsible for the process that got rid of such prayer from our liturgies. With this in mind, I asked whether he would mind being interviewed for this article. The following is a transcript of the interview (Burrows 2019) I did with him on the evening of October 8, 2019:

> **Hanley:** So, you were involved in the LAC when the prayers for rain and the ceasing of rain were removed from the prayer book?
> **Burrows:** That is true, yes.
> **Hanley:** And what was the thinking behind that?
> **Burrows:** Well, the older prayer books, particularly the Irish Revision of 1878, did include prayers for rain and for, I think what was called, temperate weather. They always struck me—although I'm sure they were used to the latest faithfulness and, as it were, integrity, they always struck me as being somewhat amusing, that one was really beseeching God to sort of turn the taps on, and the other was beseeching God, when the taps had been on for too long, to turn them off again. And they did this in the context of mellifluous phrases, like asking for temperate rain and showers. They were in the occasional prayer section, and, when it came to the revision of the contents of that section, obviously, we recognized that intercessions now tend to be much more fluid and flexible and locally based and often in the intercessor's own words. But in so far there are still a few fixed/occasional prayers, we decided to delete the prayers for weather, as I think they really weren't faithful to the worldview that we felt that Christians might or even should have today.
>
> For example, I've always been reticent myself in this area, although I was a country priest for a long time, and I know that the strong feelings of those who work on the land in relation to these matters, and I know too that you have a very strong spirituality, which I wouldn't wish to diminish for a moment. But having said that, first of all—and this hit me during the opening of the Ploughing[2] recently when Annie May McHugh was sort of thanking the clergy for praying for good weather on the lovely week of weather, which the Ploughing was, as if we had done it! But it struck me, first of all, that I don't think the relationship between God and the world in terms of what we know about physics and cosmology and the checks and balances of the laws of creation that order things. I don't think they really operate in the context that God makes casual interventions in order to cause dry weather in one corner of Carlow for the Ploughing! And, equally if God were to do that, I would find it very hard to put my faith in so mercurial or capricious a God who because of the, as it were, the entreaties of a few people, again, in one corner of County Carlow, would cause them to have lovely weather for the Ploughing while

tolerating dramatic weather events, flooding and suffering in other parts, particularly of the developing world. So, it would seem to me, very unjust for a loving God to intervene so capriciously in these matters.

So, I was never happy praying either that God would turn the taps on or turn the taps off. But, more recently, I have discovered a way where we might pray about the weather if not so explicitly for the weather which I might outline to you?

Hanley: Yes, please do.

Burrows: Well, it always seems to me that when you're saying your prayers, you're not actually twisting God's arm to do what God wouldn't otherwise do. That would be a strange view of prayer. Obviously, great strength derives from prayer, and the people who are prayed for know that. But a large part of prayer is aligning our wills with the will of God. So that. . .

Hanley: Is that not exactly what the Roman Catholic Church says?

Burrows: Let me continue my thoughts. . . Aligning our wills with the will of God so that we don't just throw things at God for his intervention and action and think we've done our duty. As well as placing them into the loving arms of God, we also commit ourselves in so far as lies in us to be the part of the answer to our own prayers, so we are looking for the strength and the grace of God that he'll help us do and strengthen us to do what we are praying will come about. And so, when you say "amen" at the end of a prayer "so be it," you're really saying, "yes please, this is what I want," and as I cast the matter into the care of God, I'm also committing myself to working with God to make it happen. So, there's no point in asking God, for example, to feed the hungry children of the third world unless we're prepared to say "amen" to our prayers and by our actions and by our generosity and by our practical measures to take some prayerful and active part in strengthening the poor.

So, to come back to where I was in relation to the weather, I think we now know, as never before, that dramatic weather events, particularly in developing countries, are caused by our actions, by our decisions, by our omissions, and by our folly as we warm up the world and the ice caps melt, the sea levels rise, and that disproportionately affects the poor who live in shanty towns near the coast—we know all that! And so, we've suddenly realized, yes, it's actually right to pray about the weather in those contexts and for the deliverance of people from those kind of dramatic weather events, and it's right to pray about it precisely because we need God's help and God's strength to equip us and motivate us to become the answers to our own prayers and, when we say our "amen," actually doing something about it. Whether through our willingness to stop being negative about carbon tax or our willingness to reduce our own emissions or to do all the things we know will make some small contribution to the deceleration of global warming. So, we become the answers to our own prayers in relation to the weather, and we recognize that actions, including prayerful actions, in one part of the world affect what is actually experienced in another.

Hanley: In a way, is that not reflected in the story of Noah, in its interpretation has been more or less turned upon its head, in the Victorian age was a tale of how God is in control of everything and now has become a morality tale of our duty to look after the global climate?

Burrows: Well, there would be nothing wrong with that particular interpretation, would there? It's more useful than, as it were, literally believing that two of every animal went into the ark and stayed there for forty days, delightful as that story is, inspiring as it is for children, you've got to move them on, and it always seemed to me that I couldn't understand when I was a child two things about that: first of all, if God had created everything in the first place, why he couldn't have just started again? Having wiped the whole lot out, it seemed to be very troublesome to have to put two of all of them in an ark! But they were all going to have an incestuous relationship anyway, that was going to cause problems for their future descendants! And also, why did God bother saving one or two of all the nasty things, like the snakes? Though, of course, we now know that they are necessary for the eco-balance so, you know, you could go on and on about it . . . . You know, there is another lovely example that I quoted recently that I came across in a sermon, which being a lover of Celtic spirituality you will know about, Life of Columba—there's a lovely passage where Columba's in the middle of a terrible storm off the western isles of Scotland, and all the monks in the boat with him, who think rather like the disciples with Jesus that they are all going to perish, cry out to him to pray, and he refuses to do so, but he says, "the person who will pray for us is my old friend Abbot Calice over in Aghaboe." So, by some kind of telepathy, Calice realizes that he was meant to do this, and so, he rushes from the refectory, leaving one of his shoes behind, and he prays madly for Columba, and so, the storm ceases. Now, to me, that's a rather, as it were, eccentric, charming, and having an eco-graphical early realization that behavior and prayer for behavior in one part of the world influences weather in another.

Hanley: The prayers for rain or the ceasing of rain seem to be very much what Nadia Bolz Weber would call "the gumball Jesus version" of you know, put out your prayer and getting your request at the other end. I suppose I would have thought of prayer as being a seeking out of the light of God's grace in any situation and that striving in prayer should be a striving to see that. It seems to me that in rural communities that understanding of prayer, as in praying for the light of Christ to shine from where it is, seems very easy in land-based prayers and in things which are down here. But weather seems to come from the heavens and therefore, still remains in a kind of old-fashioned view of God.

Burrows: The waters above the dome . . . God can close the trap doors of the dome if we're getting too much of a shower coming down. . . .

Hanley: I just wonder whether there is something to do with. . . People are very happy to have a modern theological understanding of things they can understand but still use maybe an old-fashioned God of the gaps or superstitious God when it's something that they don't.

Burrows: I hate the God-of-the-gaps concept because if you want to, as it were, destroy people's sense ultimately of the presence of the power of God, you know, resort to the God of gaps. There's a hymn that really annoys me—and children sing it a lot at harvest time in our schools and so on, I'm sure with the best of intentions. It's called "Who Put the Colors in the Rainbow." Again, it's a sort of authentic. . . rather primitive. . . primitive in the true sense, kind of acknowledgement that God is at the heart of creation, and somehow cosmology echoes the glory of the creator. But having sort of sung its way through various wonderful lyrics about who put the colors in the rainbow, the salt into the sea, the stripes on the tiger, and all the rest of it, it kind of comes to the great conclusion, it surely can't be chance! But we know there are lots of things that in the past people said, "surely they can't be chance," and then, the more they learned about the genetics, they knew there was a perfectly good scientific reason, and, as it were, they displaced God. And so, I'm never keen to get children to sing things where, as it were, they are playing into the hands of the possibility that God will be readily displaced by their greater scientific knowledge.

Hanley: Which brings me to the, maybe one of the last things I like to ask you in that, you said before, as have many people, "if you need to know the theology of an Anglican, read their prayer book." If you want to know the folklore of Anglican congregations, should you read their hymnal?

Burrows: Well, I don't know if you'll find many prayers about the weather in the hymnal, but I agree with your essential sentiment, and of course, by not having prayers about the weather in the official liturgy, it doesn't mean you are forbidden to do it. It means that bishops like me might think it was a bad idea, but I'm not going to forbid anyone to do it on the grounds that sometimes things that make no sense may still be helpful. And that's a paradoxical reality in many aspects of religion. It's in the nature of religion that we sometimes do senseless things on the face of it and yet find deep and real comfort in it. So, if praying about things that I'm not inclined to pray about truly helps some people, so be it. But I hope that the people who like praying for the weather, good Protestants that they are, will be equally happy that I pray about some things that are optional on the Anglican spectrum but that I find helpful.

...

Hanley: Is it then that particularly in our understanding of the weather, in a rural sort of folklore context, that it is one of the few things that a modern farmer feels is outside their control in a world where everything is very controlled, by regulations from the EU or by modern farming practices, which are scientifically pinned down, so therefore this one element which is outside of their control, is something that they need to take to their place of religiosity?

Burrows: I accept that, there are in all our lives, things that, however much science we know, remain mysterious, and sometimes mysterious things—again using "mystery" in a good sense—are things that we want to associate with or place in the heart of God. Perhaps no harm if we are praying about these, to pray that these things actually will remain acts of God. Because I would much prefer the weather sort of remained what the insurers would call "an act of God" than, for instance, Mr. Trump decid[ing] that he could drop nuclear bombs to kind of stop dramatic weather events, as he told us he would soon be able to do the other day! So, you know, best to pray that some things remain in God's hands, but having said that, I wouldn't want... I'm not mixing my metaphors now, I'm actually being consistent, I'm happy to pray that some things remain in God's hands, but I'm not going to myself pray in a way that suggests that God will act capriciously simply to please me by giving good weather for the Ploughing.

Hanley: There was one other thing that came into my mind that I was just going to tag on. It was about control and the idea of mystery that, should we in fact be encouraging people, as you have tagged the sense of mystery with things that are beyond our control? Coming from a farming background, not a farming background but a fishing background, there's much more sense that you can't control everything. Maybe, is praying for weather in churches where there are a lot of farmers in some ways a good thing because it reminds them that there are things beyond their control, and that's ok?

Burrows: Find appropriate words to do it. When I was in West Cork, I had a very small number of people who fished, including one family for a living, and as you say, when you compare fisher persons to farmers—and we pray often in one breath about the harvest of the sea and the harvest of the land, but actually, despite all the challenges of the weather and working on the land, I think those who work at sea fishing have probably a much more developed sense of, as it were, the rawness of the elements than any of the rest of us ever have. They don't necessarily pray about it in ways that we might be arguing about this. They simply acknowledge it. The human predicament in the midst of raging torrents is a very, as it were, graphic description of, in a way, the best of metaphors for the human predicament. The waters have come up to my neck, I often feel like.

Hanley: I just think about the time my cousin went down, well his boat went down, and he died. There was diver sent down to the boat who died. One of the people who was saved off that boat was saved quite seemingly miraculously, and when the diver went down to investigate the boat and was lost, I think they were looking for one body. It wasn't my cousin's boat, but there were two boats that went down at the same time, and the comment among lots of people was "the sea wanted one more." *And there was a fatalistic, in a good way there is an acceptance of the sea*, but there's also a fatalistic streak that goes with that.

Burrows: Yeah, I remember again in West Cork, there was an old man that I used to know when I was working down there who had worked his whole life on the Cape Clear ferry, and his son had died. He was a very young fisherman at the age of eighteen or nineteen, he couldn't swim—that was, of course, was considered bad luck, to be able to swim. He's commemorated on a memorial to those from the island who died at sea, in the north harbor. But he died at the age of nineteen while out at sea jumping from one fishing boat to another, but the boats moved, and he went between them, and he was drowned. The old man used to say, forty years later, kind of, "ah, the Lord wanted him." And there are, of course, sometimes when it is not pastorally wise or sensitive or appropriate to, as it were, take people on in such views. People are perfectly entitled to the views they wish to hold that they find helpful, which have sustained them in stresses the rest of us probably could never experience or imagine. So, I am conscious that perhaps this brings the sort of conversation to an end. There can be a certain arrogance for people like me saying it's terribly inappropriate to pray about the weather. I don't wish to question the integrity of those who do it, but I just think that they might think a bit harder about why they do it, about how the church officially should or should not do it, and if we are to bring the weather into intercession, perhaps great pressing issues of climate change provide a new context in which we should pray about it.

Theological View from a Well-Grazed Hill

This next extract is from an interview I conducted with a Mr. Dermot Finley (2019), who is a sheep farmer in the parish in which I serve. It took place on October 17, 2019. This second interview is very much representative of every discussion I had with farmers under the age of about fifty-five. Members of the community older than this seem to have a view more aligned with that found in Semple's book, *The Rector Who Wouldn't Pray for Rain*: that is, a belief that God is in control of the weather in a very direct sense.[3]

> **Hanley:** I wanted to talk to you about the idea of praying for weather.... Would you, as a farmer, expect your clergymen to pray for weather in a church service?
>
> **Finley:** Well, I gave it some thought, and I actually feel farmers, as you know, in their nature, are very hard pleased, so in actual fact, there's really no weather to suit all farmers because there are so many different [people]; when one is looking for rain, others want dry weather. So, to pray for good weather... It's difficult to see just what exactly that is, for the different people. I'm just not sure, you know, whether it should be more to help farmers due to the weather, that could nearly be a better thing. To pray for good weather might be actually the right thing, you know, because some farmers will require rain, others want dry weather, so I don't know what to think about it really now.
> ...
>
> **Hanley:** For other things, if you are praying for God's help, would you pray for (a) just to have strength to bear the situation as it pertains to yourself right now; (b) ask God to shine a light on your way forward so you know what to do; or (c) ask God to provide a particular set of circumstances? Which of those would be your attitude to prayer?

Finley: The way I feel, it's probably not really the answer you're looking for. I feel everything belongs to God, so I believe that really whatever comes, comes, you know. I could give an example now. I lost a lot of lambs one time, but I didn't get that upset because I never feel that I actually own those lambs. I feel everything is belonging to God, and that's being honest about it. So, the tragedy of it all didn't really upset me. So, I'm inclined to leave everything, whatever comes in his control. It's not really the answer you want in a sort of way.

Hanley: Well, it's not a question of what answer I want. I'm trying to do a couple of interviews with different people to get a variety of different answers to actually kind of feel where people are at. . .

Finley: But funnily enough, the weather is not something I'm inclined to really. . . I don't really do want. . . whatever comes, I take it as it comes. On my list of prayers, I don't actually put it high on the list now.

Hanley: That's fine. That's more or less it.

Finley: But that's the way I feel, that's really the reason why: because I believe a lot of it is whatever comes, comes, and He's after dictating that, and all you can do is your best.

Hanley: So, your regular prayer is. . . ?

Finley: My regular prayer is mostly probably starting with my family, and then maybe it would be someone who is after losing something, and I realize what they are going through. . . It generally features around people now more so than. . . It would be mostly people.

Hanley: People and their. . . Would you say then your prayer is for their health and mental strength rather than prayer for their physical. . . ?

Finley: Yes, it would. It would.

There is a sense of evolving folk belief, a faith that is growing and maturing in the general populace, in seeing God as a micro- rather than a macroclimatologist. One would like to think that this reflects the change in theology we present in the prayer book that hymn writers have not yet, but will eventually, catch up with. However, one young farmer recently made a very keen observation. With all the developments in modern farming since the 1980s, farmers in the part of the country in which I now find myself ministering are far less dependent on weather than their fathers and grandfathers were. There are various technological advancements that mitigate the effects of bad weather. This, in turn, results today in a mental shift that bad weather is not a divine retribution to be suffered but a challenge to be overcome, a mindset that can be seen very clearly in the interview above. By giving people technology, which sometimes mitigates the effect of bad weather, there is an expectation that, whatever the weather, they should be able to get done what they have to do.

It does occur to me that maybe an older folk belief is at work here. This new ideal feeds into the Protestant work ethic, the idea that to work is to pray and to work hard is to pray properly.

The term "husbandry" might be used more for the mental attitude of farmers of this later generation. There is less the idea of the world as a piece of clockwork kit, a machine, and more an attitude of the organic nature of creation. The farm, the climate, and creation are animals needed to be cared for, nurtured, and passed onto the next generation. They have their own life and can seemingly express their own will. They are in our care more than under our command, things of value out of which we can make a living and can have great skill with and understanding of, but which are never entirely predictable. In my experience, farmers of this younger generation tend to talk about their farms more like they talk about a prize bull rather than how they would talk about a favorite tractor.

Noah, the Environmental Hero

Bishop Burrows's concerns may be pastoral but are also undoubtedly for theological consistency and rightness. Mr. Finley's view is more grounded in everyday experience.

To round out my reflections, I thought I would seek the thoughts of someone whose frame of reference lay between the eternal and the immediate. Reverend David White is a fellow priest in Cashel Ferns and Ossory but was, for some time, a horticulturalist and is a prominent member of the ecumenical environmentalist group Eco Congregation. Reverend White (2020, n.p., *sic* throughout) responds:

> Part of the problem when it comes to this whole area, is our inadequate understanding of our place in the kingdom of God. It's easy to be distracted and to think of the kingdom as a territory or an almost unreachable place. This has led some to speak of the reign of God rather than the kingdom.
>
> In any case, the reign of God at its essence, is about a way of living, individually and communally, that allows God to be at the center so that God infuses every dimension of life.
>
> When we say that we are in the kingdom of God, we mean that we are confident in God's love; are open to God's initiatives; seek God's guidance in all our choices and decisions which all allows for our lives to become a generous response, in love to the love that God holds for each of us and all of creation.

The implications of this include allowing God to reign in our own attitudes, words, actions and personal relationships. But also includes allowing God reign in our social structures, institutions, decision making processes by which we live in solidarity with each other and all of creation.

God has hopes and desires for the world and for us and provides whatever is necessary to allow them to reach fulfilment. Our true destiny and happiness lie in helping to create the kind of world that offers a framework within which we all can have the life and the destiny which God desires.

So when we are asked the question "Will you pray for rain?" we need to remind ourselves that the key to finding the will of God in the kingdom lies in our use of the gift of freedom. Through this gift of freedom we are invited to cooperate with God in creating a world that enables us, individually and collectively, to become the people who God desires us to be. Therefore, our prayers should reflect this desire to see people and all of creation flourish, to demand that our societal structures respond effectively to issues like climate breakdown and that collectively and individually we work for the kingdom, which at its heart is about allowing God to infuse every dimension of life.

In some sense, it would seem that the folklore of weather in the church has tracked along behind the development of the church's theology of creation.

One way in which we can trace how folk belief has changed in the Church of Ireland in particular is through Sunday-school teaching. Sunday-school lessons tend to focus on favorite biblical stories with a nice picture to color and a heartfelt moral to take home. I notice in my ministry that there are some stories that used to be well known, like the story of Shadrach, Meshach, and Abednego, that are now unknown to the younger members of the congregation, and there are some stories that still form the bedrock of the theological education of our youth.

The story of Noah is one of those; however, its teachings have changed quite significantly. What we tend to see now are pictures of a happy Noah saving all the lovely creatures, and sometimes you might even find some potted plants dotted around the place on the ark. The disobedience of humanity in the story has become a cipher for humanity's willful destruction of God's gift of nature. Noah is the environmental hero who listens to God and saves the creatures and plants. The destruction and the rains are seen as a direct consequence of the sinful actions of most of humanity, of which they were warned but scoffed at the warning. They had forgotten God's commandment to Adam to look after creation. The rainbow is a sign of the beauty of creation and how God wants to make all things beautiful in their time. The idea is that nature gives back what we put in, that if we look after our climate, our climate will look after us. It is also about the long-term payback for righteousness, that we are called as Christians to look after the planet which, while it may not give us the microclimate for our barbecue today, it will gift the next generation the world in which they will live—no cheap grace and no instant karma.

This stands in stark contrast to the lessons that this story used to be mined for. The flood used to be very much depicted as the act of a rightfully angry god on a very sinful humanity. In this view, the moral to be drawn is about the need to listen to God and to behave well. It suggests that everything belongs to God, and God can do with it as God pleases. The rainbow becomes a sign that, once justice is served, God will return to being a benevolent creator. Noah's mastery over the animals is far more a part of this interpretation of the story in which the underlying themes are control and obedience than care and beauty. I should, at this point, say that I believe that control and care and obedience and beauty are all important things in society. One can see, though, that when we pray at a family service having passed on this new understanding of the Noah narrative to children, our prayers flow more in the direction of "Dear God, help us to look after the wonderful world you have given us so that are weather will be clement" rather than "Dear God, please bless Mr. Murphy's potato field with rain and his maze field with sunshine."

Contemporary Christianity takes a far less mechanical view of God's creation. God is no longer a benevolent mechanic who has engineered a perfect universe, a universe that we have to drive with occasional help from the manufacturer, asking for assistance with the loose screws that caused it to rattle and must be tightened up and requiring occasional spiritual oiling from above. We have developed instead a theology in line with our use of technology, one that sees creation as something organic and sees God's presence in that creation as an atomic grace, a boson of blessing, to be discovered—something that is woven into its fabric instead of relying on a reach from outside. It is a creation that we know we are part of, not apart from, in and through which we can meet the divine. So, while we may still have an instinct to plead on Mr. Murphy's behalf, our reflections will lead us on to see the blessings and possibilities in whatever weather Mr. Murphy's various crops are blessed with.

Notes

1. William Paley was an English clergyman who was also a lecturer in moral philosophy at Cambridge University for some time and was well known in the late eighteenth and early nineteenth century for his work in the field of natural theology.

2. The "Ploughing" refers to the National Ploughing Championship, which is an annual event combining a ploughing contest, an outdoor agricultural show, and a trade fair. It is

held in different locations in September every year attracting around three hundred thousand visitors.

3. It is this sense that one finds in about forty-seven of the seven hundred hymns in the church hymnal. At this point, I should probably say that there are a handful, about five or six, hymns in the hymnal that do mention weather but not in the same "God-directing-it" way as the forty-seven mentioned above. These hymns are either very recent or, in one or two cases, very old. The majority of the forty-seven hymns portraying God as a cosmic weatherman are, in fact, Victorian or slightly older in origin. The bulk of the Church of Ireland hymnal dates from between 1750 to 1900.

References

Burrows, Michael. 2019. Interview with author. October 8, 2019.
Clarke, Richard. 2000. *And Is It True? Truth, God and No-Man's Land.* Dublin, Ireland: Dominican Publications.
Eliade, Mircea. 1987. *The Sacred and the Profane.* New York: Harcourt Brace and Jovanovich.
Finley, Dermot. 2019. Interview with author. October 17, 2019.
Semple, Patrick. 2007. *The Rector Who Wouldn't Pray for Rain: A Memoir.* Cork, Ireland: Mercier.
White, David. 2020. E-mail to author. February 6, 2020.

Section II

TEXT

Introduction

THE ROMANCE OF THE WEATHER

The thunder burst at once with frightful loudness from various quarters of the heavens. I remained, while the storm lasted, watching its progress with curiosity and delight.
—**MARY SHELLEY**, *Frankenstein*

Billy's at 44 north, 56 west and heading straight into meteorological hell.
—**SEBASTIAN JUNGER**, *The Perfect Storm: A True Story of Men against the Sea*

Mary Shelley wrote *Frankenstein* during the "year without a summer" or "eighteen hundred and froze to death." The eruption of Mount Tambora in Indonesia left a layer of ash in the atmosphere that impacted weather patterns around the globe. That European summer proved unrelentingly cloudy and cold. Crops were impacted for years to follow. It was the weather, so the story goes, that led Shelley and her friends to spend their time writing rather than hiking or boating. The weather outside was frightful and so, too, were the stories—in which the weather was also frightful. Throughout the novel, storms, ice, and cold surround the monster, and moments of calm only occur when the plot itself is calm. Bill Phillips argues that an "eco-critical" reading of *Frankenstein* reveals the impact of the natural world on Shelley's story. With its gothic, romantic heart, it not only ponders the effects of the misapplication of science but also documents a climate in crisis, suffering from the ash of Tambora, beset by storms, at the end of the Little Ice Age (Phillips 2006).

Two hundred years have passed since the publication of *Frankenstein*, but the climate is again in crisis, and the artistic response seems once again to be gothic and romantic. Since the late 1990s, there has been a steady increase in a genre of texts that could fall roughly under the label of weather stories. The genre includes fiction and nonfiction, books and films, that center on cataclysmic weather events and seek to tell the tale of those who, well, weathered them. Some are realist; some are speculative.¹ Sometimes the protagonists even live. Unlike *Frankenstein*, in which the weather reflects the action and mood, the weather of contemporary weather stories is the driver of the action—it is inevitable, it is inescapable, no matter how good the science or technology, it is ultimately unknowable, and it is overwhelming. The weather in these narratives is dramatic, transcendent, disturbing, uncanny—in other words, gothic. Contemporary literature may have embraced the postmodern and beyond, but weather stories strike a profoundly romantic chord.

George R. Stewart's 1941 novel *Storm* established the mold for weather stories to follow, ushering in decades ahead of its time both the eco-novel and the naming of storms. The best-selling novel charts the progress of a storm nicknamed Maria (pronounced with a hard *i*)² from its formation in the Pacific through its battering of the California coast as it is held in place for several days by a polar inversion. Stewart's work reveals the tensions between old and new methods of meteorology, transfixed to the page in a moment of technological change with the advent of radar, but the storm itself forms its main character. This centering of the storm, described in human terms and named "like a baby," led a recent review by critic Nathaniel Rich (2021, n.p.) to call the novel "the second strangest book ever written about a storm."³ What so bothers Rich lies in Stewart's exploration of relationships, but relationships in which one partner is not human and thus remains forever just out of reach of human understanding. From a folkloristic point of view, what makes Stewart's book remarkable is that it does exactly what we all do when we narrate our encounters with the weather: we give it the attributes and agency of humans but with neither humanity's fallibility nor its interiority.

Like Frankenstein's monster, the weather cannot ultimately be controlled for long or fully known or understood. Even in its natural state, extreme weather simultaneously evokes the horror of humankind's profound animalism and its creations run amok. Holly Bailey's *The Mercy of the Sky* (2015), a nonfiction account of the Oklahoma City tornadoes in 2013, turns the weather itself into a cyborg. The tornadoes are at once pornographic and mechanical. On one page, Bailey (2015: 177–78) describes the storm as "a tempestuous belly dancer on a mission to seduce and destroy the land below," a resident drawn to it "like a snake to a charmer." Later, it emits "a ghastly combination of the whooshing, high-pitched sound of a whining jet engine and the rattling, metallic rumble of a howling freight train speeding out of control" (Bailey 2015, 196).

The wake of these storms reveals the uncanny, a world that is both strange and ordinary. Storms feel familiar, but big storms can leave the landscape unfamiliar and those in it unsure how to move in this newly remade world. Bailey (2015, 171) describes the tornado-struck city as "leveled by an atomic bomb" such that "the streets didn't look like streets anymore. It was just rubble as far as the eye could see, and when that was gone, it looked like the surface of the moon—empty and barren." But it is the end of *The Mercy of the Sky* that is most gothic. As the principal of a school destroyed by the tornado looks up at the blue sky, Bailey (2015, 251) describes, "She had never seen the sky so beautiful. It betrayed no hints of the horror that had been unleashed only an hour or so before." Building from weather to climate, Amitav Ghosh (2006, 8) argues that the horror of climate change "lies precisely in the fact that in these encounters we recognize something we had turned away from: that is to say the presence and proximity of nonhuman interlocutors." In our stories about weather, we are called to account by a powerful and distinctly nonhuman entity. Romanticism offers a known narrative structure for the exploration and containment of that entity.

Jed Mayer suggests that *Frankenstein* is less science fiction or gothic than it is "weird," described by a "malign or particular suspension of those fixed Laws of Nature" (H. P. Lovecraft quoted in Mayer 2018, 229). The weird tale, said to have been at its height in the first decades of the twentieth century as industrialism gave way to war, is marked by a sense that "there is no stable status quo but a horror underlying the everyday" (China Miéville quoted in Mayer 2018, 231). In its sense of inescapable horror, Mayer argues, climate change fits that sense of a status quo underwritten with horror. The romantic provides audiences of weather stories a kind of meteorological epistemology. That weather stories tend so strongly toward romanticism and the gothic suggests a deep sense of the unknowability of the weather, its monstrosity, and its inevitability. Weather stories provide a way to navigate the sense of pending horror as we begin to witness the effects of climate change. And perhaps it is this sense of horror that drives the growing genre of weather narratives as audiences seek both a reckoning with climate change and some precedent of where this particular weirdness may lead. The narratives themselves offer a wide range of alternative endings—utter destruction, hope and rebuilding, a dystopia, a clean slate, or a new world.

The chapters that follow demonstrate how the weather is used as a way to not only set the scene but also to make meaning. They engage directly and indirectly with the larger genre of weather stories both in literature and in vernacular narratives. In the first four of these chapters, the weather is both actor and atmosphere, pushing and changing the primary characters and the worlds they inhabit. In the last, the weather becomes a witness to a life. First, Hannah Chapple raises the issue of how climate change disrupts vernacular weatherlore in her chapter "'The World of Sensible Seasons Had Come Undone': Climate Change and Regional Folklore in Barbara Kingsolver's *Flight Behavior*." While the novel has long been considered important in climate-change discourse, Chapple argues that it documents the breakdown of local community ties in the face of changing weather conditions, tearing asunder the protagonist's marriage and connections to the land itself. In Chapple's analysis, Kingsolver's weather acts much like Maria in *Storm*, always present but never quite *there* either, impacting the lives of people unable to control their own fates. Taking a more epistemological approach, Christine Hoffmann's "Early Modern Special Snowflakes" explores the use of "frozen words" in François Rabelais and Edmund Spenser. These words describe the weather but halt the flow of discourse, much as the contemporary epithet "snowflake" carries an accusation of stifling free expression.

While the Gulf Coast is defined by its hurricanes, it is often the Black women of the region that bear the brunt of those storms, as Jennifer Morrison and Shelley Ingram argue in "Mothering the Storm: Black Girlhood and Communal Care in Literature of Katrina." Through readings of Jesmyn Ward's *Salvage the Bones* (2011), Kiese Laymon's *Long Division* (2013), and Sherri L. Smith's *Orleans* (2013), Morrison and Ingram look not only at how racial, gendered, and historical traumas find themselves revealed in the aftermath of hurricanes but also how these storms can provide generative power for moments of maternal and communal care.

The last two chapters combine weather stories and the folklore of weather. James I. Deutsch continues this close reading of weather imagery and symbolism into the music in "'You Don't Need a Weatherman': Bob Dylan's Windlore." Dylan uses wind imagery often to suggest political and cultural change in songs like "Blowin' in the Wind" and also to imply hardy survival, like in "Girl from the North Country." The winds, of change and trouble, permeate Dylan's conception of himself and his journey through life. Taking these literary explorations back into folklore, Lena Marander-Eklund, in "'I'll Never Forget the Thunderstorm of 1960, I Think It Was': Storm Stories," documents how personal narratives of near misses from the weather encode existential fears of the weather. These are ultimately stories of survival and resilience in a dangerous world, but they also harken to the romance of the weather. The weather in the texts discussed here, over a span of five hundred years, traces the complex relationships between people and nature. Weather is at times symbolic, overpowering, unavoidable, fickle, and incomprehensible, but in each case, the representation of the weather helps us, as readers, make sense of, well, the weather and reminds us that we live and can survive in a dynamically transformational world.

Notes

1. Speculative fiction about the weather sometimes falls under the genre of cli-fi (climate fiction).

2. Pop culturally, *Storm* is said to be the inspiration for the song "They Call the Wind Maria" from Alan Jay Lerner and Frederick Loewe's film *Paint Your Wagon* (1951). Meteorologically, the novel presaged the tradition of naming tropical storms with human names, which began officially in 1953. See Liz Skilton 2018 for a thorough review of the impact of Stewart's novel.

3. Rich reviewed *Storm* as part of a current series by the *New York Review of Books* revisiting older texts. He claims that the strangest book written about a storm is Henry Darger's memoir, in which 4,878 of its roughly five thousand pages are devoted to a storm's devastation of Illinois (Rich 2021).

References

Bailey, Holly. 2015. *The Mercy of the Sky*. New York: Penguin.
Ghosh, Amitav. 2006. *The Great Derangement: Climate Change and the Unthinkable*. Chicago: University of Chicago Press.
Junger, Sebastian. 1997. *The Perfect Storm: A True Story of Men against the Sea*. New York: HarperCollins.
Mayer, Jed. 2018. "The Weird Ecologies of Mary Shelley's *Frankenstein*." *Science Fiction Studies* 45, no. 2: 229–43. https://doi.org/10.5621/sciefictstud.45.2.0229.
Phillips, Bill. 2006. "Frankenstein and Mary Shelley's 'Wet Ungenial Summer.'" *Atlantis* 28, no. 2: 59–68. http://www.jstor.org/stable/41055247.
Rich, Nathaniel. 2021. "George R. Stewart's Perfect 'Storm.'" *New York Review of Books*, July 28, 2021. https://www.nybooks.com/daily/2021/07/28/george-r-stewarts-perfect-storm/.
Shelley, Mary. [1818] 2003. *Frankenstein*. London and New York: Penguin Classics.
Skilton, Liz. 2018. *Tempest: Hurricane Naming and American Culture*. Baton Rouge: Louisiana State University Press.
Stewart, George R. 1941. *Storm*. New York: Random House.

Chapter 6

"THE WORLD OF SENSIBLE SEASONS HAD COME UNDONE"

Climate Change and Regional Folklore in Barbara Kingsolver's *Flight Behavior*

Hannah Chapple

In the Finger Lakes region of western New York, as in many corn-growing regions in the northern United States, early summer often finds locals sharing a familiar refrain: "knee-high by the Fourth of July." Growing—and more pervasively eating—sweet corn is an important part of local culture. My family has lived in the area for generations, and it is a rare drive through the country during which my parents, aunts, uncles, or grandmother do not offer the saying. However, in recent years, it has become more common to adapt it into an assessment of whether or not the corn will, in fact, reach this all-important standard: "Looks like we'll be knee-high by the Fourth of July after all," or more recently, "not sure we'll be knee-high by the Fourth of July this year."

In the summer of 2019, during which local farmers casually bemoaned the particularly wet growing season, there was a local corn crisis. By July Fourth, the corn rows stretching along either side of the roads featured sprouts that barely passed ankle level. Come the usual harvest time, our family, with many in and around the town of Bloomfield, watched White's Farm Market, waiting for the emergence of their hand-painted plywood sign featuring a half-shucked ear of corn that signaled the local corn grown on their farm was ready to purchase. But the sign neither emerged when expected nor the next week. When we asked when the corn would be ready, the staff answered that it would be a few weeks yet, that the wet growing season significantly slowed the crop: too much water, not enough corn. They started bringing in corn from southern Pennsylvania to tide over residents until their crop could meet demand, but for the local corn faithful, like my mother, raised on a dairy farm in the nearby town of Lima and known to put down no fewer than five fresh ears at dinner, it was no substitute, not even worth purchasing. Eventually, inquiries were frequent enough that White's posted a sign out front on white printer paper with estimated dates, a few weeks off, that the corn might be ready.

But the delay wasn't just at White's. Each family in the area has their favorite source for fresh corn, and all were delayed. As I stopped for dinner at my grandmother's house or my father and stepmother's house, discussing the various local farms and expected dates of the harvest became an almost nightly event. "Knee-high by the Fourth of July," a saying so certain in predicting the status of corn growth that it includes no conditional—not "knee-high by the Fourth of July and the corn will be ready by August," simply the marker that my great grandfather taught my mother to expect, as happened in so many families—had proven false. The resulting discussions while waiting on the harvest began to touch on the topic of climate change. Some relatives casually and certainly attributed the increased rainfall and resulting harvest delay to climate change, while more conservative family members dismissed these claims, stating it was just a wet year. The conversations were brief, with both sides unwilling to listen to the other.

Finally, the debate quieted when the long-awaited plywood sign with its shining green and yellow beacon appeared; we were too busy eating the corn to keep talking about it. But with climate scientists in the area unequivocally attributing the increased rainfall and resulting impacts to anthropogenic climate change, it seemed that, in our little corner of western New York, climate change was having a noticeable impact on how this environmentally dependent farming community was used to living our lives (Sherwood 2019). If things keep going this way, perhaps the saying in our region will have to change with it, becoming conditional: "if winter is dry, knee-high by the Fourth of July."

The impact of climate change on a formerly reliable element of regional culture as well as the stubborn silence permeating the resulting climate-change debate were a microcosm of the larger impacts and debate currently raging on a national and even global scale. For my family, a large part of our regional identity had been noticeably impacted. In his text *Mapping the Invisible Landscape: Folklore, Writing, and the Sense of Place*, Kent Ryden (1993, 56) argues that a symbiosis exists between lore and landscape. The question this raises, in the context of the change in my regional weather and its resulting impact on folklore—the seeming disruption of that symbiosis, is as follows: What happens to lore when the place changes, when the familiar landscape shifts or even dissolves? When anthropogenic climate change makes recognizable places and landscapes unrecognizable, the connection between place and lore can be disrupted, displaced. With this in mind, in order to evaluate potentially shifting folklore as a result of environmental change, it is important to bring to our understanding of regional folklore a focus of the biological environment: the bioregion.

Bridging Folklore and Bioregionalism in the Time of Climate Change

The ideology of region in the study of folklore has, to this point, been largely and understandably anthropocentric.[1] Henry Glassie (1969, 34), in his *Pattern in the Material Folk Culture of the Eastern United States*, describes "region" as "a section of a geographical whole established by an analysis of comparable material found throughout the whole." This construction creates regional bounds determined by human activity, by the dividing lines where shared material folk culture begins and ends. Indeed, in *Passing the Time in Ballymenone*, Glassie (1992, 326) asserts, "as the old scientists read the wondrous world as the product of divine planning, our task is to read the world for evidence of the existence of man." He describes determining his region of study in Ballymenone by the activity of the people who dwell within it, explaining that he would eventually "be able to drape the net of their social motion over a map of their place" (Glassie 1992, 13). The relationship between folklore and region for Glassie appears to treat the regional environment as the backdrop upon which region and folklore are negotiated, as though the "region" of regional folklore lives not on the very ground and in the very climate of the region but rather in the mind of the folklorist.

This is not to say that the environment has had no role in shaping folklore in Glassie's regions, only that his approach to delineating and documenting regional folklore casts this influence as primarily having already happened. The environment was a part of determining how the people came to live in their region but, having helped determine their way of life, it has done its work. As a result, there is not much room to discuss the impact of regional climate and weather and the impact shifts in those forces could have on the local folklore. Therefore, I propose that Glassie's construction of "region" can no longer hold under the current and continuing influence of anthropogenic climate change, through which the environment changes rapidly, and as a result, the folklore-place relationship is corrupted. In the time of anthropogenic climate change, when the environment itself is shifting and changing, when it can disrupt and displace folklore, region can no longer be viewed as this static plain across which folklore moves and upon which it can be mapped. Instead, this definition of "region" that relies upon anthropocentric activity must be replaced by one that places primacy on the environment, replacing anthropocentrism with bioregionalism.

I do not believe that bioregionalism is inherently at odds with the work of folklorists. In fact, the two fields function much more disparately than the foci of their study requires. Anand Prahlad (2019, ix) argues, in his forward to *Implied Nowhere: Absence in Folklore Studies*, that "all too often, texts reflect a field that imagines itself in a vacuum, disconnected from disciplines whose materials, upon close inspection, overlap that of folkloristics." Bioregionalism is one such discipline. In *LifePlace: Bioregional Thought and Practice*, Robert Thayer (2003) describes "bioregionalism" as both "a body of theory and technique with great significance to the nature of community life, public citizenship, personal lifestyle, regional planning, ecosystem management, and education." This foundational chapter represents a central and important tenet of bioregionalism: that it is both a body of theory as well as a critical approach. Bioregionalism as a critical approach is widely applied in fields as disparate as ecology, architecture, and economics. To do so in folklore studies requires that regional folklore be understood through the lens of the bioregion.

To understand the aims of such an approach, it is necessary to understand the concept of "bioregion" as it functions within the field, as well as the ways bioregion comes to bear on human culture and on the study of that culture and its products. Thayer (2003, 3) defines a "bioregion" as "literally and etymologically a 'life-place'—a unique region definable by natural (rather than political) boundaries with a geographic, climatic, hydrological, and ecological character capable of supporting unique human communities." Each unique bioregion is "distinguishable from other areas by attributes of flora, fauna, water, climate, soils, landforms, and the human settlements and cultures those attributes have given rise to" (Sale 1983). Of central importance in these descriptions is the directionality and temporality of influence between bioregion and human inhabitant. Human culture is inseparable from the bioregion, the culture has taken the shape it holds specifically because of the bioregion in which it is situated, and the bioregion continues to influence the culture actively. Humans are not, in fact, simply inhabitants of their bioregion—the human culture is one facet that makes up the bioregion, no more determining or essential than any other, except perhaps in the scale of damage it could cause. Peter Berg and Raymond Dasmann (2014, 66) explain that bioregion "refers both to a geographical terrain and a terrain of consciousness—to a place and the ideas that have developed about how to live in that place. Within a bioregion the conditions that influence life are similar and these in turn have influenced human occupancy." Reading cultural products and representations of bioregion centers the relationship between the human and the region. The region is not passive, not a place upon which human culture is enacted. The bioregion and the culture of its inhabitants are inextricable. Ryden (1993, 201) asserts, in his analysis of bioregional literature, that the very concept of bioregion "implies a particular shared ethical, communal, and political awareness among the people who live within it." The very beliefs of the inhabitants arise as a result of the bioregion and its influence on their culture.

In attempting to explore the implications of bioregional theory for the humanities, bioregional scholars have in many ways been engaging with folklore without recognizing it as such. One example of a bioregional text that is unknowingly engaging with folklore is Christine Cusick's (2012) "Mapping Placelore: Tim Robinson's Ambulation and Articulation of Connemara as Bioregion." Cusick's (2012, 135) essay defines Tim Robinson's work representing Connemara through maps, poetry, and collections of local narrative as bioregional in that it "offers us an enactment of how humans might more purposefully dwell in the places that sustain them." Despite frequently using the term "placelore" to represent material Robinson documented from locals in his walks through the region, Cusick's essay discusses regional folklore without engaging with folklore as a field. Though Robinson is perhaps equally unaware of the context of his work within folklore, he describes performing fieldwork. He explains that "when people realize that I really do want to know something—a scrap of an old song, a decayed placename, a half-forgotten anecdote—they respond by taking the item seriously, perhaps for the first time, having been persuaded by modernity that it is obsolete nonsense" (quoted in Cusick 2010, 142).

Despite the importance that Robinson and those he interviews place upon the placelore he collects, Cusick seems to wrestle with valuing local lore in the context of Robinson's larger project. She explains that "Robinson has given value to what Chet Bowers calls 'low-status knowledge'" and that "Robison's attentiveness to these forms of knowledge, and his commitment to their transcription, is evidenced in the way he carefully weaves local lore with natural history, cultural artifact with sensory perception, granting to these different forms of knowing an epistemological equivalence" (Cusick 2012, 142). Cusick's description of Robinson's fieldwork and treatment of the folklore Robinson collects demonstrates the perhaps too-familiar discussion of folklore without awareness of folklore as a field and represents a missed opportunity to discuss the bioregional imagination at work on regional folklore. It is time, with anthropogenic climate change acting daily on the bioregion and upon regional folklore, to lean into the vacuum, to build a bridge between bioregionalism and regional folklore. It is time to strengthen the work of both by valuing rather than remaining blind to the nexus between them, by exploring what can be revealed through a bioregional folklore.

Kingsolver's *Flight Behavior* as Bioregional Text

To explore the potential implications of a bioregional folklore and its merit for tracking potential folkloric disruption resulting from anthropogenic climate change, I wish to examine a fictional text in which the impact of climate change on folklore is extant and which makes a significant statement on the group dynamics at play in the discussion of climate change: Barbara Kingsolver's (2012) novel *Flight Behavior*. Kingsolver's novel explores local impacts of the global problem of anthropogenic climate change. The Turnbow Farm in rural Tennessee struggles to thrive due to the impacts of climate change. In addition, in a speculative turn, the farm becomes the nesting ground for a millions-strong monarch butterfly population whose flight behavior has changed when their traditional habitat in the mountains of Mexico is destroyed by a landslide. These bioregional concerns become both the backdrop and impetus for a life-altering season for Dellarobia Turnbow, who struggles with the realities of her life and yearns for romance, for escape, demonstrating repeated flight behaviors of her own. The monarchs bring outsiders to the Turnbow's farm, specifically biologist Ovid Byron and his team of graduate students. Dellarobia takes a job with Ovid to help document what he presents as the inevitable downfall of a monarch population that is beyond saving. The resulting interaction brings the two groups, bioregional insiders and outsiders, to chafe against each other as the circumstances of their meeting bring beliefs regarding social codes and climate change into debate.

Kingsolver's novel is representative of a bioregion in southern Appalachia, with an adherence to bioregional concerns of what it means to live in place in some degree of harmony with the bioregion. What makes it bioregionalism is its interest in lives that are "stitched into the physical environment. It is not their 'setting,' and they are not its 'inhabitants.' Rather, life . . . is a constant process of relationship and negotiation among phenomena" (Blair 2012, 165). Given that memetic narrative is suited to and relies on human subjects with the gift of speech and storytelling, the fate of the humans in the story is not dramatically more important than that of the monarch population. Instead, each population's struggle is a single phenomenon currently at work in the bioregion. *Flight Behavior* is "a story of movement" on the global scale, a novel that "links local environmental issues with global ones" (Blair 2012, 167). Reading Kingsolver's text through a bioregional lens situates the story in the personal, in the immediate bioregion. But rather than disrupting the connection to the global, it bears out global issues on the bioregional scale and thus represents bioregions as a part of the interconnected global.

The structure of the narrative, as well as the introduction of an "outsider" group to play against the local residents, intensifies the importance of a bioregional reading. This becomes clear early in the novel, when monarch specialist Ovid explains to Dellarobia the central scientific question of the monarchs' arrival, mainly wondering why "a major portion of the monarch population that has overwintered in Mexico since God set it loose there, as you say, would instead aggregate in the southern Appalachians, for the first time in recorded history, on the farm of the family Turnbow" (Kingsolver 2012, 122). In one sentence, Kingsolver encapsulates the bioregional and personal focus through which she filters a global catastrophe: she filters a large area (the cisatlantic monarch range) into a single foreign area (Mexico), then into the southern Appalachian bioregion, then to a single-family farm, and finally to the woman who lives there, who first experiences the monarchs as a personal kind of magic. In this way, though the arrival of the outsiders could very easily broaden the scope of the texts beyond the bounds of bioregion, Kingsolver structures the narrative instead to work from the outside in, to examine the ways these bigger changes and issues impact the bioregional home. The change is presented as having profound personal importance before being expanded to the scope of the bioregion. By reading the novel's constructed folklore through a bioregional lens, we have the opportunity to understand regional folklore as born of and situated in, inextricable from, the local bioregion. This allows exploration of the impact of climate change's potential disruption on bioregional folklore, as well as investigating interactions between folk groups, one of which is of the bioregion in question.

Reading Bioregional Folklore's Disruption in *Flight Behavior*

In *Flight Behavior*, Kingsolver addresses the impact of anthropogenic climate change on the folklore of the novel's rural Tennessee community. Kingsolver first creates the folklore of her fictional community and then disrupts that folklore as a way to represent the displacements caused by climate change in the world at large. Kingsolver uses the tension created by this displacement to fictionalize the debate between climate-change believers and climate-change deniers, then offers an intervention into the silence between the groups to complicate the assumptions that frame the debate. She uses this intervention to investigate one way in which the impacts of climate change could actually reinforce climate-change denial in an environmentally dependent community by displacing folklore and thus disrupting identity. This intervention employs folkloric motifs traditionally used to "other" social groups in literature to instead complicate the dichotomy between two folk groups usually represented as "folk" and "elite" culture.

Weather in *Flight Behavior*, weather unusual for and negatively impacting the region, is central to the drama of the text and unequivocally presented as the result of climate change's impact on the regional environment. The narrator describes the unusual fall weather as a part of a larger disrupted system, in which autumn found that "the leafless pear trees . . . had lately started trying to bloom again, bizarrely, little pimply outbursts of blossom breaking out on the faces of the trees. Summer's heat had never really arrived, nor the cold in its turn, and everything living now seemed to yearn for sun with the anguish of the unloved. The world of sensible seasons had come undone" (Kingsolver 2012, 49). Kingsolver makes clear, through the speculative thread of a threatened monarch population appearing in Tennessee and the arrival of the scientists who follow to study them, that this weather is the result of anthropogenic climate change and that the negative impacts are far-reaching and dire. In the earliest pages of the novel, a single fallen tree serves as a synecdoche for the regional damage and a metaphor for the cultural loss of this Appalachian region. The single tree then rapidly devolves into an entire hillside of timber lost to the rains:

> The tree was intact, not cut or broken by wind. What a waste. After maybe centuries of survival it had simply let go of the ground, the wide fist of its root mass ripped up and resting naked above a clay gash in the wooded mountainside. . . . After so much rain upon rain this was happening all over the county. . . . The ground took water until it was nothing but soft sponge, and the trees fell out of it. Near Great Lick a whole hillside of mature timber had plummeted together, making a landslide of splintered trunks, rock and rill. (Kingsolver 2012, 5)

In the face of such destruction, the reaction of the residents makes clear how unprecedented, how unrecognizable the world had become: "People were shocked, even men like her father-in-law who tended to meet any terrible news with 'That's nothing,' claiming already to have seen everything in creation. But they'd never seen this, and had come to confessing it" (Kingsolver 2012, 5). As Ryden (1993) posits is possible regarding the risk of threatened placelore, the essential and tenuous relationship between environment and stories has been disrupted. The regional environment—the world around them—has become incompatible with their stories. As a result, the shifts in the bioregion are effectively silencing the community storytellers.

The unusual weather continues to change the face of the landscape not only for the community at large but also for the family farm at the center of the text. On the Turnbow farm, where Cub Turnbow grew up and now lives with his wife Dellarobia, Dellarobia observes the wreckage in the aftermath of a storm, from which "the drainage gully down the center of the pasture had swollen into a persistent, gushing creek. . . . No creek had ever run here, in any year Cub could remember, and now a series of waterfalls climbed the hill like a staircase" (Kingsolver 2012, 124). The known shape of the land has been violated, become unknown—the land literally divided. As the novel progresses, the disrupted migratory pattern of the monarch population is paralleled with the disrupted folklore of the human population. In an environmentally dependent agricultural community, when regional weather becomes unrecognizable, the occupational folklore of the farm is particularly vulnerable. In the case of the Turnbow family, their shared knowledge of husbandry and harvest are thoroughly disrupted, in some cases to great detriment.

Dellarobia notes one such displacement in which knowledge of hay harvest proves unreliable for the new conditions. Despite the fact that they should have a hayloft packed to the brim with the several seasonal harvests, the loft is not even half filled. She laments that "they'd lost the late summer cutting because three consecutive rainless days were needed for cutting, raking, and baling a hay crop. All the farmers they knew had leaned into the forecasts like gamblers banking on a straight flush: some took the risk, mowed hay that got rained on, and lost. Others waited, and also lost" (Kingsolver 2012, 36). The community knowledge of how many hay cuttings are viable and what conditions can be safely expected during certain times of the year does not hold up to the new world. Though there was some variation in response to the unusual weather—waiting for a rainless forecast or taking the risk that the forecasted rain would not arrive, the variety actually signals the consistency of the folk knowledge: three dry days are necessary, and the two approaches had the same goal of somehow finding the three dry days that did not arrive. Whether they chose to rely on their knowledge and experience and risk the cutting or risk the wait, the farmers did not get the results they needed.

The animal agriculture upon which the local farms make their living depends upon hay to feed the animals and having to purchase hay from out of state cuts into the already narrow profit margins of the struggling area. This import damages the self-sufficiency of the bioregion's agricultural economy. The implications of the crop loss are far-reaching, especially once the moisture from continued rainfall causes even the early harvested hay to mold. The Turnbows, who raise sheep, suffer from the loss but fare better than their neighbors. Dellarobia and Cub "threw out some hay every morning, bales purchased out of Oklahoma for a harrowing price because this farm's meager hay crop had molded, along with all the other hay within a hundred miles. Their cattle-raising neighbors were losing a fortune on hay this winter, with little choice but to start selling off calves for nothing" (Kingsolver 2012, 332). While it hurts the Turnbows, their herd simply eats less than cattle do, and their loss is lessened as a result. Meanwhile, for their neighbors, it has become environmentally and economically unstable to raise cattle in a community for whom that has been a traditional and sustained income source, persisting for so long that even the Turnbow's sheep farm was a former dairy.

Like with the harvest, husbandry expertise is displaced by the new weather pattern. The sheep are first moved away from the bottomland where they are usually kept, an area that is normally good for them because "the grass down there is good. But not this year" (Kingsolver 2012, 129). But even the move to a part of the farm that has never been used for sheep pasturing does not solve the problem. Dellarobia laments the muck-covered flock later in the text, reflecting that "they were pastured here for the higher ground, but after the past week of torrents any former notion of high ground was called into doubt" (Kingsolver 2012, 331). In the end, there is no land on the farm that is dry enough to keep the herd from hoof rot. Whether for hay harvest or cow and sheep husbandry, the regional knowledge of farming practices proves fruitless in the face of climate change.

Other income sources, too, are put in jeopardy by the changing weather. Though the area is not one with infrastructure based upon timber trade, timber is a traditional income source. Every few generations, standing timber can be harvested for a profit to supplement the farm. However, this income source is threatened by the new weather. When the rain is so relentless that standing timber is falling on entire hillsides, cutting timber now leaves property open to dangerous mudslides. Dellarobia tries to explain this to Cub when he tells her the senior Turnbows are planning to log the property for much-needed supplemental income. Dellarobia asks:

> Have you looked at that mountain since they finished logging it out? It's a trash pile. Nothing but mud and splinters . . . it looks like they blew up bombs all over it. Then all these rains started and the whole mountain is sliding into the road. They have road crews out there blading the muck out of the way. I bet I've seen that six times since July. (Kingsolver 2012, 40)

In this environmentally dependent community, it has been critical to have the option for logging the property in order to support the farm in hard years. But when the weather that makes it a hard year for the husbandry and harvest also makes it dangerous to log, the farm is in real jeopardy, and traditional income sources are eliminated in a landslide of their own.

In addition to the occupational lore at the heart of surviving and thriving in an agricultural community, the rules around weather and play upon which the community operates are also called into question. When December finally arrives, the Turnbow's farm—the world—does not look like it should, as "the front yard became a flat, grassy pool. Dellarobia couldn't let the kids play out there unless they wanted to pull on their rubber boots and splat around on it. She would have considered putting them in their swimsuits, if it were just a hair warmer, so they could run around as they did in summertime under the sprinkler. But this was winter, the dead of it" (Kingsolver 2012, 123). The world outside is closer to summer weather but just barely too cold for kids to splash around in swimsuits. Indeed, the rules about weather-appropriate play make the obvious option—allowing the kids to splash around in their swimsuits—feel like a violation Dellarobia cannot bring herself to commit. It has not even been cold enough outside yet for Dellarobia to purchase a jacket for their toddler, and her longing to take the kids outside to play in the snow remains unanswered. Instead, without tradition in place to dictate how the children should interact with their environment in this state, Dellarobia and her children are stranded inside the house, transforming the house to a lifeboat of sorts that is also penetrated by the water seeping under door and window frames. Her children have to play inside, and special attention is paid to that play, especially to the way the unsuitability of her daughter Cordelia's toys mirrors the unsuitability of the season. Through this isolation, the relationship the children have with their natural environment is curtailed, just as that environment becomes inhospitable for their parents and for the animals they raise. Kingsolver presents a circumstance in which the loss of rules governing the relationship with the environment can curtail or cut off that relationship, can sever the ties between region and dweller. The dissolution is rampant, the disruption and displacement of folklore directly linked to living in the bioregion—rules by which the community earns a living, operates their farms, and even entertains and teaches their children as dictated by the weather—penetrates all facets of life.

Replacement Folklore and Practice in *Flight Behavior*

With the community folklore rapidly being displaced, it is perhaps unsurprising that new folklore and practice emerge, even within the small temporal space of the text, to supplement the community relationship with their drowning world. One such practice is a unified community response to the weather; when considering strange weather events, some are worth commenting on and lamenting and some are not. In February, a fog rolls over the region, the only fog mentioned in the book and one Dellarobia describes as "strange," yet "in a winter so persistently deviant as this one, most people were sick of weather talk and greeted the latest act without a salute" (Kingsolver 2012, 310). However, not long afterward when another deluge of rain appears, Dellarobia observes that "this last week of rain had stacked up more layers of crazy on folks who had lost whole harvests and the better part of their minds to a year of drizzle. Water torture, they were calling it" (Kingsolver 2012, 336). The rain, the new weather that has been actively changing their environment and directly changing the way they live their lives, is met with universal chagrin despite the fact that the earlier and more singular fog was largely ignored. Despite its singularity, there is a comfort to be found in the fog that obscures a landscape so thoroughly transformed by the rains and mudslides of recent months: the residents cannot see the changed land and thus receive a break from having to confront its difference from their more familiar image of their region. In this way, the yearning to retreat into the comfort of the known and unchanged rises to the surface.

With this latest rain, the response to which is set significantly in contrast to the strange fog, Dellarobia shares a new legend that has emerged in the community: "this morning she'd heard about a man in Henshaw who walked outside and unloaded his Smith & Wesson into his old horse, claiming he'd seen a vision of it drowning in mud. The vision was familiar to most by now" (Kingsolver 2012, 336). The town, Henshaw, is not mentioned before this moment in the text, and yet, its location is not described in detail. In a text that takes such pains to situate the locations of events and their relationship to each other, this gives the impression that Henshaw is close enough to be known and local, indeed close enough to share the regional weather pattern, but far enough away that the Turnbows do not personally know the parties in question, as would be the case of someone in their hometown of Feathertown. In this way, this legend follows Suzi Jones's (1976, 110) hypothesis for the ways in which folklore can become regionalized, situated in regional proximity and with shared consciousness resulting from the shared environment. The legend is attributed to a safe but not-too-distant location that is accessible to and allows the legend to function for most human residents of the bioregion who have been experiencing significantly changing weather patterns.

Another source of new or replacement folklore at work in the text, one that satisfies the yearning for consistency revealed by the fog, is climate-change-denial discourse. It is extant in the text that climate-change denial functions as folklore in the local community, so much so that Kingsolver introduces a folklorist character to vocalize the fact.[2] And it makes sense that, under the circumstances of bioregional disruption, climate-change-denial discourse could function this way. Though Ryden (1993, 95) argues that "the sense of place can outlast the place itself," the community in *Flight Behavior* finds itself unable to thrive when their folk knowledge of regional weather cannot be depended upon. Disruption of community folk knowledge of their region has displaced their sense of place. If, as Ryden posits, sense of place and sense of self are—for locals—indivisible, the disruption of place sense necessarily disrupts the sense of self. It is easier to understand, then, why the community would effectively replace the function of folk knowledge with alternative discourses of authority. That displacement creates a gulf in the relationship of folk to region, one that cannot simply be replaced by any source but whose replacement must have some bearing on the natural environment, on the bioregion. It is thus that the impacts of climate change have left a gulf in regional folklore that necessitates a substitute, one in which the closed system of climate-change-denial discourse is uniquely appropriate to fill.

The conflict between local folk relationship and external discourse in the novel's use of climate-change denial is highlighted during an exchange between Dellarobia and Cub when he responds to his wife's use of the term "climate change" with: "Al Gore can come toast his buns on this" (Kingsolver 2012, 260). Dellarobia recognizes the phrase, one that relegates discussion of local weather to national politics rather than the home region: "It was Johnny Midgeon's line on the radio, every time a winter storm came through" (Kingsolver 2012, 260). Dellarobia presses back, asking him about the conditions their family has actually experienced and connecting that experience to generations of local history. She asks, "but what about all the rain we had last year? All those trees falling out of the ground, after they'd stood a hundred years. The weather's turned weird, Cub. Did you ever see a year like we've had?'" (Kingsolver 2012, 260). Rather than following Dellarobia's direction and relocalizing in the debate, Cub answers her by referring again to the greater and distant discourse of the climate-change debate, saying simply "they don't call it 'global weirding'" (Kingsolver 2012, 261). Because climate-change-denial discourse has become naturalized as the way residents think about the environment, even discussions of local weather, of their bioregional conditions, do not engage with lived experience but rather repeat the words of external authority.

This serves as a mirror for the revelation earlier in the book that the change in weather had silenced the community's storytellers. Their own stories no longer work, so rather than staying silent, they have adopted this external source of authority. It is in this moment that a bioregional reading of the novel's constructed folklore is uniquely generative. Because it is not that climate-change denial is functioning as folklore that is particularly significant, it is the origin of that folklore and the reason for its adoption, the way that bioregional lore is displaced. The danger of this displacement, in bioregional terms, is that the discourse is external, inherited from mass media and news sources. Localized and environment-based knowledge, gathered and transmitted over generations and natural to the bioregion in which it was developed, is replaced with external voices of authority that are not of the region and that further disrupt the folk's relation to the environment. This discourse has been so thoroughly adopted, however, that its use is represented in the text as central to signaling membership in the regional folk group.

Group Dynamics and Climate-Change Discourse

Perhaps most significant in Kingsolver's use of bioregional folklore is the way she targets the group dynamics at play around the climate-change debate. It is Dellarobia's very violation of the practice of climate-change denial that sets her apart as a liminal member of her folk group, the one positioned between herself and the members of the "elite-culture" group of academics who come to their farm to study butterflies or rally to save them and raise awareness. While folklore is often employed in fiction to reinforce cultural hierarchy between elite and folk culture, in which elite culture is educated and correct and folk culture is quaint and "other," Kingsolver's text challenges this hierarchy. And while it would have been easy for Kingsolver to use Dellarobia's positioning to create a kind of awakening tale in which the uneducated folk learn about science and see the error of their ways, the way Kingsolver plays out group dynamics resists this reading.

Dellarobia observes the challenge of the debate playing itself out on her family farm: "There were two worlds here, behaving as if their own was all that mattered. With such reluctance to converse with one with the other. Practically without a common language" (Kingsolver 2012, 153). She communicates the two sides to Ovid in a way that yields primacy to the economic realities at play for each group, an element that Ovid does not include in his own comparison, when she explains to him that "the teams get picked, and then the beliefs get handed around . . . team camo, we get the right to bear arms and John Deere and the canning jars and tough love and taking care of our own. The other side wears I don't know what, something expensive. They get recycling and population control and lattes and as many second chances as anybody wants" (Kingsolver 2012, 321). It takes time for Ovid to actually hear Dellarobia's description of her group, to get over his frustration with what he describes as their refusal to accept fact. Perhaps unsurprisingly, Ovid had not considered anything other than scientific fact versus fiction in his imagining of the two groups. However, as she and Ovid have gotten to know one another and spoken about the challenges of the debate, Dellarobia refers frequently to the economic realities of life for her family as central to the way they view climate science. She asserts that worrying about immediate needs makes it hard to worry about something as far away as extinction. Despite Ovid's initial dismissal that anything other than the world-ending eventualities of climate change should be primary in understanding one's place in the world, Dellarobia eventually communicates to Ovid that having the time and resources to actively engage with the realities of climate change is as much a luxury as the graduate students' fancy field equipment. Despite their initial dismissal of one another's understandings of climate change, learning more of the beliefs and realities of one another's folk groups builds enough of a bridge that Dellarobia and Ovid can now discuss climate change in good faith.

Despite the way Dellarobia portrays her own group, Kingsolver uses this economic reality to make a significant move in challenging the expected cultural hierarchy. During a critical exchange, a Mr. Akins comes to town to preach the gospel of sustainability to the uneducated local yokels, explaining that the stereotypes of the area are exactly "why I come to places like this, instead of Portland or San Francisco. You people need to get on board, the same as everyone else. If not more so" (Kingsolver 2012, 315). He walks Dellarobia through the sustainability pledge he is asking the locals to sign, and the exchange becomes a farcical landslide of elite-culture obliviousness and blind superiority. His assumptions about the recklessness of the southern Appalachian people are belied when he talks Dellarobia through the behaviors he wants Dellarobia and her neighbors to perform. For each promise Mr. Akins asks her to make, he is surprised to find that his attempts at environmental preservation fall short of Dellarobia's and her family's self-denial, such as when he asks her to "bring your own Tupperware to a restaurant for leftovers, as often as possible" and Dellarobia responds: "I've not eaten at a restaurant in over two years" (Kingsolver 2012, 325). She continues to assert that her family would not think of paying for bottled water or forgetting to take gas mileage into account when driving and that her husband's truck is so well maintained that it is on its third engine (Kingsolver 2012, 328–29). When Mr. Akins suggests that she agree to fly less, Dellarobia reiterates with notable disdain: "fly *less*" (Kingsolver 2012, 329). Dellarobia is confronted with an outsider who has come to her community to educate the locals on the importance of sustainability and reducing their carbon footprint, only to find that the lifestyle in this place—in large part determined by the economic challenges of their bioregion—is already one with a lower carbon footprint than that of Mr. Akins, who traveled across the country to educate them. Rather than assigning all the right thinking to the elite group of Mr. Akins and visiting scientists, Kingsolver attributes viable knowledge and sustainable practices to the citizens of southern Appalachia despite the prejudices against them. In this, she takes the specific claim of many of the outsiders in the text—the unwillingness to consider the beliefs of the other side—and turns it against those who most frequently wield it.

It is not only in these exchanges across group divides that Kingsolver resists the elite-folk hierarchy, showing instead that they are each their own folk group defined by different practices. In a conversation between Dellarobia and Cub, Cub surprises his wife with his astuteness about the relationship between science and farming and the credence he lends to the value of husbandry knowledge. After she describes the scientist's work to Cub, he responds, "I guess it's in knowing what to measure . . . same in farming" (Kingsolver 2012, 258). Dellarobia is surprised and concedes: "it was astute. Someone on this farm had to check the inner eyelids of the ewes and lambs every week, watching for anemia by degrees as an indication of parasite load. They monitored the hayfield for the right proportion of seedhead to stem. They bred and culled the sheep based on meat yields and staple lengths of the fleece" (Kingsolver

2012, 258–59). Dellarobia had fallen prey to the authoritative thrall of academic culture and relegated any useful knowledge to the side of the scientists, but Cub—the very character most devoted to climate-change denial—opens her eyes to the validity of knowledge and practices of their own family, which do not need validation from an exterior source. The problem then is not, as each group had asserted, the beliefs of one group or the other; it is the very silence between the two groups that reinforces the hierarchy both have internalized but that the practices of each belie under closer inspection.

After all, Dellarobia argued early in the text of elites that "they would never come see what Tennessee was like, any more than she would get a degree in science and figure out the climate things Dr. Byron described" (Kingsolver 2012, 166). However, throughout the text, that is precisely what happens: many of those elites have come to Tennessee, and Dellarobia has decided to go to school to get a science degree. This is not a redemption of Dellarobia from the dregs of the folk nor of Ovid from the ivory tower. Through the interaction between groups, the redemptive move is not a change from one group to the other. It is the interaction itself because, while the negative connotations of anthropogenic climate change and its far-reaching negative impacts are kept at the forefront by centering news reports of damaging weather events simultaneously all around the world, those negative impacts are also unifying in their breadth of influence. These groups with completely different perspectives, whose beliefs regarding climate change are so strongly definitive of group membership that identity is threatened when they are challenged, are able to interact, to discuss, and to learn from one another. Neither has the right answer, and the voices of both are required to understand the full picture. Kingsolver presents a world in which no opposing groups must remain fully divided—even those whose beliefs have centered around their mutual rejection can perhaps come to understand one another.

Conclusion: The Necessity of Bioregional Folklore in the Time of Climate Change

This chapter imagines a functional nexus between folklore and bioregionalism, and it explores what a reading of regional folklore that centers bioregional thinking and employs a bioregional approach might accomplish. In an age in which anthropogenic climate change is altering the face of region and the relationship of human cultures to their regions—both of which are incorporated into a bioregional approach, employing a bioregional folklore strengthens the abilities of each approach to address the times in which we humans now do, think, make, and believe. What is at stake is understanding how we might live among one another in this changing world—as well as how we might challenge false hierarchies and bridge group divides in order to face it. Through my reading of Kingsolver's work, I've tried to show that folklore and bioregional literary criticism can complement each other to make important statements about what it means to live in place when the regions upon which we've built and through which we come to understand and sustain our folklore become unrecognizable and how that experience can come to displace and shape folk belief and practice.

Notes

1. For a comprehensive evaluation of how concepts of "region" have been constructed and explored in folklore studies, see Allen 1990.

2. The folklorist, who comes to study folk-art representations of the monarchs, describes the use of climate-change-denial discourse as a "cargo cult," stating that "it's become fully identified with the icons of local culture, so it's no longer up for discussion" (Kingsolver 2012, 395).

References

Allen, Barbara. 1990. "Regional Studies in American Folklore Scholarship." In *Sense of Place: American Regional Cultures*, edited by Barbara Allen and Thomas J. Schlereth, 1–13. Lexington: University Press of Kentucky.

Berg, Peter, and Raymond Dasmann. 2014. "Reinhabiting California." In *The Biosphere and the Bioregion*, edited by Cheryll Glotfelty and Eve Quesnel, 65–70. New York: Routledge.

Blair, Ruth. 2012. "Figures of Life: Beverley Farmer's *The Seal Woman* as Australian Bioregional Novel." In *The Bioregional Imagination: Literature, Ecology, and Place*, edited by Tom Lynch, Cheryll Glotfelty, and Karla Armbruster, 164–80. Athens: University of Georgia Press.

Cusick, Christine. 2010. "Mindful Paths: An Interview with Tim Robinson." In *Out of the Earth: Ecocritical Readings of Irish Texts*, edited by Christine Cusick, 205–11. Cork: Cork University Press.

Cusick, Christine. 2012. "Mapping Placelore: Tim Robinson's Ambulation and Articulation of Connemara as Bioregion." In *The Bioregional Imagination: Literature, Ecology, and Place*, edited by Tom Lynch, Cheryll Glotfelty, and Karla Armbruster, 135–49. Athens: University of Georgia Press.

Glassie, Henry. 1969. *Pattern in the Material Folk Culture of the Eastern United States*. Philadelphia: University of Pennsylvania Press.

Glassie, Henry. 1992. *Passing the Time in Ballymenone: Culture and History of an Ulster Community*. Philadelphia: University of Pennsylvania Press.

Jones, Suzi. 1976. "Regionalization: A Rhetorical Strategy." *Journal of the Folklore Institute* 13, no. 1: 105–20.

Kingsolver, Barbara. 2012. *Flight Behavior*. New York: Harper Perennial.

Prahlad, Anand. 2019. "Foreword." In *Implied Nowhere: Absence in Folklore Studies*, by Shelley Ingram, Willow G. Mullins, and Todd Richardson, ix–xiii. Jackson: University Press of Mississippi.

Ryden, Kent C. 1993. *Mapping the Invisible Landscape: Folklore, Writing, and the Sense of Place*. Iowa City: University of Iowa Press.

Sale, Kirkpatrick. 1983. "Mother of All: An Introduction to Bioregionalism." Lecture at Third Annual E. F. Schumacher Lectures, Schumacher Center for a New Economics, Mount Holyoak College, South Hadley, MA, October 1983. https://centerforneweconomics.org/publications/mother-of-all-an-introduction-to-bioregionalism.

Sherwood, Julie. 2019. "Straight Talk on Climate Change." *Canandaigua Daily Messenger*, October 6, 2019. https://www.mpnnow.com/news/20191006/straight-talk-on-climate-change.

Thayer, Robert. 2003. *LifePlace: Bioregional Thought and Practice*. Kindle ed. Berkeley: University of California Press.

Chapter 7

EARLY MODERN SPECIAL SNOWFLAKES

Christine Hoffmann

Liberals, millennials, Mitch McConnell, Milo Yiannopoulos, Donald Trump, British army recruits, white supremacists, Theresa May, Merrick Garland, Robert De Niro, Yale University, and an autistic child: all, according to various news and social media publications between 2016 and 2021, are "snowflakes." As *The Guardian* pointed out in late 2016, "there's really no comeback to it: in calling someone a snowflake, you are not just shutting down their opinion, but telling them off for being offended that you are doing so. And if you, the snowflake, are offended, you are simply proving that you're a snowflake. It's a handcuff of an insult and nobody has the key" (Nicholson 2016, n.p.). Commentators may not resolve the equivocality of the "snowflake" insult, but by historicizing and ecologizing the term, we can better understand its fleeting significance. In this chapter, I examine the snowflake as a radical moniker for a critically posthumanist ethos whose roots may be traced to the premodern era. Rather than signal self-importance, the snowflake ethos can nurture skepticism about humankind's attachment to egoistic narratives of preservation.

On the surface, rhetorical malfeasance is the common intellectual shortcoming for which snowflakes are targeted; their viewpoints are dismissed as wrongheaded annoyances but with the potential to obstruct the free flow of ideas. A draft addition to the *Oxford English Dictionary* entry for "snowflake, *n.*" records the late twentieth century as the earliest use of the word to refer to "a person, esp. a child, regarded as having a unique personality and potential." But the entry introduces the word as "usually derogatory and potentially offensive," suggesting its primary use in the twenty-first century for mocking characterization of those who consider themselves "entitled to special treatment or consideration" (*Oxford English Dictionary* 2019, n.p). Many descriptive labels exist for the wrongheaded, but the word "snowflake" is special for the very reason that it is associated with specialness.[1] To attach this label to someone with whom you disagree is to suggest that what is much faultier than their defective rhetorical position is their anxious need to see their special perspective acknowledged and esteemed. Such is the crude irony of the label, which summons up the concept of nature-made distinctiveness only to lampoon it as another in a line of indistinguishable examples of identity politics run amok. What a shame to be so special, the tough talkers suggest about each delicate snowflake; to be so vulnerable to the momentum of stronger forces; so unsavable from the too-late-to-prevent disasters of their own unlucky natures. As iconically adrift from consideration as the starving polar bear on the ice floe, the snowflake is positioned to ponder the use of having one's specialness acknowledged when the status of specialness has been so attenuated, its congruency with capability rewritten as vulnerability preceding collapse.

When the label "snowflake" enters an argument, then, it shifts the debate from ascertaining particular right-versus-wrong positions to judging general fitness to survive and thrive in conversation and beyond. Most attempts to defend or reclaim the snowflake put notable stress on the term's environmental associations as opposed to its rhetorical ones—as in George Takei's (2017, n.p.) tweet: Snowflakes "are beautiful and unique, but in large numbers become an unstoppable avalanche that will bury you." A sign-turned-meme from the January 2017 Women's March in Washington DC reads, "Damn right we're snowflakes: Winter is Coming" (Barnes 2017, n.p.). "Snowflakes are not fragile," Laurie Penny (2017, n.p.) wrote in a piece for *The Baffler* around the same time. "They pack down into glaciers that outlive civilizations and carve out mountains; they form snowbanks that sweep away whole towns. Millions of snowflakes together can make an avalanche, a hurricane, a killing frost."

Assuming an eco-phobic appropriation of nonhuman destructive power in order to prevail temporarily in games of ideological one-upmanship is clearly one response to the scattershot range of this insult.[2] But it's a too-easy turn of the tables: reversing the implied menace in the snowflake insult to declare the threatened snowflake as, all along, the naturally more threatening threat. To identify with such forces of destruction, to hinge empowerment on the capacity to ruin, is problematic in a time of climate crisis, and it disregards the opportunity to contemplate how such circumscribed articulations of empowerment correspond to the forms of assertion promoted by capitalist culture. I offer an alternative interpretation of the snowflake insult that suggests—despite the kneejerk eco-phobia in evidence—a readiness among human beings to recognize the fruitlessness of asserting human advancement through appeals to nature's power, given the unscalable, uncontrollable ecology in which more and more humans are noting their inopportune position.[3] The snowflake insult invites us to reckon with our increasing vulnerability within an environment that we have made increasingly vulnerable and with the disempowerment that is the ironic result of our powerful influence on the conditions of the territories we inhabit.

I move to (re)claim the snowflake label not as a call to aggressive or destructive action but as a character trait and a constructive, critically posthumanist ethos. One of the editors of *Posthuman Shakespeare* specify this critical attitude as "a strategic move away from anthropocentric premises: the human can no longer be taken for granted; humanity as a universal value is no longer self-legitimating" (Herbrechter 2012, 5). Such a move involves recognizing rather than dismissing the value and sophistication of the analogical thinking that inspires a statement like Takei's above; to claim correspondence between human and snowflake is to credit the correspondence between human and nonhuman natures. It is to embrace a structure of thought that Michel Foucault (1994, 21) assigns to the pre-Enlightenment, Renaissance mind, which imagined all the stuff of creation "linked together like a chain." Several early modern critics insist that "such correspondences are how the world is" (Egan 2006, 26), that a premodern model positing sympathetic connections between all things "begin[s] to look like a helpful anticipation of what we increasingly and indispensably understand as ecological networks" (Watson 2011, 51).[4] Claiming the snowflake label is part of a larger project to claim "the concept of a N/nature to which we already belong" and to "restor[e] the notion of nature as a multivalent, complex, interrelated living system" (Bruckner 2011, 20 and 30).

Language, naturally, is part of this reclamation project. As Rebecca Bushnell (2011, 212) asserts, nature, "past or present, does not 'speak' for itself, in an unmediated way." Humans substitute for the illegibility of the more-than-human world terms we understand—that's why we carve lover's initials into trees, assign family values to animal mates, or project emasculating effects of hyperconsumerism onto snowflakes. The latter example remains a favorite script for individuals fond of quoting the "real talk" of antihero Tyler Durden from Chuck Palahniuk's novel *Fight Club* (1996): "You are not special," Brad Pitt intones in the film adaptation; "You're not a beautiful and unique snowflake. You're the same decaying organic matter as everything else"[5] (Fincher 1999, n.p.). Despite such pathetic fallacies, snowflakes remain "perplexingly nature/culture," a phenomenon we continue to examine through "diverse mediating structures"—films, tweets, memes, urban-dictionary definitions, even those paper cut-outs we all made in school—and through discourses, the limits of which may "only be detected in [their] failure" (Bowerbank 2004, 6). Often, "we get 'nature' wrong," but we keep going, "reinventing nature collectively, as we go along, working out the snags" (Bowerbank 2004, 6). Bushnell (2011, 213) adopts a tone of careful encouragement when she defends language as "an act of connection as well as mastery" that "both alienates us from nature and binds us to it." Dialogue with nature is possible, and dialogue "implies equality and mutual respect" rather than hierarchy, forges "a bond . . . which will give us the knowledge and the strength we need to foster change" (Bushnell 2011, 217).

Lately, however, I find myself freezing up at the assumption that the work of eco-criticism must be cooperative to be generative. The implied quid pro quo of Bushnell's dialogue—in exchange for participation, knowledge is transferred from one party to another—acknowledges the inescapability of language as a mediating system but also limits that system to productive outcomes or at least privileges measurable exchange in deference to the assumption that the aim of eco-criticism is to serve as a guide for humankind to untangle its errors. But why assume every snag should be untangled? Along with Vicki Kirby (2008, 234), I ask, "What happens if nature is neither lacking nor primordial, but rather a plenitude of possibilities, a cacophony of convers(at)ion?" That we do *not* form productive bonds with the nature to which "we already belong" is a realization within ecological thought as perplexing as it is undeniable. Even if/when we avoid the trap of speaking for nature and instead dialogue with it, we should anticipate and investigate those conversations that are unproductive, that end abruptly, that retreat into silence, that snag, that freeze.

Here, I investigate the snowflake insult as the tentative embrace of a subject position disencumbered of the privileges assumed to accompany participation in or cooperation with cultural hegemony. Rather than rewrite snowflakes' vulnerability as unrecognized or stifled strength, I want to recognize their light touches for what they are: special but forgettable, captivating but ultimately inexpressive or otherwise unaccountable according to the usual social metrics that associate survivability with measurable, preservable consequence. That this former "word of the year" has faded in prominence and popularity is itself evidence of the kind of inconsequence I want to explore. The snowflake ethos accepts accountability without impact and, rather than take for granted the "metaphysical commitment to 'living on' after we die" (Colebrook 2008, 58), supposes an uncharted aftermath of expendability. Such an ethos is "critically posthumanist" in that it takes seriously "and, to a certain extent, literally" the "posthuman challenge" to go beyond the limitations of humanist perspectives and look out for "the possible advent of a new 'episteme,' in which the human again becomes a radically open category" (Herbrechter 2012, 11 and 15). The humanized incarnations of snowflakes in the premodern literature that I discuss below represent the material realization of unmeasurable impact, and as such, they retrieve for readers a language beyond eco-phobic appropriation. We should explore the possibilities of this nonhuman identification not because a collective acknowledgement of collective vulnerability will provide a means of escape from the material conditions of the Anthropocene, but because revising the narratives of survivability that humans like to tell themselves—and expanding them so as to feature characters who disrupt the associations between specialness, impact, and self-preservation—means reckoning more honestly with the human as threat and threatened, as the figure whose methods of preservation against erasure have ironically proved profitable enough to inscribe erasure as obligatory.

Rabelais's Frozen Words

In order to understand the relationship between an ethos based on specialness disconnected from consequence and the rhetorical significance of the snowflake insult, let us consider the rhetorical weight carried by this cold criticism by putting it in conversation with premodern narratives that make use of a familiar trope: the frozen words. In the *Moralia* [c. 100 AD], Plutarch tells of a city where words freeze in the winter and thaw in the summer. Baldassare Castiglione's *Book of the Courtier* [1528] relates a similar tale about two groups of traders whose business is delayed when words spoken by the buyers on one side of a river freeze in place before they make their way across. In *Le Quart Livre* [1548] of François Rabelais's Gargantua and Pantagruel series, the author devotes two chapters of his Odyssean parody to an encounter at sea with frozen words.[6] In chapter 56, Pantagruel "cast[s] fistfuls of frozen Words on to the deck," where the shipmates warm them with their hands: "they melted like snow, and we actually heard them but did not understand them" (Rabelais 2006, 829).

> I saw many sharp Words, and bloodthirsty Words too . . . ; there were dreadful Words, and others unpleasant to behold. When they had all melted together we heard: *Hing, hing, hing, hing: hisse; hickory, dickory, dock; brededing, brededac, frr, frrr, frrr, bou, bou, bou, bou, bou, bou, bou, bou. Ong, ong, ong, ong, ouououong; Gog, magog* and who-knows-what other barbarous words; and the pilot said that they were vocables from battles joined and from horses neighing at the moment of the charge, and then we heard other ones, fat ones which made sounds when they melted, some of drum or fife. . . . Believe you me, they provided us with some excellent sport. (Rabelais 2006, 830)

Initially, the cacophonous manner in which the words become audible renders them ugly to the narrator's eyes/ears. Perhaps his discovery of the words' context—a violent battle—colors his reaction. Or perhaps it is the limits on this contextual knowledge that leave the narrator cold, the way the reification of language is made so conspicuous in this episode that words' symbolic meanings are lost as their "literal meanings become suddenly and absurdly volatile" (Anderson 1996, 19–20). Indeed, an argument erupts when Panurge takes Frere Jean "literally at his word when he was least expecting it" (Rabelais 2006, 830). Judith Anderson (1996, 35–36) reads Rabelais's episode as evidence of "an awareness of the way *sententiae* operate as templates of meaning, freeze language, and appear to solidify it" as a "kind of thing" to be employed when the occasion arises for its proper, profitable, and/or pleasurable employment, but by exaggerating the frigidity of the discourse in this chapter, Rabelais invites readers to ask whether there is a meaningful distinction between the absurd language games played in this novel and regular, respectable humanist practice.

As "word things," frozen utterances reflect the human(ist) tendency to nominalize and collect language through a process that involves recording select phrases as commonplace. Ann Blair (1992, 541) describes this "quintessentially humanist method of reading and storing information," in which "one selects passages of interest for the rhetorical turns of phrase . . . or the factual information they contain; one then copies them out in a . . . commonplace book . . . to facilitate later retrieval and use." "Commonplace," then, is a designation that grants both exceptionalism and accessibility. The frozen-words trope calls attention to a difficulty embedded in this humanist practice in that preserved words, dwelling in an unkairotic landscape of irrelevance and/or inopportunity, often prove to be obstacles in the free flow of discourse rather than facilitators of its smooth, accumulative progression toward learning. As Eric MacPhail (2014, 73–74) remarks on the commonplace tradition, though their "infinite adaptability was prized," proverbial sayings could be versatile to the point of cancellation, because "the same saying [could] be used to argue on both sides of the question." This perhaps explains why nothing comes from Panurge and Frere Jean's verbal sparring and why the episode with the frozen words abruptly ends without revisiting the high-minded explanation Pantagruel suggests upon the travelers' approach: "that there are several worlds so touching each other as to form an equilateral triangle at the core and centre of which lay . . . the Manor of Truth, wherein dwell the Words, the Ideas, the exemplars and portraits of all things, past and future" (Rabelais 2006, 827). He and Panurge anticipate the meaningfulness of this encounter by associating the frozen words not only with the Manor of Truth but also with Gideon's fleece, Homer's poetry, Plato's philosophy, Orpheus's song, and Moses's Law (Rabelais 2006, 827–29). Yet none of the characters essay the "philosophical inquiry" that Pantagruel initially advises (Rabelais 2006, 828), and when the narrator tries "to preserve a few" of the words, Pantagruel stops him, "saying that it was madness to pickle something which is never lacking and always to hand as are gullet-words amongst all good and merry Pantagruelists" (Rabelais 2006, 830). It is madness to preserve words already conveniently accessible, says Pantagruel, suggesting how familiar this supposedly exceptional encounter with frozen words has felt, how "seemingly natural [a] tendency" it is "to hypostasize language in this way" and thus give "the impression that the concept really exists, . . . that it has 'weight' and 'substance'" (Anderson 1996, 22). Without explicitly rejecting the possibility that here dwell the Platonic forms, the conclusion of these chapters suggests that nothing more exemplary resides in this episode than cold words delivered in a cold style. This matters because, in the premodern era, coldness takes on particular meanings when applied to rhetorical situations. Aristotle's (2000, 88–89) *Rhetorica* (fourth century BCE) defines frigidity of style as a matter of excess: "strange words," "obscure" metaphors, and "unseasonable" epithets. Demetrius (n.d., 119)—writing in the first century CE—states later that the proper temperature of a speech may be determined, in large part, by a speaker's sense of decorum, and he draws a comparison between coldness and imposture: "The impostor boasts, facts notwithstanding, that qualities belong to him which do not. In like manner, . . . the writer who invests trifles with pomp resembles one who gives himself airs about trifles." Demetrius (n.d., 126) goes on to claim that "the most frigid of all figures is hyperbole" because "it suggests something impossible"—a move occasionally appropriate for comedy and poetry but generally out of place.

What, then, is the takeaway from Rabelais's frozen words? It's one thing to demonstrate "the remarkable mobility and fluidity of Renaissance commonplaces in humanist fiction" (MacPhail 2014, 74) by imagining them as maneuverable, material objects. It's another to suggest that even the most literally outlandish words ("frrr," "bou," "ong") may be taken for templates with substance and authority in the world. Rabelais's travelers prepare for the possibility of entering a Manor of Truth, essentially the dwelling place for the justification behind humanist practice, where the concept of preservation is preserved as much as any individual, exchangeable truths. The coldness of Rabelais's environment, however, conjures skepticism about the correspondence between preservation and value and thus skepticism of a humanist rhetorical tradition in which the accumulation of words as cultural capital signal one's socially responsible contribution to civilization's progressive momentum. If gullet words are special enough to guarantee their preservation and (over)accessibility, of what value is a specialness grown so common? If we think of Rabelais's frozen words as "vaulted offerings to futurity" and of the commonplace-making methodology as a vault designed to "enciphe[r] the power to ensure continuity by summoning a conjectural future against which it has been built" (Bruyére 2013, 66), then in these chapters, Rabelais dramatizes the moment of the vault's opening, the words' thawing. But instead of validating the preservation process by narrating a scene of epiphanic revelation or even a modest, sensible exchange between communicants, he treats readers to a noisy equivocation. The frozen words' unseasonable coldness marks them as inappropriate to occasion even as their state of preservation marks them as most worthy of use. The result is that the question of what is worth preserving proves far less significant than the larger question, What is preservation worth?

The anxiety provoked by such questions is arguably what develops along with the awareness that "with respect to language we are late," that "something is given ... before we are in the condition of agreeing to it" (Lyotard 1993, 103 and 105). Rabelais's story illustrates this lateness with respect to language; it dramatizes the crew's introduction to the frozen words as a (re)introduction to the most outlandish regions of language *and* to the uncanny notion that these regions are reachable even if not relatable, that the scope of our "psychic ecology" (Lyotard 1993, 105) is more dense than vast. The realm of the frozen words is not the distant Manor of Truth but a well-trod zone of hyperbole, where unarticulated or even inarticulable trifles have the weight of truth without the instructive impact, where language has been preserved without being adjusted for the "optimal performance [that] predominates everywhere" in logocentric knowledge systems (Lyotard 1993, 98). The frozen words are at once special and dysfunctional; they form a "stuttering" literary language that fails to operate "as the instrument of a willing organism" or "the indication of a fulfilling sense," instead standing as evidence of how often life "produces singularities that may remain detached, unlived and devoid of relationality" (Colebrook 2008, 75). The anxiety produced in the frozen-words episode stems from (1) awareness of an expanded scope of humanist discourse that accommodates "the discourse of the secluded, of the thing that has not become public, that has not become communicational, that has not become systemic" (Lyotard 1993, 105); and (2) awareness that the promiscuous preservation of language that is not optimal, perhaps not even communicational, is already regular practice and can result in confusing overlaps between the outlandish and the commonplace, the special and the inconsequential, the wasteful and the profitable—categories that we rely on to orient not only our arguments but also ourselves.

Indeed, the frozen-words chapters point to their own promiscuity within *Le Quart Livre*'s wider quest narrative: Panurge's search for the Oracle of Bacbuc, final consultant on his question of whether he ought to marry (book 3 sees Panurge visit several prophets and advisors, all of whom tell him that if he marries, his wife will beat, rob, and cuckold him). When Pantagruel assures the narrator that he need not go to the trouble of preserving any of the frozen words, Rabelais gives readers simultaneous permission to dismiss the preserved content of this novel, which might be held as lightly as the gullet words melting in the narrator's hands. Rabelais thus swerves from typical usage of the frozen-words trope as a reminder of the existence of cultural codes and the process of cultural encoding—i.e., the "fixing of perception" in language that then seems to operate independently, beyond the language user, to ease the incorporation of subjective experiences "into existing categories" (Anderson 1996, 21 and 23) and thus confirm an intelligible system of manners and morals. Instead of using the moment of the thaw to confirm that the arrested moment "is only meaningful when inhabited by story" (Garrett-Petts and Lawrence 1998, 174), Rabelais illustrates Pantagruel's willingness to exempt himself from the obligation to tender every encounter as a contribution to a grander account. The frozen words have as little impact on him as the countless prophecies confirming his future marriage of misfortune have on Panurge. Thus, by their casual handling of language that might be truth and might be nothing, Rabelais's characters propose not exactly a break from a humanist tradition that treats encounters with preserved language as revelatory, but rather "a shift in the kind of attention revelation demands, an attenuation of its stakes" (François 2008, 54). They act in defiance of the humanist conviction that, in the course of one's learning, to encounter something preserved is to encounter something worth preserving and that to act in the interest of its continued preservation is obligatory (François 2008, 30). Instead, the frozen words are received as special but nonimpactful, and preservation is treated as a mysterious fact of nature, a phenomenon of an unpredictable rhetorical climate through which the characters sail.

Such rogue preservation puts in question a tradition that relies on a direct correlation between revelation and what humanist texts, like Giovanni Pico's ([1496] 1948, 225) *Oration on the Dignity of Man*, present as the singular opportunity divinely granted to humankind to intensively "observe whatever is in the world" and "with freedom of choice . . . fashion [ourselves] in whatever shape [we] shalt prefer." Rabelais's frozen words thaw into a future in which they cannot be incorporated into such human-centered processes of meaning making; their moment of revelation is also a moment of dubiousness for humanism's underlying assumption that revelation is the prerogative of humans. Pantagruel's failure to endorse the links between preservability and continuity instead exposes the stacked assumptions in a humanist philosophy that assumes the inseparability of being and impact and insists on preservation as the sole means of measuring value. If we take away anything from these chapters, it should be the chastening reminder that the matters humans so carefully preserve may thaw into a future not at all continuous with what they presently know.

"The Snow-Child"

Rabelais's frozen words offer cold comfort: that language *can* be preserved, just not apart from the dynamic activity of a posthuman ecology in which nature "punches back at humans and the machines they construct to explore it in ways we cannot predict" (Alaimo and Hekman 2008, 7). As physicist Karen Barad (2007, 185) remarks, "we don't obtain knowledge by standing outside the world; we know because we are *of* the world." Knowledge plus the words we use to articulate it are real as weather, and words matter not as forecasts of continuous value systems but as actors in an onto-epistemological cosmos in which "an inherent difference between . . . mind and body, matter and discourse" is never assumed (Barad 2007, 185). Rabelais's frozen words illustrate perfectly this mutual implication of language and matter. Along with the words, what melts before the narrator's eyes is the belief that ontology and epistemology are separate and isolatable. What becomes graspable as a result of this loss is a new "accountability for what materializes, for what comes to be" (Barad 2007, 361)—in this case, a language that is itself unaccountable, a secluded discourse unencumbered by the burden of continuity and hinting at the vitality of a posthuman *world without us*. Rabelais's chapters are frustrating because the frozen words are neither incorporated as contributors to prevailing ideologies nor unequivocally rejected as valueless. But I venture that the more frustrating scenario would be one in which Pantagruel insisted on total incorporation of every word, down to the last "frr" and "ong."

Such a scenario of stubborn incorporation plays out in the eleventh-century fable "The Snow-Child." In this "amusing" (according to the narrator) tale, an unfaithful wife tells her husband that the child she bore during his long absence was conceived innocently: "once, stricken with thirst in the Alps, I quenched my thirst with snow. So, pregnant from that, alas, I gave birth to this son" (Ziolkowski 2009, 259). The cuckold is suspicious but waits five years before taking the snow-child on a voyage, selling him to a trader and returning home with his own fantastic explanation: "A storm arose and a raging wind drove us, too tired to resist, onto sandy shoals," after which "the sun scorched us all terribly, and that child of yours melted." The narrator tidily concludes the story: "thus fraud overcame fraud; for the child whom snow engendered quite rightly melted under the sun" (Ziolkowski 2009, 259–60).

Undoubtedly this is a story about the violence of reification; the husband reifies his wife's lie, reduces her flesh-and-blood child to her fantastic account of a "natural" conception. The "amusing" part of this story, apparently, comes from seeing the cheating wife taken at her word, her epistemology taken as ontology. As for the moral lesson, if it leaves us cold, it's because we recognize the narrator's impossible-to-credit insistence on the "quite rightness" of an ending that tries to trade a child and melt it too. To accept the story's "rightness" is to accept the narrator's frigid humor, taking him at his word when he describes the human victim of vengeful human justice at the center of this story as a nonhuman enactor of natural processes at the end of it—a snow-child quite rightly, naturally, and not at all hyperbolically, melts in the sun. As W. F. Garrett-Petts and Donald Lawrence (1998, 174) remark of the frozen-words trope, "the interruption [by] the temporarily frozen moment only heightens our anticipation of the flow" of the narrative. This is distressingly true of "The

Snow-Child," for which it is inescapably "right" to understand, from the title alone, that the natural conclusion of this fable and the doom of its title character are equivalent. His mother's falsehood made flesh, the snow-child drives the fable's plot and is doomed by it.

It's not that "The Snow-Child" is elegant but unjust—it is unjust because it is so elegant, because the snow-child and all surrounding figures are so tidily incorporated into its moral, *because* the recognition of the snow-child's special distinction in the story inscribes his sacrifice to our anticipation of his/the story's end. The narrator calls this justice, and he's right insofar as the momentum of the story justifies the snow-child's erasure; the structure of the joke inscribes it. Here, annihilation is no less serious for being primarily discursive. Here, doom exposes itself as part of the grammar of narrative continuity.

Remembering both this fable and Foucault's (1972, 229) advice to "conceive discourse as a violence that we do to things," we can continue to entertain interpretations of the snowflake insult's entry into current debates as an elegant imposition less about ascertaining right/wrong positions than about judging and discursively shaping participants' fitness to stand up amidst the rhetorical currents flowing around them. To use the snowflake insult is to narrate a fable that reifies an opponent's specialness in order to frame it and them as vulnerable—easily swept away from hearing or view. "All they can do is melt," as one writer put it, reflecting on the weaponization of the snowflake insult and referring to its primary target: "drifting young people with little prospect of the prosperity or security of postwar generations" (Green 2016, n.p.). Each instance of the insult's use allows the insulter to assert that specialness is akin to weakness; but given the insult's indiscriminate application, we are approaching a generalized correlation between specialness and vulnerability. If anyone who weighs in on debates about how humans shape culture can be labeled a special snowflake, then anyone is quite rightly doomed to get what's coming to them/us: acknowledgement indistinguishable from deletion. We may, like the cheating wife in the fable, deploy the term "snowflake" to identify anomalies that we wish at the same time to dismiss as inconsequential, thus preserving our security and self-interest. But once everyone is a snowflake, we will find ourselves, like the snow-child, confined to a narrative that coalesces that which stands out with that which cannot stand up. Such is the nature of knowledge-making practices: an identification constructed as an ideological tag is also a material enactment of vulnerability; the extensive employment of the snowflake label alerts us to the extent of our individual vulnerabilities in a time of social, economic, and environmental crisis. And it alerts us to the more comprehensive vulnerability of cultural hegemony, which since the early modern period has treated human ingenuity as the species' special feature, as the thing that somehow both breaks ground and smooths the way for the continuation of identical processes of production that, more and more, we experience as unsustainable.

On the inevitable overlap of cognition and materialism, Gerard Passannante (2017, 459) is severe in his assessment that "materialism does not free the mind from its fictions . . . but stuns and paralyzes it with its own," and "sometimes merely entertaining an idea is enough to find oneself entranced by it." What is significant about the snowflake insult is just this exposure of the idea of humankind's special distinction as an entrancing material fiction entertained to a point of no return. My argument for (re)claiming the snowflake insult is not a call to accept dead ends as destiny or to imitate the snow-child's silent acquiescence to disappearance within a dooming narrative. My final premodern snowflake example illustrates a richer response to the threat posed by the use and abuse of preservation narratives as well as to the opportunities granted by survival stories that distinguish vitality from impact and duration.

Spenser's Florimell(s)

Elizabethan poet Edmund Spenser took seriously his contemporary Philip Sidney's ([1595] 2002, 105) reminder that "poesy may not only be abused, but that being abused, by the reason of his sweet charming force, it can do more hurt than any other army of words." Sidney was not boasting out of turn of the power of his craft, not in an age when "allusions to the power of poesy often insist[ed] upon its superior ability to impress itself on the world," with *impression* understood to be both ornamental and phenomenal (Mann 2017, 238). Early modern writers made a habit of "pictur[ing] the hidden operations of perception and cognition" by "equating the human interior with other material elements of the natural world" (Mann 2017, 239). Frequently, the poetic voice of *The Faerie Queene* [1590]—Spenser's epic endeavor to represent Christian and courtly virtues in allegorical figures—interrupts his verse to lament that characters must suffer the trials he himself has thought up for them. The constantly imperiled Lady Florimell melts the poet's heart when he thinks, "how causelesse of her owne accord / This gentle Damzell, whom I write upon, / Should plonged be in such affliction" (Spenser 2001, III.viii.1).

One outcome of Florimell's being constantly afflicted by rapacious men—even as she is tormented by the irony that the one man she desires, Marinell, has sworn off women entirely—is the creation of her double. In canto 8 of book 3, a witch conjures a snowy clone of Florimell to console her love-stuck son, from whom the original Florimell has fled. In addition to "purest snow" and "virgin wex," the witch installs "two burning lamps" in place of eyes, "golden wyre" for hair, "and in the stead / Of life, she put . . . A wicked [male] Spright yfraught with fawning guile" and skilled in "all the wyles of wemens wits. . . . That who so then her saw, would surely say, / It was her selfe, whom it did imitate, / Or fayrer then her selfe" (Spenser 2001, III.viii.6–8). False Florimell is not only mistaken for the original at several points in Spenser's epic, but also, and confusingly, compared with her, for example, in book 4, when "all were glad there *Florimell* to see; / Yet thought that *Florimell* was not so faire as shee" (Spenser 2001, IV.v.14).

The original Florimell appears in the Book of Chastity, and it is a hard-won virtue for her, given the frequency with which she is in danger of ravishment; given the implication in the verses quoted above that her admirers would be universally unable to authenticate the essentiality of her virtue even if they were looking at the "real" thing; and given the running theme throughout *The Faerie Queene* of characters failing to distinguish between the simulated and the real. Florimell is often read as a representation of "the elusiveness of true beauty and transcendent ideals in the world," ideals that Kenneth Borris (2009, 250) argues Spenser ultimately makes legible by "allow[ing] for skeptical counternarratives about them . . . in the context of a conclusion that valorizes" them. When the two Florimells finally come face-to-face in book 5 and "Th'enchaunted Damzell vanisht into nought / Her snowy substance melted as with heat" (Spenser 2001, V.iii.24–25), readers should understand that the false Florimell's inauthentic virtue *quite rightly* melts once in proximity to the original Florimell's true radiance. Furthermore, "Florimell's vindication symbolically applies to all that she represents . . . and to her creator's poetic" (Borris 2009, 250), which I take to mean it invites readers to emphasize or deemphasize the material or immaterial components of the text's figured virtues and their simulations, depending on which move better serves the epic's valorization of an ideal (regardless of whether the clarity of the definition of that ideal is enhanced or diminished as a result). All of which is to say that the false Florimell's fate is inscribed as baldly as the snow-child's, if not quite as neatly; though she is animated by an assemblage of materials, it is her snowiness that is retroactively elevated as her defining feature, the better to explain her disappearance as a natural response to the "blazing" qualities of her double—the original Florimell is, in earlier cantos, compared to or associated with blazing suns, stars, and comets, while "the impostor's bodily substance is snow from 'a shady glade'" (Spenser 2001, III.viii.6), no match for the original's vaporizing brightness (Borris 2009, 247).[7] Of the lamps, the golden wires, even the male spright who/that is part of the Florimell assemblage, nothing more is said.

A further inconsistency arises when the original Florimell's enchanted girdle, her lost garment that appears in earlier cantos as a kind of chastity detector that no unchaste female character can successfully don, appears in the spot where the false Florimell vanishes. Unlike the snow-child, the false Florimell leaves behind a visible remainder that prevents her complete subsumption to a moralizing lesson/punchline. In book 4, the false Florimell tries to secure the girdle around her waist, but with every attempt to fasten it, "it loos'd / and fell away, as feeling secret blame" (Spenser 2001, IV.v.16). Spenser (2001, V.iii.28) reiterates later, "Full many ladies had assayd, / About their middles that fair belt to knit ... Yet it to none of all their loynes would fit." Yet stanza 24 of the same canto states that "th'emptie girdle" the false Florimell leaves behind "about her wast was wrought." This isn't the only narrative inconsistency in the six complete books of *The Faerie Queene*, of course, and some readers dismiss the girdle's description as another of Spenser's lapses; however, this particular lapse "is not a matter of the poet forgetting what he has written over the course of two books, but, apparently, forgetting over the course of five stanzas" (Yearling 2005, 137). Regardless, it is unmistakable that the text here gestures to the false Florimell's capacious ecology, to the unpredictable vitality of her "transcorporeal" nature,[8] even as it draws the reader's attention to her devitalization as a doomed-to-be-vaporized-mockery damsel of snow.

False Florimell is a troublemaker. She's summoned into existence as the "other" whose imposture confirms the legitimacy and moral solidity of the figure from whom she was copied. But once again, we have an author who, by exaggerating the coldness of his creation, invites skepticism about the broader legitimacy of a value system in which the dysfunctional or nonoptimal performances of cold characters are readily accommodated as constitutive of the same values they simultaneously pervert. False Florimell's disappearance disorients because, as we recall from "The Snow-Child," it's a cheap resolution that tries to instantiate a figure and melt it too. Once again, questions arise about the preservation process; if the only way to preserve the original Florimell's value is to doom her double to disappearance, that suggests a miscalculation in the process itself, and it leaves open the question of what the preservation of Florimell's chastity is worth when the falsified version of it seems to function just as well right up until the moment it melts, if not beyond. As Robert Tate (2014, 201) points out, "one cannot describe False Florimell's vanishing act as iconoclastic" because "iconoclasm implies the agency of an image-breaker" and the false Florimell melts "for no apparent reason beyond the juxtaposition itself ... her only residuum being the witnesses' 'wonderment' (V.iii.26)." Tate goes on to argue that this "wonderment" promises to have a material effect within the psychic ecology of every witness. For Marinell, in particular, the false Florimell's disappearance threatens to linger as trauma, in that he "los[es] his sense of what constitutes purity" as he confronts the likelihood that his failure to distinguish between the false and original is "because he has been projecting an idealized image upon Florimell from the beginning" (Tate 2014, 211).

I agree with Tate's (2014, 201–2) observations that it is unwise to leave the scene of the false Florimell's disappearance "under the illusion that there is nothing more to track"; that readers must pursue Florimell "beyond the threshold of visibility"; that we must "start asking, with seriousness, 'What of her became?'" Indeed, to continue to track Florimell postdisappearance is one way to take as seriously as Spenser does the force of poetry and to escape "the philosophical view behind thinking that objects are one thing and relations . . . another" (Morton 2013, 73). However, I hesitate to go so far as to claim, as Tate (2014, 213) does, that "False Florimell does not leave . . . has never left," but rather "crosses back into the 'liquid ayre' that surrounds and cycles through her perceivers" because "it is of this air, their breath, their words that she is 'framed.'" While this interpretation rightly acknowledges the onto-epistemological vitality of language- and knowledge-making practices, it also imposes on the false Florimell a relationship of cyclical cooperation with the community that has just expelled her.

Rather than rewrite Florimell's alterity as rarefied embodiment, I suggest we dwell on the possibility that one answer to the question "What of her became?" really is "nothing." Such is the reading that takes most seriously the entanglement of preservation and erasure in the composition of humanism and hegemony to which Spenser's *Faerie Queene* contributes. False Florimell came into being, passed as a figure of virtue for most of her existence, and at her melting point, confronts how little that means. The functionality of her nature neither saves her nor guarantees her preservation in any coherent record of events: "what of her became, none understood" (Spenser 2001, V.iii.26). She melts away like one of Rabelais's frozen words, and as with those words, we are not obliged to find her absence revelatory.

But Spenser also invites us to leave his question, "What of her became?" open. False Florimell's proximity to the girdle proves she should not be investigated "as (solely) the ground for the action of something else" (Alaimo and Hekman 2008, 257), but neither must she be investigated as a temporarily dematerialized agent whose crossings back and forth between liquid air and solid ground must ultimately assume a legible social function. That the false Florimell survives as untraceable matter is another possibility. For one thing, it may better explain why Marinell is dismayed to the point of crisis by her disappearance: because *nothing* suggests itself to be made of it. Mightn't he find the false Florimell's reasonless vanishing even more unsettling than the relatively coherent explanation that she was a trick and he a dupe? Marinell would be in good company—with Una and Redcrosse, with Arthur, with several heroes of Spenser's epic who find themselves repeatedly unmasking the same villains[9]—if all he had to worry about was being confronted again by the same test of perception the next time some errant knight stumbles on the witch's recipe for Florimell: take ten snowflakes, one evil fairy, two cups liquid air, and mix well. It seems to me, however, that Florimell survives in this text not as pop-up antagonist but as a figure of alterity, devoid of the relationality and functionality that would permit her to be "framed" again as either villain or damsel. She is matter that remains unaccountable, that refuses "to be granted meaning by thought" (Colebrook 2008, 56). In place of a benevolent discourse that would speak *for* her in order to consolidate or undo her otherness, we can contemplate the false Florimell as she is: "unknowable alterity, an excess, which elides comparison and exchange" but which deserves our attention regardless (Birla 2010, 96–97). Claire Colebrook (2008, 56) suggests that "as long as the 'life' of vital matter is deemed to be creative, productive, and intensive, then we remain caught in an age-old moral resistance to those aspects of life that remain without relation" and we reinforce the binaries "that privileg[e] act and production over inertia and passivity." So even though it feels cold to conclude that the false Florimell's disappearance in *The Faerie Queene* makes no representable impact on the story, it is important to linger in this coldness. Here, we might find the "space to thrive" for which Stacy Alaimo and Susan Hekman (2008, 259) suggest eco-critical readings must advocate—space expansive enough to accommodate ways of being that are "too complex to be predicted in advance" and too challenging to our commonplace assumptions about vitality, duration, and worth.

Early modern writers lived and wrote during what we now describe as the Little Ice Age, a decades-long cold snap that played havoc with expectations of weather and climate. The particular climatological volatility of this era does not map seamlessly onto our own environmental crises, yet it is instructive to examine the ways premodern artists "internalize[d] [Nature's] unpredictability" and created "fictions of stability" as "a compensatory rhetoric to project an idealized vision of Nature onto an ever-changing and often threatening world" (Markley 2008, 132, 140, and 134, respectively). Preservation narratives were and still remain a large part of this compensatory rhetoric, but fictions of nonpreservation may play a larger role in future works of both art and oratory. The climate crisis is a phenomenal and an epistemological problem, having to do with particular "threat[s] to the reproduction of life" made against "multiple subjectivities," but also having to do with the fact that, alongside these experiences of "disarray," the world is "not politically unpredictable enough" (Berlant and Stewart 2012, n.p.). To turn to figures such as the false Florimell is to turn toward a politically unpredictable space in which humans can imagine themselves as a species not entitled to preservation. We can grant the horror of that thought while also granting the horror of the contrary: the clinging belief that, more than literally anything else, humankind is entitled to endure. False Florimell invites the dread/wonderment that existence is imaginable and that accountability for being is obligatory even when the assumption of life's extension is absent. "Our desires change as we see ourselves differently," and if we do the work "of reenvisioning ourselves as other than those entitled to help," to stay, to endure (Cornell 2010, 113), imagining instead what might constitute *being* inside a "politics beyond the lived" (Colebrook 2008, 81), how might we learn to treat each other and change each other?[10]

Conclusion

Survival narratives are often more concerned with "defin[ing] zones of interest in life, rather than ways 'to keep death—or the wrong kind of death—at bay'" (Bruyére 2013, 73), meaning they act less as practical survival guides than as containers preserving the hierarchies of value in which humans are presently invested and away from which we have trouble straying. Lauren Berlant (2008, 23) agrees that although fantasy "always expresses a desire for continuity with a better world than yet exists," the implied endorsement of that better world is "expressed in an attachment to objects that do already exist and therefore misrepresent inevitably the desires that bring people to them." Enter snowflakes, who may not provide the keys to escape attachment to predictable hierarchies of value and the preservation-for-some ideologies they secure, but who can alert us to ways we might, from within the cuff, express, validate, and make commonplace certain "zones of interest in life" that don't get recognized, let alone endorsed, as inhabitable. With the material feminists cited here, I am interested in the possibilities that emerge from a "literal 'contact zone' between human corporeality and more-than-human nature," in the snowflake's transcorporeality as "a place where corporeal theories and environmental theories meet and mingle in productive ways" (Alaimo and Hekman 2008, 238). I am interested in willingly entering the snowflake zone as a place where specialness is credited but not exploited, where agency and self-distinction are managed via enigmatic methods that are only self-defeating if we accept "the modern ideology of improvement that cannot admit the waste of unexploited powers" (François 2008, 22).

I venture that the fleeting popularity of the snowflake insult is connected to the apprehension of unlearning the normative expectation that self-worth is realized by *making the most of* what a self has to offer and that claiming the insult for rhetorical history and material feminist practice can be part of "undoing a historically contingent affect world that offers the apparent security of identities and things in lieu of just and equitable social relations" (Radway 2012, 343). Snowflakes are not redeemer figures, nor is redemption a matter of humans finally promoting a narrative of self-interest compatible with nonhuman vitality, but adding the option of inconsequential, nonimpactful agency to stories of survivability is a rhetorical exercise with nontheoretical implications. It is difficult to shift "the normative presumption favoring the articulation of human potential" as the crucial action for the development of self-regard and the advancement of civil equilibrium, but since this presumption coincides with the hegemonic insistence that value must be determined transactionally and that development of resources is estimable up to, or even past, the point of their depletion, I see a need for more stories in which characters participate casually, if at all, in the preservation and exchange of social capital, in which they "accept the risk of going unnoticed" (François 2008, 22). I do realize the perversity of suggesting indifference to duration as part of an ethical praxis given the resurgence of fascist politics across the globe and the embrace of systematic dehumanization as a strategic response to refugee crises and, in the United States specifically, police brutality. Nor am I confident that inserting this conversation about snowflakes into long-running debates about critical theory's attachment to socially legible expressivity as the mark and means of subjectivity makes it any less callous.[11] The threat of extermination is more real for some human beings than others; that has to be said. But for those of us with the privilege to take duration for granted—those with enough social or political capital to count on the modest extension of their current livelihoods, imagining our own nonpreservability is work worth doing, even when it is the least we can do. That means recognizing that preservation itself is troublesome, that when it is the linchpin of humanist theorizing, humanitarian aid, and environmental activism, it limits the extent of the change in the human-nature dynamic we say we are open to fostering.

What happens if we delay the critical impulse to "give in to the rehumanization reflex" and instead face the foundation-altering prospects of a posthumanist ethos (Herbrechter 2012, 53)? After mattering to such an extreme that human beings exist in an era named after our own influence, we may find the fantasy position of *not mattering* to be valuable for the narratives we imagine about our future and our present survivability—valuable not necessarily as one-size-fits-all models of imitation for day-to-day resistance against populist frenzy, but as part of a continuing critical dialogue on posthuman ethics. We must explore the limitations of parochial narratives that imagine survival from a strictly human point of view, even as those of us who participate in the dialogue acknowledge that ruminating on the climatological drama of the imperiled human species is not an invitation to deem irrelevant the sociological perils of minoritarian groups and their daily struggles for visibility.

Notes

1. The association between snowflakes and childhood continues to hold water for writers eager to malign millennials born between 1981 and 1996 as members of "generation snowflake," even as the word "snowflake" resists confinement to a specific age group (*Oxford English Dictionary* 2019).

In 2019, *Merriam-Webster* likewise added new definitions for "snowflake (noun)": "a. someone regarded or treated as unique or special" and "b. someone who is overly sensitive." Both uses are further categorized as "informal + usually disparaging" (*Merriam-Webster* n.d.).

2. Simon Estok (2018, 1) defines "eco-phobia" as "a uniquely human psychological condition that prompts antipathy toward nature" and can be expressed as "fear, contempt, indifference, or lack of mindfulness (or some combination of these) toward the natural environment."

3. It is unscalable due to the existence of what Timothy Morton (2013, 124 and 108, respectively) describes as "hyperobjects"—"fatal substances" such as Styrofoam or plutonium "that will outlast [humans] and their descendants beyond any meaningful limit of self-interest" and thus signal "the end of the human dream that reality is significant for [humans] alone."

4. Early Modern philosopher Johannes Kepler ([1611] 1966, 33), for example, does not by the standards of modern science account for the crystalline structure of the snowflake in his essay "On the Six-Cornered Snowflake," but he certainly observes the correspondence between its hexagonal shape and "the ordered shapes of plants and of numerical constants" before concluding that this mathematical structure is part of "a formative faculty in the body of the Earth, and its carrier is vapour as the human soul is the carrier of spirit."

5. See also Palahniuk 2017.

6. W. F. Garrett-Petts and Donald Lawrence (1998, 144) discuss several more recent variants, primarily from Canada, where they point out that, "by reason of climate and geography, the . . . literary and visual arts scene has been especially involved in the use of northern tropes and interarts strategies." Environmental artist Nicole Dextras (2019, n.p.), for example, constructs outdoor "social typography" installations that consist of block letters made from ice, spelling words such as "legacy," "consume," and "resource." The eight-foot-tall blocks are installed in various environments and photographed as they melt.

7. For example, the initial description of Florimell in book 3 is as follows:

> All as a blazing starre doth farre outcast
> His hearie beames, and flaming lockes dispredd,
> At sight whereof the people stand aghast:
> But the sage wisard telles, as he has redd,
> That it importunes death and dolefull dreryhedd. (Spenser 2001, III.i.16)

8. Stacy Alaimo and Susan Hekman (2008, 238) define "transcorporeality" as "the timespace where human corporeality, in all its material fleshiness, is inseparable from 'nature' or 'environment.'"

9. Duessa and Archimago, for example, are two antagonists who show up repeatedly, in various disguises, to torment Spenser's valorous heroes.

10. Thousands of words into this chapter, I irresponsibly recall Sharon O'Dair's (2008, 19 and 22, respectively) argument that the "responsible way to make historicist scholarship of the early modern period eco-critically active today is not to do it at all," that we "promote sustainability within the profession" by essentially adopting "a no-growth policy."

11. The "unwillingness" among practitioners of critical theory "to allow anyone to go uncounted" is a perspective Anne-Lise François (2008, 30) scrutinizes in *Open Secrets: The Literature of Uncounted Experience*, which was crucial to the generation of this chapter, as were the deconstructive challenges posed in Gayatri Spivak's (2010, 56) "Can the Subaltern

Speak?" regarding the disappearing figure of the third-world woman; caught between "two contending versions of freedom"—tradition and development—"the constitution of the female subject in *life* is the place of the différend." Can we mark this place of disappearance, Spivak (2010, 61) asks, "with something other than silence and nonexistence?"

References

Alaimo, Stacy, and Susan Hekman, eds. 2008. *Material Feminisms*. Bloomington: Indiana University Press.

Anderson, Judith H. 1996. *Words That Matter: Linguistic Perception in Renaissance English*. Stanford: Stanford University Press.

Aristotle. 2000. *Rhetoric*. South Bend: Infomotions. https://public.ebookcentral.proquest.com/choice/publicfullrecord.aspx?p=3314386.

Barad, Karen Michelle. 2007. *Meeting the Universe Halfway: Quantum Physics and the Entanglement of Matter and Meaning*. Durham: Duke University Press.

Barnes, Sara. 2017. "31 of the Most Creative Protest Signs from the Global Women's March." My Modern Met, January 23, 2017. https://mymodernmet.com/womens-marches-signs/.

Berlant, Lauren. 2008. *The Female Complaint: The Unfinished Business of Sentimentality in American Culture*. Durham: Duke University Press.

Berlant, Lauren, and Kathleen Stewart. 2012. "Forms of Attachment: Affect at the Limits of the Political." Lecture at ICI Berlin, Berlin, Germany, July 9, 2012. http://www.ici-berlin.org/docu/formsof-attachment.

Birla, Ritu. 2010. "Postcolonial Studies: Now That's History." In *Can the Subaltern Speak? Reflections on the History of an Idea*, edited by Rosalind C. Morris, 87–99. New York: Columbia University Press.

Blair, Ann. 1992. "Humanist Methods in Natural Philosophy: The Commonplace Book." *Journal of the History of Ideas* 53, no. 4: 541–51.

Borris, Kenneth. 2009. "Platonism and Spenser's Poetic: Idealized Imitation, Merlin's Mirror, and the Florimells." *Spenser Studies: A Renaissance Poetry Annual* 24, no. 1: 209–68. https://doi.org/10.7756/spst.024.007.209-268.

Bowerbank, Sylvia. 2004. *Speaking for Nature: Women and Ecologies of Early Modern England*. Baltimore: Johns Hopkins University Press.

Bruckner, Lynne. 2011. "N/nature and the Difference 'She' Makes." In *Ecofeminist Approaches to Early Modernity*, edited by Jennifer Munroe and Rebecca Laroche, 15–35. New York: Palgrave.

Bruyére, Vincent. 2013. "Paroles En L'air: Climate Change and the Science of Fables." *Diacritics* 41, no. 3: 60–79. https://doi.org/10.1353/dia.2013.0014.

Bushnell, Rebecca. 2011. "Afterword." In *Ecofeminist Approaches to Early Modernity*, edited by Jennifer Munroe and Rebecca Laroche, 211–17. New York: Palgrave.

Colebrook, Claire. 2008. "On Not Becoming Man: The Materialist Politics of Unactualized Potential." In *Material Feminisms*, edited by Stacy Alaimo and Susan Hekman, 52–84. Bloomington: Indiana University Press.

Cornell, Drucilla. 2010. "The Ethical Affirmation of Human Rights." In *Can the Subaltern Speak? Reflections on the History of an Idea*, edited by Rosalind C. Morris, 100–114. New York: Columbia University Press.

Demetrius. n.d. *On Style*. Translated by W. Rhys Roberts. Attalus. Accessed September 26, 2019. http://www.attalus.org/info/demetrius.html.
Dextras, Nicole. 2019. "Social Typography." https://nicoledextras.com/portfolio/social-typography/.
Egan, Gabriel. 2006. *Green Shakespeare*. London: Routledge.
Estok, Simon. 2018. *The Ecophobia Hypothesis*. London: Routledge.
Fincher, David, director. 1999. *Fight Club*. Los Angeles, CA: 20th Century Studios.
Foucault, Michel. 1972. *The Archaeology of Knowledge*. Translated by Alan Sheridan. New York: Pantheon Books.
Foucault, Michel. 1994. *The Order of Things: An Archaeology of the Human Sciences*. New York: Vintage Books.
François, Anne-Lise. 2008. *Open Secrets: The Literature of Uncounted Experience*. Stanford: Stanford University Press.
Garrett-Petts, W. F., and Donald Lawrence. 1998. "Thawing the Frozen Image/Word: Vernacular Postmodern Aesthetics." *Mosaic: An Interdisciplinary Critical Journal* 31, no. 1: 143–78.
Green, Miranda. 2016. "Year in a Word: Snowflake." *Financial Times*, December 21, 2016. https://www.ft.com/content/65708d48-c394-11e6-9bca-2b93a6856354.
Herbrechter, Stefan. 2012. "Introduction—Shakespeare Ever After." In *Posthumanist Shakespeares*, edited by Stefan Herbrechter and Ivan Callus, 1–19. New York: Palgrave.
Kepler, Johannes. [1611] 1966. *The Six-Cornered Snowflake*. Oxford University Press.
Kirby, Vicky. 2008. "Natural Convers(at)ions: Or, What If Culture Was Really Nature All Along?" In *Material Feminisms*, edited by Stacy Alaimo and Susan Hekman, 214–36. Bloomington: Indiana University Press.
Lyotard, Jean-François. 1993. "Oikos." In *Political Writings*, translated by Bill Readings and Kevin Paul Geiman, 96–107. Minneapolis: University of Minnesota Press.
MacPhail, Eric. 2014. *Dancing Around the Well: The Circulation of Commonplaces in Renaissance Humanism*. Leiden: Brill.
Mann, Jenny. 2017. "The Orphic Physics of Early Modern Eloquence." In *The Palgrave Handbook of Early Modern Literature and Science*, edited by Howard Marchitello and Evelyn Tribble, 231–56. New York: Palgrave.
Markley, Robert. 2008. "Summer's Lease: Shakespeare in the Little Ice Age." In *Early Modern Ecostudies: From the Florentine Codex to Shakespeare*, edited by Thomas Hallock, Evo Kamps, and Karen Raber, 131–42. New York: Palgrave.
Merriam-Webster. n.d. s.v. "snowflake." Last modified May 24, 2019. https://www.merriam-webster.com/dictionary/snowflake.
Morton, Timothy. 2013. *Hyperobjects: Philosophy and Ecology after the End of the World*. Minneapolis: University of Minnesota Press.
Nicholson, Rebecca. 2016. "'Poor Little Snowflake'—The Defining Insult of 2016." *The Guardian*, November 28, 2016. www.theguardian.com/science/2016/nov/28/snowflake-insult-disdain-young-people.
O'Dair, Sharon. 2008. "Slow Shakespeare: An Eco-Critique of 'Method' in Early Modern Literary Studies." In *Early Modern Ecostudies: From the Florentine Codex to Shakespeare*, edited by Thomas Hallock, Evo Kamps, and Karen Raber, 11–30. New York: Palgrave.

Oxford English Dictionary. 2019. s.v. "snowflake, *n*." Last modified June 2019. www.oed.com/view/Entry/183512.

Palahniuk, Chuck. 2017. "On the Origins of 'Snowflake.'" *Entertainment Weekly*, November 17, 2017. https://ew.com/books/2017/11/17/chuck-palahniuk-snowflake-insult.

Passannante, Gerard. 2017. "On Catastrophic Materialism." *Modern Language Quarterly* 78, no. 4: 443–64. https://doi.org/10.1215/00267929-4198220.

Penny, Laurie. 2017. "Meltdown of the Phantom Snowflakes." *The Baffler*, January 4, 2017. https://thebaffler.com/war-of-nerves/meltdown-phantom-snowflakes-penny.

Pico, Giovanni. [1496] 1948. "Oration on the Dignity of Man." Translated by Elizabeth Forbes. In *The Renaissance Philosophy of Man*, edited by Ernst Cassirer, Paul Oskar Kristeller, and John Herman Randall Jr., 223–56. Chicago: University of Chicago Press.

Rabelais, François. 2006. *Gargantua and Pantagruel*. Translated by M. A. Screech. London: Penguin.

Radway, Janice A. 2012. "Cultivating a Desire to Become 'Not-Something': Lauren Berlant, the Idioms of the Ordinary, and the Kinetic Temporality of the 'Nearly Utopian.'" *Communication and Critical Cultural Studies* 9, no. 4: 337–45.

Sidney, Philip. [1595] 2002. *An Apology for Poetry, or the Defence of Poesy*. Edited by Geoffrey Shepherd and R. W. Maslen. Manchester, UK: Manchester University Press.

Spenser, Edmund. 2001. *The Faerie Queene*. Edited by A. C. Hamilton, Hiroshi Yamashita, and Toshiyuki Suzuki. London: Pearson.

Spivak, Gayatri Chakravorty. 2010. *Can the Subaltern Speak? Reflections on the History of an Idea*. Edited by Rosalind C. Morris. New York: Columbia University Press.

Takei, George (@GeorgeTakei). 2017. "The thing about 'snowflakes' is." Twitter, January 21, 2017. https://twitter.com/GeorgeTakei/status/822922284256100353?ref_src=twsrc%5Etfw.

Tate, Robert W. 2014. "Haunted by Beautified Beauty: Tracking the Images of Spenser's Florimell(s)." *Spenser Studies: A Renaissance Poetry Annual* 29, no. 1: 197–218. https://doi.org/10.7756/spst.029.009.197-218.

Watson, Robert. 2011. "The Ecology of Self in *Midsummer Night's Dream*." In *Ecocritical Shakespeare*, edited by Lynne Bruckner and Dan Brayton, 33–56. Farnham: Ashgate.

Yearling, Rebecca. 2005. "Florimell's Girdle: Reconfiguring Chastity in the Faerie Queene." *Spenser Studies: A Renaissance Poetry Annual* 20, no. 1: 137–44. https://doi.org/10.1086/SPSv20p137.

Ziolkowski, Jan M., ed. and trans. 2009. "C. 'Modus Liebinc,' or 'The Song to the Liebo Tune' (*Carmina Cantabrigiensia* 14; ATU 1362 'The Snow-Child.'" In *Fairy Tales from before Fairy Tales: The Medieval Latin Past of Wonderful Lies*, by Ziolkowski, 257–61. Ann Arbor: University of Michigan Press.

Chapter 8

MOTHERING THE STORM

Black Girlhood and Communal Care in Literature of Katrina

Jennifer Morrison and Shelley Ingram

Contemporary writers like Jesmyn Ward, Kiese Laymon, and Sherri L. Smith have turned to the Gulf South as a site of creativity and as a place from which to interrogate the African American experience. These writers engage the geographical space of the Gulf South to reveal how it allows for and suppresses the voices of young Black women. Central to this geography is the hurricane. Much has been written about Hurricane Katrina and its aftermath; folklorists, journalists, historians, and social scientists have sought to make sense of the devastation wrought by the storm, generation defining in its devastation and the trauma that ensued. The storm quickly made its way into American letters, as writers of all genres worked their way through this national tragedy. In fact, Samantha Pinto and Jewel Pereyra (2019, 1) suggest, "In the storm's material and symbolic wake . . . an entire movement of black arts, literatures, and cultures emerges—a 'post-Katrina' periodization." For how can you talk about Black life in the Gulf South without talking about the storm?

Carl Lindahl writes that "nothing is more important to trauma survivors than ownership of their own stories." He says that this is particularly true of stories of Katrina. Folks of the Gulf South "felt profoundly misrepresented by media accounts"—they "wanted their stories back" (Lindahl 2012, 153). In this chapter, we investigate novels by Black writers set in the Gulf South that take their stories back, articulating experiences of identity and trauma in relation to storms in order to, as Kathleen McKittrick (2006, x) argues, "make visible social lives which are often displaced, rendered ungeographic." We look to three novels that connect the lives of young Black teenage women in the American Gulf South to the hurricane: Ward's *Salvage the Bones* (2011b), Laymon's *Long Division* (2013), and Smith's *Orleans* (2013). Each novel links the storm to particular types of folk narrative, including Greek mythology, hip-hop, and sacred ritual, as the Gulf South hurricane functions as a powerful metaphor for the forces that seek to erase Black subjects physically and psychically. However, we argue that these novels also suggest additional relationships between subject and storm that create a powerful sense of place and belonging, generative of Black subjectivity and creativity in the Gulf South after devastating trauma: maternal and communal care.

As Melissa V. Harris-Perry (2014, 15) says of Katrina, "television news and popular magazines used images of desperate, frightened African American women to dramatize the tragedy facing residents as they battled the aftermath of the hurricane with little official assistance." It was through the filter of these images in the wake of the disaster that conversations about race, community, family, and gender began. Hurricane Katrina as an event exposed who the most vulnerable citizens of the Gulf South were—those who did not have the protection of money or access to escape the most vicious aspects of a hurricane, as storms like Katrina serve to "make visible the architecture of anti-blackness" (Leader-Picone 2019, 72). Toni Morrison muses in her novel *The Bluest Eye* (1970) that some ground is hostile for some seeds, that there are places where they cannot be nurtured, where they cannot grow. If this is true, how able are the young women in *Long Division*, *Salvage the Bones*, and *Orleans* to make a life and imagine a future in which they are happy and fulfilled? In other words, how fertile is the proverbial ground in the Gulf South for Black women? Will they thrive? Or will they drown, overcome by both climatological and ideological disasters born of antiblackness? We look to the work of these Black writers with deep ties to the Gulf South for some answers, as they use the symbolic power of the hurricane to explore both the tension and the profound affinity between Black girlhood, mothering, and community.

Ward's *Salvage the Bones*

To have control over your narrative, to take back your story, *is* to have control over how you define yourself and your destiny. Ward (2011a, n.p.) describes the terrifying and awe-inspiring experience of being caught in Hurricane Katrina by saying the following of their physical vulnerability during the storm: "It reduced us to improbable metaphor." The novel is written from the point of view of Esch Batiste, a young, devastatingly poor, and pregnant Black girl, as she and her family of men move about the world in the twelve days leading up to and following the landfall of Katrina on the Mississippi Gulf Coast. Ward draws on the folk knowledge of those who have lived their lives waiting for storms. In doing so, she connects four women, four mothers, through time and space: Esch, the goddess Medea, the dog China, and the hurricane Katrina.

Esch is isolated in a sea of men in rural Mississippi. She lost her mother as a child, and almost all the interactions in the text are between her and her three brothers, her father, and male family friends. She is also pregnant, and the father is Manny, one of her brother's closest confidants. The relationship between Esch and Manny is intensely physical yet emotionally distant, and he eventually refuses to acknowledge the child as his. Esch herself juggles several feminine roles: mother to her younger brother, confidante and housekeeper for her father, and sex object to more than one of her brothers' friends. She is also only fifteen. On the surface, it appears that Esch's life will take a familiar track, that she is that statistic trotted out in racist political discourse—a poor Black pregnant teenager. Yet while she is obviously undereducated, she is intelligent and intensely curious.

Esch narrates the events of her life by drawing analogies to Greek myth, as Ward invokes this mode of storytelling to establish it as part of her own mythic and literary heritage and of the cultural and literary heritage of her characters. Esch understands herself and knows herself through the lens of Greek mythology. As a young Black girl in the early days of one of the most devastating disasters of early twenty-first-century America, she relies on this "timeless" form of storytelling to describe the things that happen to her. In fact, Esch rejects contemporary narratives of Black femininity and youth in favor of ancient myth. But this mythology, the story and culture she pulls from, could not have imagined someone like her: a poor Black pregnant teenager who almost dies in a hurricane.

Esch is utilitarian and practical in her engagement with folk narrative. Her favorite book is Edith Hamilton's *Mythology* (1942); it is the only book we see her read in *Salvage the Bones*, and it spurs her imagination and provides narratives through which she can understand herself and her world. It is clear in the way she observes her surroundings that she is not only constantly evaluating them but imagining—creating—them as well. She observes a dog fight, a particularly violent and masculine, though communally satisfying, event:

> There are around ten dogs here, around fifteen boys. I am the only girl. . . . The dogs are brown and tan, black and white, striped brindle, red earth. None of them is white as China. She glows in the sun of the clearing, her ears up, her tail cocked. The dogs nap, pace, bark, strain against the leash, and lean out into the clearing where they will fight, trying to get into the sun, to feel it on their wet black noses. They will all match today, one dog against another. The boys have been drawn by gossip of the fight between Kilo and Boss to the clearing like the Argonauts were to Jason at the start of the adventure. (Ward 2011b, 160)

Esch, being the only girl and pregnant at that, is weighed down by her gender in this moment. She is about to watch a dog fight, a ritualized, masculine community event. Judging by the makeup of the crowd there, she, as a woman, is out of place in this space. The way she chooses to ease her anxiety is to return to her narrative crutch, her way of understanding the world that is both familiar and foreign. She references the story of Jason and the Argonauts, a text that is, like the dogfight, deeply masculine, yet the shadow of the powerful female character Medea hangs over the tale like a cloud. And perhaps because of her own secret connection to female power, Esch summons this particular tale as a way of asserting herself in the proceedings. The use of colloquially crass language, the "low" culture in contrast to the "high" culture of Greek mythology, also suggests Esch's cultural hybridity and her ability to move between her real life and her fantasy life. As the two blur and converge, her narrative elevates her in a moment where she is clearly being ignored by all of those around her. Because she cannot participate in the proceedings, all it seems she can do is watch.

But her observation, her narrative creation, even though it is internal, *does* allow her to take some role in the masculine ritual of the dog fight, and it emphasizes her interiority, suggesting that there is more to her than meets the eye. Her interiority is evocative of all those young Black girls and women in the aftermath of Katrina who were not allowed to speak, who were consumed and projected onto without being able to respond. At this moment, though, Esch can participate through the mythos of Medea and, by extension, though the actions of China, even while standing outside of the community. Like the women in the aftermath of the storm, Esch is not a silent, mute body, but a thinking, observing human being. She is not passive, but active, even if it is only in her mind.

Esch understands the force of desire through the stories of the Greeks. As she says, "I'd let the boys have [sex with me] because for a moment, I was Psyche or Eurydice or Daphne" (Ward 2011b, 17). But she comes to understand betrayal and rage through Medea. For example, when Esch realizes Manny, the father of her unborn child, does not love her, she slips into Medea's identity to channel her anger: "This is Medea wielding the knife. This is Medea cutting. I rake my fingernails across his face" (Ward 2011b, 204). As the novel moves in action toward the coming of Katrina, Esch's identification with Medea only intensifies. Esch thinks back to the last storm she experienced with her mother, the woman whose absent presence hangs over so much of the novel. The storm, called "Elaine" in the book, most likely references Hurricane Georges, which hit the Mississippi Gulf Coast in 1998. However, Ward gives it a woman's name, alluding to a previous storm, Hurricane Elena, which hit the coast in 1985: "Mama had talked back to Elaine. Talked over the storm. Pulled us in in the midst of it, kept us safe. This secret that is no longer a secret in my body: Will I keep it safe? If I could speak to this storm, spell it harmless like Medea, would this baby, the size of my fingernail, my pinkie fingernail, maybe, hear?" (Ward 2011b, 219). Esch turns to Medea as a source of identity as she is contemplating her pregnancy, her own journey into motherhood, and the approaching hurricane. As Esch thinks of her mother, who is no longer with her, she attempts to tap into some source of power to keep herself and her child safe. She does not look to Medea as a model of how to be a parent; rather, she looks to her because Medea knows how to survive. As the storm hits, Esch transfers the identification of Medea to herself and then to the hurricane. In her fear and sorrow over what may be coming, Esch observes: "Jason has remarried, and Medea is wailing. *An exile, oh God oh God, alone*" (Ward 2011b, 225).

As Hurricane Katrina gains power, Esch is humbled by the strength of the storm that will eventually overturn the entire social contract in its aftermath. The power of the storm for subversion, to make the intense class and race division of rural Mississippi obsolete (at least temporarily), inspires awe in Esch. She thinks:

> Her [Katrina] chariot was a storm so great and black the Greeks would say it was harnessed to dragons. She was the murderous mother who cut us to the bone but left us alive, left us naked and bewildered as wrinkled newborn babies, as blind puppies, as sun-starved newly hatched baby snakes. She left us a dark Gulf and salt-burned land. She left us to learn to crawl. She left us to salvage. Katrina is the mother we will remember until the next mother with large, merciless hands, committed to blood, comes. (Ward 2011b, 255)

After the devastation of Katrina, Esch and her family emerge into a world changed. They walk through the streets of their community, and though there is devastation all around, there is also help and hope and love. Her friend, Big Henry, says not to worry, that her baby "got plenty daddies" (Ward 2011b, 255). She understands that, like the epic tales of Medea, Psyche, and Eurydice, the story of Katrina will live in the minds and souls of the people of the Gulf Coast in a way that will inspire mythological horror and awe. And because of Esch's devotion to these myths as a narrative mode of self-articulation, she is almost uniquely able to process and understand the devastation that is happening and put it in a proper context.

This passage also suggests that the Greek myths are a part of Esch's life and will forever be so as she mentions the way the "Greeks" would remember Katrina. In this passage, the "Greeks" are the citizens of the Gulf Coast. Mae Gwendolyn Henderson (2000, 358) argues that the "self-inscription of black women requires disruption, rereading and rewriting the conventional and canonical stories, as well as revising the conventional generic forms that convey these stories." The narrative moves that Esch makes using Greek mythology do exactly that. They allow her to articulate her experience by using common stories and tropes that are familiar, which forces the reader to acknowledge that what seems to be an idiosyncratic tale of devastation is also a story of universal vulnerability to the natural world. Esch, however, ultimately moves beyond Medea. That story offers only destruction, and there is no rebirth or redemption possible. Katrina, though, "cut us to the bone but left us alive," reducing the land to its elements, to its metaphors, but upon which they could learn to crawl (Ward 2011b, 255). Speaking now with a communal voice, Esch has solidified her understanding of herself to a universal feminine experience, one that moves her from feeling helpless to being humbled by the potential of her own power, a power manifested through the twinned impending states of motherhood and aftermath. This power brings her into the protected circle of the community; no longer isolated in her home by the pit, Esch allows herself and her child to enter into communal care.

Laymon's *Long Division*

Laymon's 2013 novel *Long Division* is a complex story of time travel, novels within novels, and the historical trauma of white supremacy. The protagonist is Citoyen, called "City" by everyone who knows him. He's a high school freshman in 2013 who splits his time between Jackson and the fictional coastal town of Melahatchie, Mississippi. City is also the name of the protagonist in the book *Long Division* that City is given to read by his high school principal. *This* City is a high school freshman in 1985 who splits his time between Chicago and Melahatchie, who travels in time through a portal in the ground in the woods back to 1964 and forward to 2013.

In the novel within the novel *Long Division*, the 1985 City is also reading a novel called *Long Division*. All of the versions of *Long Division* we see and read follow the exploits of City and his friends: Shalaya Crump, the love of 1985 City's life; Baize Shephard, who 1985 City meets when he travels to 2013 and who goes missing in 2013 City's reality; and Evan Altshuler, a Jewish boy from 1964. We, and City, discover that the 1985 version of Shalaya and City are supposed to marry and have a child and that child is Baize Shephard. However, the current Baize is an orphan because her parents City and Shalaya died in 2005 when Hurricane Katrina hit the Mississippi Gulf Coast. The young Shalaya we come to know is then faced with a choice: get on with her regularly scheduled life, marry City, and die in Katrina, or change her own future—and by extension the future of everyone around her—and stay in 1964 with Evan to fight white supremacy and her own erasure.

One character thus receives autonomy at the expense of the other: in order for Shalaya to live, Baize must cease to exist. Shalaya decides to stay with Evan, as this choice offers her an alternative future to the one that she will face if her and City's lives go as destined. In the end, it is almost impossible for City to let her go. She is firmly entrenched in the narrative of his life as City had been so completely sure of their life together even before they met their future daughter, Baize. But what City cannot see and what Shalaya can is that their pairing, especially in the way City imagines it to be, will ultimately end not just with their own demise but with the loss of an opportunity to make their community in Melahatchie better, stronger. To choose City is to lose herself and her chance to change the world. So, her breaking out of the story that City has for her allows both City and her to have an alternative future that may let them live. It is Shalaya who understands that she must decide her own fate.

Shalaya first alerts City to the fact that they can travel through time by using a porthole in the ground of the Mississippi countryside. Once Shalaya leaves and takes control of her destiny, City's and Baize's lives are forever changed, as Baize's existence is dependent upon Shalaya being City's wife and her mother. Shalaya's fate is to die in Hurricane Katrina and leave Baize an orphan. By leaving to go through the portal with Evan, though, she makes for herself a different story. With City, her future is certain death. She knows that she and City will share a beautiful but brief life together that will result in the birth of Baize. She shoulders the responsibility that if she does not stay and fulfill her destiny, other lives will be affected. That Shalaya knows all of this and chooses to leave with Evan anyway suggests several things about her character and what *Long Division* imagines as the possibilities for Black female subjectivity. City assumes that because he loves Shalaya, she loves him back equally, that she also feels the weight and responsibility of the love that will result in the birth of another remarkable human being, Baize. However, we discover that it is entirely possible that the love between City and Shalaya—and its inevitability—may be a comfortable and convenient fiction on the part of City, our narrator.

What she does, then, is free herself from a constricting narrative that would result not only in her and City's deaths but would also orphan their only child, who would eventually be traumatized by Hurricane Katrina. When Shalaya opts out of this "story," she not only frees herself, but she frees City and Baize as well. City is no longer destined to die in the storm. As for Baize, the novel does not shy away from the sadness or the pain of having her character erased, figuratively and literally. However, we also know from Baize the pain of living through the death of her parents, through the trauma of Katrina. As she is talking to City about her dreams and aspirations, Baize says, "Man, I come out here all the time and imagine this is a beach with palm trees and mountains in the background . . . but no matter how nice I see it, the sky ain't ever quite right." When City asks her what is wrong with her sky, she replies: "Folks act like they hate the oil now more than the wind, but me, I'd kill the wind, the sky, and the water if I could." When City again asks her why, she simply says: "because it took my family" (Laymon 2013, 220–21).

Baize chooses hip-hop as a means for articulating this experience of trauma. Her "Storm Rhymes" express both her fury and pain and recognize the futility of trying to escape the circumstances of life as a Black family in the Gulf South:

My big fat beautiful mouth
was born right here in the South
where Ma and Daddy, they went swimming,
trying to find a way out

But Katrina was hummin
and my folks got to runnin.
Ears open for God but she
ain't telling them nothin. (Laymon 2013, 69)

Hip-hop is an avenue through which a person can perform bravado, escapism, and critique. When Baize discusses Katrina, she needs a storytelling form that will allow her to address her trauma with some degree of emotional protection and safety. Hip-hop has historically allowed young Black folks to address traumatic aspects of their experiences, like police brutality, extreme poverty, and urban violence, while also engaging in fantasy and imagination. Because Katrina is so traumatic, hip-hop as a genre enables Baize to address her trauma without leaving herself completely vulnerable. She is able to access power in identity and expression by engaging the narrative, mainly because of hip-hop's closeness to her own verbal expressions. Laymon (2015) notes in his essay "Da Art of Storytellin'" that hip-hop is the product of the voice of African American youth, that it is a type of authentic culture and expression. The artists are not speaking in someone else's language or values; it is uniquely theirs. This suggests that hip-hop can function as a salve to the inherent break or violence of engaging traditional forms of expression. For Baize to use hip-hop to articulate the pain that the violence of Katrina has brought her is also a move to access some of the power that was taken from her when she experienced the storm.

The flexible temporality of the text thus allows mother and daughter to meet while they are in the same place in their lives—young, Black, poor, and restless in Mississippi—as a way of complicating monolithic representations of the experience of Black girlhood. Furthermore, the pit in the ground makes a "distinct allusion to the underground railroad," which situates the novel "within the history of black resistance" (Leong 2021, 72). For Shalaya, to exert herself and her future is an important, and resistant, political act. She resists her destiny as a wife and a mother, one that would ultimately lead to her death, to embrace her role as a community activist because, by staying with Ethan, Shalaya gets to "save the future in a special way," even if that choice means erasing "the special child they never really knew" (Laymon 2013, 244). This choice is painful, but what *Long Division* has taught the young people in this book is that "special change, the kind that lasts, hurts" (Laymon 2013, 246). When 1985 City emerges from the time-travel pit into his own version of 2013, we understand that Melahatchie is a different place than the Melahatchie of 2013 City. There is a community center, a grocery co-op, homes with large gardens, and a neighborhood with sidewalks where "black folks and Mexican folks of all ages" were "talking and laughing out loud" (Laymon 2013, 255). This laughter signals a joy that Shalaya made possible when she chose her own path away from the storm.

As future victims of Katrina, the novel's narrative suggests that Shalaya and Baize are both bodies lost to the ecological and sociopolitical disaster that the hurricane wrought. For Shalaya to decide to escape that narrative signals a power and autonomy that her doomed future self was denied. This is remarkable because of the way that young Black women are marginalized in Mississippi. While people like Shalaya, working class or poor, Black, and rural, are not necessarily destined for a life of economic struggle, lack of access, and poor health, the odds are most definitely stacked against them. Shalaya lives in an area of the country where her health, educational, and economic options are limited. While Baize's erasure, her metaphorical abortion, is emotional and tragic, Shalaya, in the end, does not hesitate in her choice. The time machine, that pit in the earth of rural Mississippi, offers her a way out. She takes it.

Smith's *Orleans*

In the near-future world of Smith's young-adult novel *Orleans*, Katrina was the first of a litany of storms that devastated the Gulf South. In this dystopian version of the region, massive hurricanes and other ecological disasters pave the way for Delta Fever, a deadly disease that kills people of a certain blood type. This disease divides the region, now simply called "the Delta," along blood lines, and the tribes that arise out of this new system of classification both offer the people of Orleans protection and set up inter-tribal conflict. Meanwhile, the US government eventually decides that the Gulf South cannot be saved, walling the states up and severing them from the rest of the country—a reverse secession in which the Gulf states are rejected and left to fend, and die, for themselves.

The novel begins some fifty years after Hurricane Katrina, and it follows the story of Fen, a young Black girl who fights to protect the child, Baby Girl, she has been entrusted with, the daughter of her tribal leader and surrogate mother, Lydia. Lydia died in childbirth, leaving Fen to care for the girl. At first, it is a matter of simple survival as Fen and Baby Girl's tribe has been wiped out, leaving no place or group for protection. Soon, though, Fen knows that she wants more for the child than day-to-day survival and decides that she must ferry Lydia's child out of the South and into what she can only hope will be a better life. She has to get her over the wall.

Along the way, Fen meets Daniel, who is the novel's other point-of-view character. Daniel is a scientist from the "Outer States" who has discovered a cure for Delta Fever—but it is a cure with a steep price as it kills not only the fever but also the host. Understanding the immense threat to life in the South such a cure holds if the US government gets its hands on it, he decides to sneak into the "Lower States" in order to perfect his serum so that what he believes are the few remaining residents of the Lower States can be cured *and* saved. In Orleans, however, he finds a thriving, if dangerous, world. He finds that people and culture, even if they have changed, have *survived*. Together, he and Fen travel through a cityscape that has merged with the swampy surroundings of a world changed by our climate crisis, as "the landscape of southeastern Louisiana, with its bayous and swamps, tropical biome, and sweltering heat offered the promise of refuge and escape" (Thomas 2019, 297). The novel ends when he is able to slip out of Orleans with Baby Girl, now named Enola, back into the outer world. But again, that cure comes with a price—Fen sacrifices herself so that Enola may live, and the novel begins and ends with a hurricane.

Smith is clear about the genesis of *Orleans*. Her mother was living in New Orleans during Katrina, deciding to ride out the storm rather than evacuate and chance being stuck in a traffic jam on the highway when the storm came through. Smith says that "[my mother] was running out of food, water, and medicine," having been abandoned by those whose job was to protect the city (Connors 2016, 38). The local and national response to the disaster was key to her envisioning of a dystopian New Orleans. Smith thus frames *Orleans* as a story about storms and our inability, or unwillingness, to take seriously our encroaching ecological tipping point. The first unnumbered, unpaginated section of the novel, titled "Before," begins with a character named Edmund Broussard climbing the Mississippi River levee across from Jackson Square in New Orleans to play his jazz. The city is facing a threat from Hurricane Ivan, but Edmund climbs the levee to play his trumpet anyway. The "TV crews loved it, the image of a lone man facing nature, refusing to bend" (Smith 2013, n.p.). Though Hurricane Ivan spares New Orleans, the next time the city isn't so lucky because the next time is Katrina.

The next page in the text is a list of hurricanes, beginning with Katrina, that bring an unprecedented string of destructive weather to the world of *Orleans*. This list includes the name, date, and category of the storm and an increasing number of dead. The list ends with a storm called "Jesus," noted as Category 6—which we find out was a category created entirely for this hurricane. It is after this unbelievably destructive storm that the fever came. But as the region was wiped clean, a new disaster was allowed to thrive in its place. The United States and FEMA decide that "for the safety of the population at large, we deem it advisable to seal off all storm-affected areas of the Gulf Coast region" (Smith 2013, n.p.).

Like *Salvage the Bones* and *Long Division*, this is a novel about Black female subjectivity and how the landscape and weather of the region has a material impact on the construction of Black girlhood, with Fen's "journey from object to subject, from powerless to powerful" (Marotta 2016, 57). And once again, we have children without mothers—both Fen and Enola, and like the other two novels, *Orleans* "resists traditional trope of the loss of the mother" to instead "link the destruction of Katrina . . . with the forging of new bonds in the wake of loss" (Leader-Picone 2019, 70). In *Orleans*, the forging of new bonds comes about with the relationship between Fen and Enola and, perhaps more importantly, between Fen and the spiritual and the natural worlds. Hurricanes, including Katrina, have become part of the mythos of Orleans.

On the first night she must care for Enola, Fen dreams of a memory of her parents. In the dream/memory, she is reading a book about the biblical flood and Noah's ark. She asks her mother "if the Flood was Hurricane Jesus, and she says no . . . she says it was a long time ago, before that. And I think that must mean Katrina, and she laughs when I say it and kisses me on the head" (Smith 2013, 59). Fen wakes from the dream to immediately take a defensive, protective position around the child. Fen then remembers that she will have to "look after Lydia," which means finding a place to bury her body. Eventually Fen walls Lydia up in a tree, bricked in by clumps of reddish clay so symbolic of the region. Fen then smooths the clay and "carves into it with [a] knife an X in the place of a cross." "In the top crook of the *X*," Fen places the number one. To the left, she scratches "an *F*, to the right a plus sign, at the bottom, an O. This means: "*One Female O Positive*. The only marker most of us get in Orleans" (Smith 2013, 64).

This scene works to establish the importance of Katrina to the region's story of itself, mythologizing the community's origin and situating the narrative of Orleans outside of a Western, Christian tradition—Noah's flood is Katrina, the now-ancient "mother we will remember," as Ward (2011b, 255) writes, and "Jesus," the most catastrophic of storms. The localized burial customs, drawn from the physical landscape of southeast Louisiana, also have their roots in one of the most visceral sets of images to arise from media portrayals of the aftermath of Katrina—the spray-painted *x*'s that served to mark the homes of the dead, fully incorporating the lore of Katrina into the rituals of the region. Perhaps most importantly, this scene also links the destruction brought by storms to the sanctity of the bond between parent and child, between protector and protected. Fen connects herself to the mothering of her past (her biological mother), present (Lydia), and future (Enola) through the myths of hurricanes and the soil of the Gulf South.

Smith continues to fold the history of storms into the fabric of ritual, sacred life in Orleans and the Delta. November first is still All Saints Day, but now the date has meaning because it marks the end of hurricane season. Fen and Daniel watch a community rite when an "All Saints krewe," decked in Mardi Gras masks and beads, rides into the streets, carrying flambeaux and chanting the names of storms past: "'Katrina, Isaiah, Lorenzo. Olga, Laura, Paloma.' Up and down, over and over, they be going faster and faster. 'Jesus, Jesus, Hay-SEUS'" (Smith 2013, 170). Fen mouths the words with them as she watches, finding a way to take part in the exuberant ritual, in which the horse riders shape themselves into a sphere and then a spiral, spinning round and round in the shape of a hurricane. As they "shout, hoot, holler" Fen says, "I got to hold my tongue not to join them out loud" (Smith 2013, 170). She takes a moment to teach Daniel about the ritual and thus about Orleans, a city that he had expected to find empty except for the dead. She begins by recounting the history of the storms and the absolute failure of the government to care for its people in the Gulf South, eventually ejecting them from the Union. She tells him:

> We no longer part of Louisiana or even the United States. We just the Delta, and we been making our own way for half a hundred years.
> So that what they be singing about. How we the Delta, how we still Orleans. . . . Somebody found an old Mardi Gras warehouse or something, and he pull out some costumes and go riding through the streets, just one man holding up a lantern, saying "We still here, we still here, thank Lord almighty, we still here." . . . And other people started, too, 'til they all been wading along, with they flashlights and torches and all kinds of things, and they start singing and dancing 'cause "this be New Orleans and that be what we do." (Smith 2013, 173)

In this way, Fen is telling Daniel, a white man from up north who knows nothing true about the Delta, about herself as well. For Fen, her identity is predicated on this neglect but also, more importantly, on the society that sprung up in defiance of that neglect. Cameron Leader-Picone (2019, 67) says of Ward and Laymon that their novels have "embrac[ed] the spaces that

African Americans have repurposed in search of a freedom still denied by the state." We see that same dynamic writ large in *Orleans* as the All Saints' Day ritual repurposes a spring religious festival—Mardi Gras—into an autumnal, meteorological one. This syncretism sanctifies the weather *and* Orleans, while preserving the spirit of the old city. And it seems to be working, as Orleans has begun to develop a "fragile ecosystem, but one that works . . . Orleans is healing itself" (Smith 2013, 265).

But while the land might be healing, this is still a community deeply stratified—and like Esch at the dogfight, Fen can only participate in the ritual silently, from the outside. She is still limited by her gender, age, and tribe, and she knows that Enola will be as well. In fact, Enola will most likely be captured and sucked dry for her newborn blood, which is "untainted" by the fever, unless Fen can get her over the wall. She knows that there "is no place safe for this child, who is not valued in this space as anything but a commodity" (Marotta 2016, 62). Melanie Marotta (2016, 58) thus reads *Orleans* as a neo-slave narrative, one in which Fen's roles of mother and leader, both of which she inherited from Lydia, will "ensure the survival of future generations." But in order for that to happen, Fen must sacrifice herself—or remake herself—for the child. Fen, just like Esch in *Salvage the Bones* and Shalaya in *Long Division*, has a choice: abandon this child and retake her place in the only community she's ever known or give herself up for Enola, Daniel, and the promise of a better world. Like both Esch and Shalaya, she decides, in the end, to save the future.

In order for both Enola and Daniel to escape, for Daniel to finish his cure for the fever, and for Enola to live a life outside of blood boundaries and bounties, they have to slip through a crack in the guarded wall that keeps the Delta segregated. The only way Fen can imagine their escape is to sacrifice herself as a distraction. She pretends to carry a child up to the wall, waving her arms and drawing the attention of the guards. Daniel looks back once to see that "her arms were raised, her face turned up, the bundle held high in the air. She rotated in a slow circle as the rain washed the mud from her skin." Shots are fired, and Fen falls down. After his and Enola's escape, Daniel thinks back to his "last glimpse of Fen swirling through the water, spinning like the wheel that turns the world" (Smith 2013, 322–23). Washed in the rain, she gives her life for the life of the child who had become "her tribe," her family. In her final moments Fen becomes the mother who sacrifices herself for her child. But more pointedly, she spins like a hurricane, like the "wheel that turns the world." For it is both mothers and storms that turn the world of *Orleans*, making possible a generative movement that, like in *Salvage the Bones* and *Long Division*, creates familial and communal love from loss, change from chaos. Fen is sanctified through her embodiment of the hurricane, becoming part of the ritual fabric of Orleans and the mythological origin story of the new world to come.

Conclusion

Because of the extensive tragedy of Hurricane Katrina, the country had to grapple with images of Black despair. And while the victims of Katrina were not all Black, Melissa Harris-Perry notes that the majority of photographs and videos coming out of Katrina were of Black women. These images forced those who were witnessing the tragedy through the lens of the news media to think about *why* these were the images that stuck, those of Black suffering and of Black pain, of Black disenfranchisement and historic and systemic racism. One of the remarkable things about a tragedy like Katrina is that it forces us as citizens to think about what it means to survive in a place where a tragedy like this can happen. Indeed, all three novels express plenty of ambivalence about their settings as they do not shy away from the bigotry, poverty, violence, and structural racism built into the foundations of the region. The way the Gulf South constructs a "home" for citizens, such as Esch, Baize, Shalaya, and Fen, seems to suggest that the novels have a pretty dark view of the possibilities this space presents for young Black women, especially those that grow up poor and rural. Is the land of the Gulf South hospitable? Is it conducive to growth? The answer seems to be "no": the young Black women suffer death and tragedy and erasure.

However, these novels complicate that answer. It is not an emphatic denial, but instead a complicated and frustrated one born out of an unfulfilled love for a place that has never really been theirs. For the young Black women of these novels, their stories are always about how their lives are shaped by and give shape to the region of the Gulf South. They find a way to remake their narrative, engaging the folklore and folk culture of the community and world around them. Esch, by thinking or imagining herself through the lens of Greek myth in times of stress or change, emerges stronger from the chaos of Katrina. Baize leans on hip-hop as a way of asserting herself, her identity as a queer woman in the South, and her trauma. Shalaya finds a way to tell a new story of herself and her community, one that exists outside of what was prescribed in *Long Division*. And Fen harnesses the ritual power of the hurricane to save her child and bring about a cure for the people of a landscape that has, in fact, begun to heal itself.

Pinto and Pereyra (2019, 4) argue that Katrina offers "an epistemological challenge for Black poetics, one that seeks historical continuities and connections even as it acknowledges the rupturing trauma of the anti-black event." The most startling thread that runs through these three novels is the deep connection the authors make between hurricanes, generative maternal power, and the sustenance of a community. As Esch says of China—but also of Katrina and Madea—in the powerful last line of *Salvage the Bones*, "She will know that

I am a mother" (Ward 2011b, 258). Esch has not only seen her community of St. Catherine's come together in a precious new way in the wake of the storm, she also has joined a community of mothers. The last scene of *Orleans* is tiny Enola, "Fen's baby girl," "waving her small fists at the weeping sky" (Smith 2013, 324). Through an embodiment of the sacred power of Katrina, Fen has passed on that power to the girl she has claimed as her own. And *Long Division*'s mother-not-mother Shalaya chooses to nurture Melahatchie, a move made possible only by the *rejection* of a prescribed biological maternalism. She remakes her community and rewrites the landscape of coastal Mississippi: where there where ditches, there are now sidewalks, gardens where there once were none. By setting these novels in the wake of Katrina, Ward, Laymon, and Smith are able to explore how the very real tragedy of the hurricane affected Black people, and young Black women in particular, while also exposing Katrina's "narrative role as both instrument of violent absence and creator of new spaces for creativity" (Leader-Picone 2019, 87). They harness the mythic, metaphorical power of storms to create, to heal fractures, to mother the self and others, and to posit a future of power and agency and resistance to white supremacy.

References

Connors, Sean P. 2016. "Dreaming, Questioning, and Trying to Find Answers: A Conversation with Sherri Smith." *SIGNAL Journal* 39, no. 1: 37–40.

Harris-Perry, Melissa V. 2014. *Sister Citizen Shame, Stereotypes, and Black Women in America*. New Haven: Yale University Press.

Henderson, Mae Gwendolyn. 2000. "Speaking in Tongues: Dialogics, Dialectics, and the Black Woman Writer's Literary Tradition." In *African American Literary Theory: A Reader*, edited by Winston Napier, 348–68. New York: New York University Press.

Laymon, Kiese. 2013. *Long Division: A Novel*. Evanston: Bolden.

Laymon, Kiese. 2015. "Da Art of Storytellin' (A Prequel)." *Oxford American*, November 19, 2015. https://main.oxfordamerican.org/magazine/item/702-da-art-of-storytellin.

Leader-Picone, Cameron. 2019. *Black and More than Black: African American Fiction in the Post Era*. Oxford: University Press of Mississippi.

Leong, Diana. 2021. "An(im)alogical Thinking: Contemporary Black Literature and the Dreaded Comparison." In *The Palgrave Handbook of Animals and Literature*, edited by Susan McHugh, Robert McKay, and John Miller, 65–78. Palgrave Studies in Animals and Literature. London: Palgrave Macmillan. https://doi.org/10.1007/978-3-030-39773-9_5.

Lindahl, Carl. 2012. "Legends of Hurricane Katrina: The Right to Be Wrong, Survivor-to-Survivor Storytelling, and Healing." *Journal of American Folklore* 125, no. 496: 139–76.

Marotta, Melanie. 2016. "Sherri L. Smith's *Orleans* and Karen Sandler's *Tankborn*: The Female Leader, the Neo-Slave Narrative, and Twenty-First Century YA Afrofuturism." *Journal of Science Fiction* 1, no. 2: 56–70.

McKittrick, Katherine. 2006. *Demonic Grounds: Black Women and the Cartographies of Struggle*. Minneapolis: University of Minnesota Press.

Pinto, Samantha, and Jewel Pereyra. 2019. "The Wake and the Work of Culture: Memorialization Practices in Post-Katrina Black Feminist Poetics." *MELUS* 44, no. 3: 1–20.

Smith, Sherri L. 2013. *Orleans*. New York: G. P. Putnam and Sons.

Thomas, Ebony Elizabeth. 2019. "Notes toward a Black Fantastic: Black Atlantic Flights beyond Afrofuturism in Young Adult Literature." *The Lion and the Unicorn* 43, no. 2: 282–301.

Ward, Jesmyn. 2011a. "National Book Award Winner Tells Tale of Katrina." *All Things Considered*. National Public Radio. Washington, DC: NPR, November 17, 2011. https://www.npr.org/2011/11/17/141285171/national-book-award-winner-tells-tale-of-katrina.

Ward, Jesmyn. 2011b. *Salvage the Bones*. New York: Bloomsbury.

Chapter 9

"YOU DON'T NEED A WEATHERMAN"

Bob Dylan's Windlore

James I. Deutsch

On May 24, 1941, the day that Robert Allen Zimmerman was born in Duluth, Minnesota, the local prevailing winds suddenly shifted from "moderate northerly" to "strong southerly" (*Duluth (MN) News Tribune* 1941a, 1; *Duluth (MN) News Tribune* 1941b, 1). What this shift may have portended is open to interpretation. According to the weatherlore of ancient Greeks, a change to southerly winds brings "ill effects . . . upon the human body. Theophrastus says that, during the prevalence of this wind, men felt more sluggish and less efficient, but the north wind made them energetic" (McCartney 1930b, 26). Similarly, for the ancient Romans, the north wind (known as Aquilo) was "the most salutary of winds," while the south wind (known as Auster) might have brought "especially destructive earthquakes" (McCartney 1930a, 13 and 15). The philosopher Plutarch—born in Greece, but later a Roman citizen—coined a proverb in "De Primo Frigido" ("On the Principle of Cold"): "If Southwind challenges North, instantly snow will appear" (Plutarch 1957, 249).

It did not snow in Minnesota on May 24, 1941, but Bob Dylan—the boy who was born that day in Duluth—has used the wind as a regularly recurring element in the lyrics of his songs. Indeed, one assessment from 2005 of Bob Dylan and weather imagery notes that the word "wind" appears in fifty-five of Dylan's 465 songs, second only to "sun" (appearing in sixty-three songs), and ahead of "rain" (forty songs), "sky" (thirty-six songs), "cloud" (twenty-three songs), and other words related to the weather (Robock 2005, 484). This article cites only a small percentage of those fifty-five songs—a number that has increased since 2005—by focusing on some of his best-known compositions. Much more than a meteorological phenomenon, the wind in Dylan's lyrics becomes a metaphorical commentary on his life, his art, and his identity. Dylan's appreciation of and fascination with the wind—blowing, chilling, hitting, howling, rotating, swirling, whipping, and whistling—constitute a corpus of informal knowledge and belief that folklorists might term "windlore."

At age six, Dylan moved with his family to Hibbing, a once-thriving, iron-mining town some seventy-five miles northwest of Duluth and just one hundred miles south of the Canadian border. This part of the country, according to meteorologists, has a "severe continental climate" that "could have instilled [in Dylan] a deep appreciation for the weather" (Brown et al. 2015, 203). It may also have instilled in Dylan an appreciation for poetry and labor radicalism—at least according to a Berkeley bookstore customer who exclaimed: "There's poetry on the *walls*. Everywhere you look. There are bars where arguments between socialists and the IWW, between Communists and Trotskyites, arguments that started a hundred years ago, are still going on. It's there—and it was there when Bob Dylan was there" (Marcus 2009, 3). Still another dimension is that, at certain times of the year, this area of northern Minnesota "turns into Winter Wonderland, and cocooned inside it, Hibbing shines and twinkles like the set for an old Perry Como Christmas Special" (Gray 2008, 312).

Dylan's personal recollections seem much less shiny and twinkling. In one of his early poems, "11 Outlined Epitaphs" (1963), the bleakness of Duluth and Hibbing are striking:

The town I was born in holds no memories
but for the honkin' foghorns
the rainy mist
an' the rocky cliffs . . .
it was a dyin' town . . .
deserted
already dead
with its old stone courthouse
decayin' in the wind
long abandoned . . .
dogs howled over the graveyard
where even the markin' stones were dead
an' there was no sound except for the wind
blowin' through the high grass . . .
south Hibbing
is where everybody came t' start their
town again, but the winds of the
north came followin' an' grew fiercer
an' the years went by
but I was young
an' so I ran
an' kept runnin'. (Dylan 1973, 101)

Only fifty lines of a poem that runs some eight hundred lines deal with Duluth and Hibbing, but Dylan's references to the wind on three of those lines seem to highlight the vernacular understanding of the wind's most disturbing effects: the courthouse is decaying in the wind; the cemetery is silent except for the blowing wind; and the north winds grow even fiercer over time. As the poem recounts, Dylan kept running, presumably in the sense that he moved east after just one academic year at the University of Minnesota in the Twin Cities, following his graduation from Hibbing High School in 1959. According to one early account, "In February 1961, Bob Dylan landed on the New York Island at the end of a zig-zaggy thumb ride across the country from South Dakota" (Turner 1962, 5). Adding more details about the weather, Dylan tells a 1985 interviewer, "I stood on the highway during a blizzard snowstorm believing in the mercy of the world and headed East, didn't have nothing but my guitar and suitcase. That was my whole world" (Crowe 1985, 10). Dylan also made brief stops in Madison and Chicago between Minneapolis and New York—but it is not clear at which point the "blizzard snowstorm" occurred.

Dylan arrived in New York City's Greenwich Village in early 1961, and as luck or fate might have it, the winter of 1960–1961 in New York was "as rugged as any on record" (Wing 1961, 1). Shortly after Dylan arrived, the city broke a record (set in 1881) for the longest stretch of below-freezing cold: sixteen consecutive days starting on January 19, 1961. Moreover, the cold and snow were enhanced by "winds up to thirty-five miles per hour" (Wing 1961, 1). Nevertheless, Dylan seemed unfazed, perhaps borrowing from the traditions of northern Minnesota, where hardy residents must weather frequent blizzards. As he recalls, "Where I came from, there was always plenty of snow so I was used to that" (Crowe 1985, 10).

However, the winter wind in New York City—like the winter wind in Minnesota—was an element that Dylan not only noticed but also incorporated into his writings, both prose and song lyrics. For instance, his published memoir opens with a scene in January 1962, shortly after he signed a contract to publish his songs: "Outside the wind was blowing," begins the sixth paragraph, with "straggling cloud wisps, snow whirling in the red lanterned streets, city types scuffling around, bundled up—salesman in rabbit fur earmuffs hawking gimmicks, chestnut vendors, steam rising out of manholes" (Dylan 2004, 4).

A short while later, "on a cold winter day near Thompson and 3rd, in a flurry of light snow when the feeble sun was filtering through the haze," Dylan spots Dave Van Ronk, one of his mentors in the Greenwich Village folk-music scene walking toward him. Significantly, Dylan uses a wind metaphor to describe the encounter: "It was like the wind was blowing him my way" (Dylan 2004, 16). Dylan's memoir contains several other instances in which the wind is howling, whistling, or whipping:

- "I wouldn't be going to see Woody today. I was sitting in Chloe's kitchen, and the wind was howling and whistling by the window" (Dylan 2004, 100);
- "Cutting across the vacant lot to a bank of field flowers where my dogs and horses were, the strangled cry of a gull came whipping through the wind" (Dylan 2004, 162);
- "I left the ice cream parlor, went back out on the sidewalk. A wet wind hit me in the face" (Dylan 2004, 192);
- "Wind whipped in the open doorway and another kicking storm was rumbling earthward" (Dylan 2004, 217);
- And "The air was bitter cold, always below zero, but the fire in my mind was never out, like a wind vane that was constantly spinning" (Dylan 2004, 26).

The phrase "like a wind vane that was constantly spinning" seems an apt metaphor for the song that promoted Dylan from folk performer to songwriter. According to Dylan-scholar Clinton Heylin (2009, 77–78), "Blowin' in the Wind" (composed during the second week of April 1962) is the song that "would change his world—nay, *the* world—fusing much of what he'd been reaching for in his foundation year."

What Dylan told his friend, Gil Turner, who in fact had first performed the song in Gerde's Folk City in April 1962 was:

There ain't too much I can say about this song except that the answer is blowing in the wind. It ain't in no book or movie or T.V. show or discussion group. Man, it's in the wind—and it's blowing in the wind. Too many of these hip people are telling me where the answer is but oh I won't believe that. I still say it's in the wind and just like a restless piece of paper it's got to come down some time. (Dylan 1962, 4)

When Peter, Paul and Mary recorded the song, it became a hit and reached millions of people. As such, it perhaps is the song that most directly links Dylan with windlore and his perceptive understanding of contemporary cultural forces. As music critic Paul Williams observes, "I don't think Dylan ever put quite as much of himself into the song again. He didn't have to. The song itself was in the wind at that point" (Williams 1990, 52).

According to Todd Harvey's (2001, 14) assessment of Dylan's formative years, "Blowin' in the Wind" had become, by late 1963, Dylan's "signature piece" and "the song most associated with the early stages of his career." However, Dylan's acoustic folk style would disappear within the next two years. As Dylan explains in his interview with *Playboy*, "It had to go that way for me. Because that's where

I started and eventually it just got back to that. I couldn't go on being the lone folkie out there, you know, strumming *Blowin' in the Wind* for three hours every night" (Rosenbaum 1978, 69). The shift came with his next album, *Bringing It All Back Home* (1965), in which his music turns resoundingly electric.

One of the album's songs, "Love Minus Zero/No Limit" (composed ca. January 1965), includes the lines "The wind howls like a hammer / The night blows cold and rainy," which suggest the wintry bleakness of Hibbing and Duluth from "11 Outlined Epitaphs" (Dylan 2016, 145). However, it is the opening song on the album *Subterranean Homesick Blues* (also composed ca. January 1965) that contains one of Dylan's most celebrated observations, "You don't need a weatherman / To know which way the wind blows" (Dylan 2016, 141). With its succinctness and common-sense wisdom, the line fits the folkloristic definition of a "proverb," i.e., "a concise, traditional statement of apparent truth with currency among the folk" (Mieder 2006, 996), and thus is a superb example of Dylan's windlore.

The line's use of weather and the wind as metaphors for what was happening in the mid-1960s inspired its adoption by a militant leftist organization, the Weathermen, later known as the Weather Underground (in keeping with greater gender neutrality). As Mark Rudd, one of the group's leaders, recalls, he and several others in the National Collective (a faction within Students for a Democratic Society or SDS) were struggling to come up with a title for the strategy paper they would present at the SDS National Convention starting on June 18, 1969, in Chicago. Then:

> Terry Robbins, who had memorized dozens of Bob Dylan songs, blurted out the line, "You don't need a weather man [*sic*] / To know which way the wind blows" from "Subterranean Homesick Blues."
> That was it! Just the perfect sort of countercultural dig at the super-straight PLers [the Progressive Labor faction of SDS], and one that made an explicit statement, too: You don't need ancient dogmas to understand the reality among us. Within days the strategy paper became known as "the Weatherman paper," and our faction of SDS became "the Weathermen." (Rudd 2009, 146)

Indeed, the line has become an example of weatherlore itself, an observation so memorable and universal that many American jurists—typically on the other side of the law from the Weather Underground—have referenced it in their legal opinions. According to one analysis, "Judges at all levels in the United States judicial system have cited Bob Dylan far more often than any other popular music artist" (Long 2011, 2). And the one line that is "most-frequently used" by judges from all of Dylan's songs is "You don't need a weatherman . . ." (Long 2011, 10).

The genesis of the line is uncertain. In a three-hour radio interview in 1991, Dylan told Elliot Mintz that he could not recall his creation of *Subterranean Homesick Blues*. "Something like that, you can't be sure how much of it was made up beforehand, how much you made up right on the spot. A lot of it could have been made up right there," Dylan explains (Burger 2018, 342). However, seeing just some of the one-hundred-thousand-plus items held by the Bob Dylan Archive at the University of Tulsa's Helmerich Center for American Research suggests that very few of Dylan's songs were "made up right on the spot." The collections demonstrate that Dylan regularly jotted down and constantly revised lyrics on whatever paper was handy—"from loose leaves and personal notebooks to hotel stationery and matchbook covers" (Bob Dylan Archive [hereafter BDA] 2017, 2). Sometimes Dylan wrote his notes with a ballpoint pen in minuscule handwriting, and sometimes he would use a typewriter to compose. As Mark Howard, an engineer and producer who worked with Dylan in the 1980s and 1990s, recalls:

> Out of everybody I've worked with, Dylan is the most dedicated and focused writer. He would *always* be working o[n] his lyrics. He'd have a piece of paper with thousands of words on it, all different ways, you couldn't *read* it, it was impossible, because there'd be words going upside-down, sideways, just words all over this page. You couldn't make heads nor tails of it. (Howard 2008, n.p., emphasis original)

Thousands of these pieces of paper with thousands of words on them are in the BDA. The materials are sporadic from the 1960s but much more comprehensive for later decades through almost to the present.

In the case of *Subterranean Homesick Blues*, the BDA holds two typescripts, neither of them dated. In one (presumably the earliest of the two), the weatherman line does not appear, though we can see the idea emerging where Dylan has typed at the bottom of the page, perhaps as an afterthought: "don't be bashful / check where the wind blows / rain flows / Keep a clean nose an[d] be careful of the plain clothes (BDA n.d.a and reproduced in Dylan 1973, 161).

The second typescript, which is titled "Look Out Kid," contains a second verse that is very close to the final version, including the lines (complete with several typing errors—*sic* throughout): "ceep a clean nose / be careful of the plain clothers / you don't need a weahter man t know which way the wind blows" (BDA n.d.b).

Certainly, there are other Dylan songs from the 1960s that refer to the wind, most often in a seemingly literal sense that reinforces traditional beliefs about the wind's presence and power in our lives. For instance, "Farewell" (composed ca. January 1963) laments how "the weather is against me and the wind blows hard / And the rain she's a-turnin' into hail" (Dylan 2016, 48). Similarly, "Girl from the North Country" (composed around the same time, ca. March 1963) describes a place "where the winds hit heavy on the borderline" and where "the snowflakes storm / When the rivers freeze" and where there are "howlin' winds" (Dylan 2016, 54). And "One More Night" (composed ca. February 1969) closely follows this theme, describing a place where "the wind blows high above the tree" (Dylan 2016, 243). "Percy's Song" (composed ca. October 1963) includes the refrain "Turn, turn to the rain / And the wind" (Dylan 2016, 105–7), but it is taken from a traditional ballad, "The Wind and the Rain," a variant itself of the Child Ballad #10, "The Two Sisters." Among his other compositions from the 1960s, only "All Along the Watchtower" (composed ca. November 1967) includes a wind that may be more metaphorical than literal: "Outside in the distance a wildcat did growl / Two riders were approaching, the wind began to howl" (Dylan 2016, 224). Given the song's allusions "to the apocalyptic sections of the Bible" and its criticism of shady "businessmen [who] drink my wine," the wind that begins to howl might signal retribution, divine or otherwise (Heylin 2009, 365–66).

Dylan's metaphorical use of wind becomes much more explicit in "Idiot Wind" (composed late 1974), described by one critic as a song "about a long-failing relationship that extends to a critique of its times" (Riley 1999, 245). When introduced in the third stanza, the wind seems to emanate some "Sweet lady":

Idiot wind, blowing every time you move your mouth
Blowing down the backroads headin' south
Idiot wind, blowing every time you move your teeth
You're an idiot, babe
It's a wonder that you still know how to breathe. (Dylan 2016, 336)

This third-person approach continues in the sixth stanza, repeating the rhyme of *teeth* and *breathe* but adding "blowing through the flowers on your tomb / Blowing through the curtains in your room." In the ninth stanza, the wind's source shifts to first-person singular but on a national scale: "blowing like a circle around my skull / From the Grand Coulee Dam to the Capitol." And in the final twelfth stanza, the wind shifts to first-person plural:

Idiot wind, blowing through the buttons of our coats
Blowing through the letters that we wrote
Idiot wind, blowing through the dust upon our shelves
We're idiots, babe
It's a wonder we can even feed ourselves. (Dylan 2016, 336)

In a 1991 interview with Paul Zollo, Dylan explains the challenges he faced to complete the song: "Yeah, you know, obviously, if you've heard both versions you realize, of course, that there could be a myriad of verses for the thing. It doesn't *stop*. It wouldn't stop. Where do you end? You could still be writing it, really. It's something that could be a work continually in progress" (Zollo 2003, 85). The manuscripts of the BDA confirm this "myriad of verses." Dylan's jottings fill multiple pages of a three-by-five-inch, spiral-bound blue notebook (originally costing nineteen cents), which could fit in a shirt pocket. For instance, before settling on "Grand Coulee Dam to the Capitol," there was "Grand Coolie Dam to Omaha" and "Grand Coolie Dam to the Mardi Gras" (BDA n.d.c). But there was never any hint that this particular "idiot wind" was akin to the hard, howling winds of "Farewell," "Girl from the North Country," or "Love Minus Zero/No Limit" from the 1960s. The "idiot wind" from the mid-1970s seems more psychologically chilling than meteorologically chilly.

How Dylan came upon the phrase "idiot wind" is a matter of speculation. The poet Weldon Kees uses it in his antiwar poem "June 1940," first published in the *Partisan Review* of 1940 and reprinted in a posthumous collection of poems in 1960: "An idiot wind is blowing; the conscience dies" (Kees 1960, 17). Another theory is that Dylan first heard the phrase from the painting teacher Norman Raeben, who used it as "one of [his] favorite terms of abuse" (Wilentz 2010, 140). Whatever its origins, "idiot wind" has since entered the broader lexicon of weatherlore to signify any type of uninformed discourse. It even appeared as part of the 2008 presidential campaign when the Republican candidates criticized Barack Obama for associating with Columbia University's Professor Rashid Khalidi. When the *Washington Post* asked Khalidi to respond "to the campaign charges against him," Khalidi "answered, via e-mail, that 'I will stick to my policy of letting this idiot wind blow over.'" The newspaper editorialized, "That's good advice for anyone still listening to the McCain campaign's increasingly reckless ad hominem attacks. Sadly, that wind is likely to keep blowing for four more days [until Election Day]" (*Washington Post* 2008, A18).

Roughly three years after composing "Idiot Wind," Dylan's life was in turmoil. According to one biographer, he "was preoccupied by a divorce and custody battle with his wife of eleven years" during the winter of 1977–1978 (Heylin 2010, 149). It was at this time when Dylan composed "I'm Cold," a song that (so far as anyone knows) was never recorded and was once thought completely lost—until jottings were found in one of the notebooks in the BDA. Meanwhile, however, the song had acquired an almost-legendary aura, thanks to the account by Steven Soles, one of the musicians who was part of Dylan's "Rolling Thunder Revue." As described by journalist Nick Hasted (2005, 66):

> In 1977, while visiting Rolling Thunder tour-mates Steven Soles and T-Bone Burnett, he [Dylan] played a set of songs too frightening to ever be heard again: like *Blood On The Tracks 2*, with the love torn out. "They were all very, very, very tough, dark, dark, dark songs," Soles told Howard Sounes. "None of them saw the light of day. They got discarded because I think they were too strong. They were the continuation of the Bob and Sara tale, on the angry side of that conflict." One of these blackest of tracks, "I'm Cold," scared Soles. "It was scathing and tough and venomous. A song that would bring a chill to your bones. That's what it did to me. T-Bone and I, when he left, our mouths were just wide open. We couldn't even believe what we'd heard."

That being said, the several references to wind in "I'm Cold," which Dylan jotted down ca. 1997–1998 in an artist's sketchbook (roughly 5½ by 8½ inches) seem neither scathing nor venomous. On one page toward the front of the notebook, Dylan writes: "Spirit of the high wind, spirit of the Green Grass, Spirit of the Garden Sun / Spirit of the . . . Now our spirit is one" (BDA n.d.d). Some ten pages later, he writes: "TEARS OF LOVE, KISS OF FIRE / Rotating – Wind / Moment of fire." And still another ten pages later, he writes: "Sea wind blowing from the loins of a man / Hot sun burning ~~in~~ thru the ~~mirrors~~ center of the land" (BDA n.d.d, strikethrough original). These references to a high wind, rotating wind, and sea wind are intriguing, especially when combined with a song title that has both a literal meteorological meaning and a metaphorical, emotional one. Nevertheless, any speculations about "I'm Cold" and a possible new direction for Dylan's windlore must remain inconclusive because the song's final lyrics have never been disclosed.

A few years later, ca. 1980, Dylan composed "Caribbean Wind." Like "Idiot Wind," his previous song with "wind" in its title, "Caribbean Wind" was constantly revised even to the point that it was never released on any Dylan album. Moreover, Dylan performed it publicly only once: on November 12, 1980, at the Warfield Theatre in San Francisco's theater district. As Dylan recalls in a 1985 interview:

> That one I couldn't quite grasp what it was about after I finished it. Sometimes you'll write something to be very inspired, and you won't quite finish it for one reason or another. Then you'll go back and try and pick it up, and the inspiration is just gone. Either you get it all, and you can leave a few little pieces to fill in, or you're trying always to finish it off. Then it's a struggle. The inspiration's gone and you can't remember why you started it in the first place. Frustration sets in. I think there's four different sets of lyrics to this, maybe I got it right, I don't know. I

had to leave it. I just dropped it. Sometimes that happens. I started it in St. Vincent when I woke up from a strange dream in the hot sun.... I was thinking about living with somebody for all the wrong reasons. (Crowe 1985, 54)

The manuscripts in the BDA confirm the constant revisions and especially the shifting winds themselves. In the final published lyrics, which according to one scholar "stand as a mess of titanic proportions" (Heylin 2010, 195), "them Caribbean winds still blow from Nassau to Mexico" (Dylan 2016, 453–54). Appearing three times in the lyrics, this refrain suggests a geographical literalness, with the winds blowing Nassau in the eastern Caribbean to Mexico in the west.

However, other versions of the lyrics found in the BDA demonstrate a much less literal approach to the Caribbean. An undated notepad associated with his *Shot of Love* album includes much more varied sources (with spellings as in the original—*sic* throughout):

And that Carrabian winds blows down
From these iron gates in China town
to the royal walls of the palace filled with tears
And that Carabian wind still swirls
On a heart of gold and the teeth of pearls. (BDA n.d.e)

Similarly, the 1980 public performance in San Francisco has the winds blowing around the world: "And that Caribbean wind blows from Trinidad to Mexico . . . / And that Caribbean wind blows hard from the Valley Coast into my backyard . . . / And that Caribbean wind still howls from Tokyo to the British Isles" (Henrik N. 2017, n.p.). Still other variants have the winds blowing "from the Port of Spain to my backyard," "from Cape Horn to Hell's Gate," and "from Mexico to Curecuo" (BDA n.d.f). Indeed, Dylan's winds here are blowing all over the map, perhaps reinforcing his aforementioned observation in the 1985 interview that he "couldn't quite grasp what [the song] was about." Although the final lyrics illustrate a literal interpretation of the "Caribbean wind" blowing from Nassau to Mexico, Dylan's attempts to create more metaphorical meanings seem to have stalled, resulting in a rare instance in which there is something happening here, but no one knows quite what it is.

A more successful attempt from the same period was "Shot of Love" (composed ca. March 1981), which became the title cut on the album, *Shot of Love* (1981). If "Caribbean Wind" leaves unresolved the literal directions or metaphorical meanings of the wind, "Shot of Love" is much less ambiguous. The manuscripts in the BDA show how Dylan was working through his ideas:

> What makes the wind want to blow tonight?
> I don't even feel like crossing the street [above the crossed-out "I got guitar trouble"] and my car aint acting [above "~~working~~"] right.
> Called home and everybody seemed to have moved away
> And my conscience is beginning [replaces "~~seems~~"] to be bothering me today
> I need a shot of love. (BDA n.d.g, *sic* throughout, strikethrough original)

Written in pencil to the right of this typewritten verse is "dog dont even wanna bite"—a line that does not appear in the song's final lyrics:

> What makes the wind wanna blow tonight?
> Don't even feel like crossing the street and my car ain't actin' right
> Called home, everybody seemed to have moved away
> My conscience is beginning to bother me today
> I need a shot of love, I need a shot of love. (Dylan 2016, 440)

For Dylan—like instances of traditional weatherlore, a blowing wind that humans cannot control seems synonymous with a litany of problems: guitar trouble, car not working, everyone moved away, troubling conscience, and a dog that doesn't even want to bite.

More negative associations of the wind prominently appear in two of Dylan's songs from the late 1980s and early 1990s. One of them, "Dignity" (composed 1988–1989), has a checkered history. As Dylan recalls in his memoir,

> I started and completed the song "Dignity" the same day I'd heard the sad news about Pistol Pete [Maravich—who died January 5, 1988]. I started writing it in the early afternoon, about the same time the morning news began to wear away and it took me the rest of the day and into the night to finish it. It's like I saw the song up in front of me and overtook it, like I saw all the characters in this song and elected to cast my fortunes with them. . . . I'd always be able to remember this song. The wind could never blow it out of my head. (Dylan 2004, 169)

However, by the time Dylan started to record "Dignity" in February 1989, the song's clarity had disappeared. Something—perhaps even the wind—had blown it out of his head: "Once we started trying to capture it, the song seemed to get caught in a stranglehold. All the chugging rhythms began imprisoning the lyrics. . . . Every performance was stealing more energy. We recorded it a lot, varying the tempos and even the keys, but it was like being cast into sudden hell.... I couldn't figure out why we weren't getting it" (Dylan 2004, 190). It was not until 1994 that the song—extensively remixed—was released on volume three of *Bob Dylan's Greatest Hits*.

The manuscripts in the BDA are not dated but demonstrate extensive revisions to "Dignity," particularly for what eventually becomes the eighth and fourteenth stanzas in the final version of the song:

> Chilly wind sharp as a razor blade
> House on fire, debts unpaid
> Gonna stand at the window, gonna ask the maid
> Have you seen dignity?
> ...
> Englishman stranded in the blackheart wind
> Combin' his hair back, his future looks thin
> Bites the bullet and he looks within
> For dignity. (Dylan 2016, 540–41)

The manuscripts that appear to be the earliest versions (presumably from 1988 or 1989) contain no references to a "chilly wind," though there are handwritten jottings that refer to "the whispering wind," and an early reference to "black wind" presumably becomes "blackheart wind"(BDA n.d.h). However, what seems most significant is how these winds seem to portend a series of calamities: "house on fire, debts unpaid" and "future looks thin," as well as the images that appear in the song's other stanzas: "Thin man lookin' at his last meal," "Somebody got murdered on New Year's Eve," "Into the valley of dry bone dreams," and "So many dead ends" (Dylan 2009, 540–41).

A second example from this same time period is "Under the Red Sky" (composed ca. March 1990). The seemingly simple lyrics follow the patterns of a nursery rhyme, opening with almost-formulaic phrasing: "There was a little boy and there was a little girl / And they lived in an alley under the red sky" (Dylan 2016, 546). The fourth stanza introduces the wind, but it's a wind that destroys the two children: "Let the wind blow low, let the wind blow high / One day the little boy and the little girl were both baked in a pie" (Dylan 2016, 546). The album's coproducer Don Fagenson (aka Don Was) claims Dylan admitted, "It's about my hometown" and about the people who were never able to leave Hibbing (quoted in Gray 2000, 692–93). Indeed, it may be true that "the rust-colored iron ore which lines the empty open-cast mines" in Hibbing is what "gives the sky a red luster" (Heylin 2010, 384). Whatever the setting, the wind seems to bring nothing but bad omens: "A bullet of lead in the middle of the head" (BDA n.d.i); "the blind horse that leads you around" and "the river [that] went dry" (Dylan 2016, 546); and most memorably two children—not "four and twenty blackbirds"—who are baked in a pie (*Mother Goose's Nursery Rhymes* 1877, 235).

One final example comes not from a Dylan song but rather from the lecture he recorded in June 2017 to confirm his acceptance of the 2016 Nobel Prize in Literature. Explaining how the themes from different books have "worked their way into many of my songs, either knowingly or unintentionally," Dylan focuses on three literary masterpieces. *Moby-Dick* "tells how different men react in different ways to the same experience." *All Quiet on the Western Front* is "a book where you lose your childhood, your faith in a meaningful world, and your concern for individuals." And "the *Odyssey* is a strange, adventurous tale of a grown man trying to get home after fighting in a war. He's on that long journey home, and it's filled with traps and pitfalls." A professor of classics at Harvard University wrote that Dylan has "long associated" himself with the Homeric hero (Thomas 2017, 256), but in his Nobel Lecture, Dylan (2017, n.p.) seems to especially relish the winds in the story: Odysseus is "trying to get back home, but he's tossed and turned by the winds. Restless winds, chilly winds, unfriendly winds. He travels far, and then he gets blown back."

In describing the significance of the *Odyssey*, Dylan observes that some of the "same things" that happened to Odysseus "have happened to you":

> You too have had drugs dropped into your wine. You too have shared a bed with the wrong woman. You too have been spellbound by magical voices, sweet voices with strange melodies. You too have come so far and have been so far blown back. And you've had close calls as well. You have angered people you should not have. And you too have rambled this country all around. And you've also felt that ill wind, the wind that blows you no good. (Dylan 2017, n.p.)

We don't need a psychiatrist to tell us that Dylan might just as easily have attributed these Odyssean trials and tribulations to the first-person "me" rather than to the second-person "you"—the drugs, the women, the magical voices, the rambling, and the ill wind that blows no good. Similarly, we don't need a weatherman to appreciate the windlore that permeates much of Dylan's best and most enduring work and that helps us make meaning of ourselves and the world in which we live.

Published Song Lyrics

"All Along the Watchtower" © 1968 by Dwarf Music; renewed 1996 by Dwarf Music.
"Caribbean Wind" © 1985 by Special Rider Music.
"Dignity" © 1991 by Special Rider Music.

"Farewell" © 1963 by Warner Bros. Inc.; renewed 1991 by Special Rider Music.
"Girl from the North Country" © 1963 by Warner Bros. Inc.; renewed 1991 by Special Rider Music.
"Idiot Wind" © 1974 by Ram's Horn Music; renewed 2002 by Ram's Horn Music.
"Love Minus Zero/No Limit" © 1965 by Warner Bros. Inc.; renewed 1993 by Special Rider Music.
"One More Night" © 1969 by Big Sky Music; renewed 1997 by Big Sky Music.
"Percy's Song" © 1964, 1966 by Warner Bros. Inc.; renewed 1992, 1994 by Special Rider Music.
"Shot of Love" © 1981 by Special Rider Music.
"Subterranean Homesick Blues" © 1965 by Warner Bros. Inc.; renewed 1993 by Special Rider Music.
"Under the Red Sky" © 1990 by Special Rider Music.

Draft Manuscripts from the Bob Dylan Archive® Collections

"Caribbean Wind" Copyright © 1985 by Special Rider Music. Additional lyrics © 2020 by Special Rider Music. From The Bob Dylan Archive® collections, American Song Archives, Tulsa, OK. Courtesy of the George Kaiser Family Foundation.

"Dignity" © 1991 by Special Rider Music. Additional lyrics © 2020 by Special Rider Music. From The Bob Dylan Archive® collections, American Song Archives, Tulsa, OK. Courtesy of the George Kaiser Family Foundation.

"Idiot Wind" © 1974 by Ram's Horn Music; renewed 2002 by Ram's Horn Music. Additional lyrics © 2020 by Special Rider Music. From The Bob Dylan Archive® collections, American Song Archives, Tulsa, OK. Courtesy of the George Kaiser Family Foundation.

"Shot of Love" Copyright © 1981 by Special Rider Music. Additional lyrics © 2020 by Special Rider Music. From The Bob Dylan Archive® collections, American Song Archives, Tulsa, OK. Courtesy of the George Kaiser Family Foundation.

"Subterranean Homesick Blues" © 1965 by Warner Bros. Inc.; renewed 1993 by Special Rider Music. Additional lyrics © 2020 by Special Rider Music. From The Bob Dylan Archive® collections, American Song Archives, Tulsa, OK. Courtesy of the George Kaiser Family Foundation.

"Under the Red Sky" Copyright © 1990 by Special Rider Music. Additional lyrics © 2020 by Special Rider Music. From The Bob Dylan Archive® collections, American Song Archives, Tulsa, OK. Courtesy of the George Kaiser Family Foundation.

Unattributed lyrics, circa 1977–78, copyright © 2020 by Bob Dylan. From The Bob Dylan Archive® collections, American Song Archives, Tulsa, OK. Courtesy of the George Kaiser Family Foundation.

References

[BDA] Bob Dylan Archive, n.d.a, Tulsa, OK, 2016.01 B34 F0304.
[BDA] Bob Dylan Archive, n.d.b, Tulsa, OK, 2016.01 B34 F0305.
[BDA] Bob Dylan Archive, n.d.c, Tulsa, OK, small notebook [15_8_smnotebook5].

[BDA] Bob Dylan Archive, n.d.d, Tulsa, OK, box 100, folder 1.
[BDA] Bob Dylan Archive, n.d.e, Tulsa, OK, box 82, folder 8.
[BDA] Bob Dylan Archive, n.d.f, Tulsa, OK, box 81, folder 8.
[BDA] Bob Dylan Archive, n.d.g, Tulsa, OK, 1016.01 B82 F0401.
[BDA] Bob Dylan Archive, n.d.h, Tulsa, OK, box 85, folders 04.03 and 04.04.
[BDA] Bob Dylan Archive, n.d.i, Tulsa, OK, box 88, folder 2.
[BDA] Bob Dylan Archive. 2017. "Finding Aids." Tulsa, OK. http://bobdylanarchive.com/wp-content/uploads/2017/03/Bob-Dylan-Archive-Public-Finding-Aid-March-2017.pdf.
Brown, Sally, Karen L. Aplin, Katie Jenkins, Sarah Mander, Claire Walsh, and Paul D. Williams. 2015. "Is There a Rhythm of the Rain? An Analysis of Weather in Popular Music." *Weather* 70, no.7 (July): 198–204.
Burger, Jeff, ed. 2018. *Dylan on Dylan: Interviews and Encounters*. Chicago: Chicago Review Press.
Crowe, Cameron. 1985. Liner notes to Bob Dylan, *Biograph*. Columbia Records, C5X 38830, vinyl.
Duluth (MN) News Tribune. 1941a. "The Weather." May 24, 1941, 1.
Duluth (MN) News Tribune. 1941b. "The Weather." May 25, 1941, 1.
Dylan, Bob. 1962. "Blowin' in the Wind." *Sing Out!* 12 (October–November): 4.
Dylan, Bob. 1973. *Writings and Drawings*. New York: Knopf.
Dylan, Bob. 2004. *Chronicles*. New York: Simon and Schuster.
Dylan, Bob. 2016. *The Lyrics: 1961–2012*. New York: Simon and Schuster.
Dylan, Bob. 2017. "Nobel Lecture." Nobel Prize, June 5, 2017. https://www.nobelprize.org/prizes/literature/2016/dylan/lecture.
Gray, Michael. 2000. *Song and Dance Man III: The Art of Bob Dylan*. New York: Continuum.
Gray, Michael. 2008. *The Bob Dylan Encyclopedia*. New York: Continuum.
Harvey, Todd. 2001. *The Formative Dylan: Transmission and Stylistic Influences, 1961–1963*. Lanham, MD: Scarecrow Press.
Hasted, Nick. 2005. "Shelter from the Storm: The Inside Story of Bob Dylan's *Blood on the Tracks*." *Uncut*, Take 92 (January): 50–66.
Henrik N. 2017. "Bob Dylan—Caribbean Wind (Extremely Rare Live Nov 12 1980)." February 16, 2017. YouTube video, 7:45. https://www.youtube.com/watch?v=IWHzEXg7OMM.
Heylin, Clinton. 2009. *Revolution in the Air: The Songs of Bob Dylan 1957–1973*. Chicago: Chicago Review Press.
Heylin, Clinton. 2010. *Still on the Road: The Songs of Bob Dylan 1974–2006*. Chicago: Chicago Review Press.
Howard, Mark. 2008. "Bob Dylan: Tell Tale Signs Special—Mark Howard!" *Uncut*, October 8, 2008. https://www.uncut.co.uk/features/bob-dylan-tell-tale-signs-special-mark-howard-37964/.
Kees, Weldon. 1960. *The Collected Poems of Weldon Kees*. Iowa City: Stone Wall Press.
Long, Alex B. 2011. "The Freewheelin' Judiciary: A Bob Dylan Anthology." *University of Tennessee Legal Studies Research Paper No. 152*, May 10, 2011. http://dx.doi.org/10.2139/ssrn.1837683.
Marcus, Greil. 2009. "Hibbing High School and 'the Mystery of Democracy.'" In *Highway 61 Revisited: Bob Dylan's Road from Minnesota to the World*, edited by Colleen J. Sheehy and Thomas Swiss, 3–14. Minneapolis: University of Minnesota Press.

McCartney, Eugene S. 1930a. "Greek and Roman Weather Lore of Winds." *Classical Weekly* 24, no. 2 (October 13): 11–16.

McCartney, Eugene S. 1930b. "Greek and Roman Weather Lore of Winds (Concluded)." *Classical Weekly* 24, no.4 (October 27): 25–29.

Mieder, Wolfgang. 2006. "Proverbs and Sayings." In *Encyclopedia of American Folklife*, edited by Simon J. Bronner, 996–99. Vol. 3. Armonk: M. E. Sharpe.

Mother Goose's Nursery Rhymes. 1877. London: George Routledge and Sons.

Plutarch. 1957. "On the Principle of Cold." In *Moralia*, translated by Harold Cherniss and W. C. Helmbold, 225–85. Loeb Classical Library ed. Vol. 12. Cambridge: Harvard University Press. http://penelope.uchicago.edu/Thayer/E/Roman/Texts/Plutarch/Moralia/De_primo_frigido*.html.

Riley, Tim. 1999. *Hard Rain: A Dylan Commentary*. Updated ed. New York: Da Capo.

Robock, Alan. 2005. "Tonight as I Stand Inside the Rain: Bob Dylan and Weather Imagery." *Bulletin of the American Meteorological Society* 86, no. 4 (April), 483–87. http://climate.envsci.rutgers.edu/pdf/DylanBAMS.pdf.

Rosenbaum, Ron. 1978. "Playboy Interview: Bob Dylan." *Playboy* 25 (March): 61–88.

Rudd, Mark. 2009. *Underground: My Life with SDS and the Weathermen*. New York: William Morrow.

Thomas, Richard F. 2017. *Why Bob Dylan Matters*. New York: Dey Street.

Turner, Gil. 1962. "Bob Dylan—A New Voice Singing New Songs." *Sing Out!* 12 (October–November): 5–10.

Washington Post. 2008. "An 'Idiot Wind': John McCain's Latest Attempt to Link Barack Obama to Extremism." Editorial, October 31, 2008, A18.

Wilentz, Sean. 2010. *Bob Dylan in America*. New York: Doubleday.

Williams, Paul. 1990. *Performing Artist: The Music of Bob Dylan, Vol. 1, 1960-1973*. Novato, Calif.: Underwood-Miller.

Wing, William G. 1961. "35-MPH Snow Caps City's Record Freeze." *New York Herald-Tribune*, February 4, 1961, 1–2.

Zollo, Paul. 2003. *Songwriters on Songwriting*. 4th ed. Cambridge, MA: Da Capo Press.

Chapter 10

"I'LL NEVER FORGET THE THUNDERSTORM OF 1960, I THINK IT WAS"

Storm Stories

Lena Marander-Eklund

It was not too long ago that I had a rather unpleasant experience with a storm. In 2013, I was traveling with my husband on a holiday to London. We flew out from Turku and changed planes in Copenhagen. The waiting area at our gate in Copenhagen Airport was filled with travelers impatiently waiting to board the plane, which was slightly delayed. The captain came out to the waiting area to let us know that we would indeed be taking off soon, but we should prepare ourselves for the stormy weather. Extra measures were being taken to secure the load, which would take some time. He also explained that the flight itself would be a bit out of the ordinary due to the storm, as the plane would be taking off at a very steep angle to get over the storm clouds as quickly as possible. We boarded the plane, which was shivering and shaking from the gusty winds battering it while it stood at the gate. Not long after we took off, a loud bang pierced the cabin. The interior lights went out momentarily (or perhaps it was just me closing my eyes tight), and I, presumably along with many others, screamed out in alarm. I was absolutely terrified. The bang turned out to be the plane flying into the clouds of the storm, which was dubbed "Sven." The rest of the flight went well, I calmed down, and we made a normal landing in London. We found out later that Copenhagen Airport had been closed just a half an hour after we took off.

We often talk about the weather. As a topic of conversation, weather is a key part of our daily interactions with other people. It is a way of starting up a conversation and establishing contact with acquaintances and strangers as well as filling the silence (Golinski 2003, 7). Talking about the weather is "phatic communication." This involves engaging with others as a way of prolonging dialogue and maintaining communication by means of formulaic expressions (Malinowski [1923] 1949, 315). Anthropologists who study weather as a cultural phenomenon consider that people react to weather through speech, but experiences with weather are also characterized by temporal factors (Strauss and Orlove 2003, 6). Many of us compare a given day's weather with the weather from the day before as well as from the year before, perhaps also moaning about the current (bad) weather. There are times when we hearken nostalgically back to the always beautiful summers of our childhood and times when we cast a worried eye toward the future of our climate. We follow the weather reports and respond to weather warnings in a variety of ways. We joke about the weather. Powerful storms or intense thunderstorms can serve as a topic of conversation. One might, for example, talk about a storm or thunderstorm from a certain year that was perhaps referred to as the "storm of the century."

However, there are also cases where the talk of weather takes a narrative form and has a more dramatic character. The story above is based on weather-related events that I personally experienced and is an example of what I call a "storm story," which deals with the experience of thunderstorms and storms. It contains many of the characteristics that constitute a narrative. Naturally, there are types of inclement weather other than thunderstorms, such as black ice, downpours, blizzards, floods, hurricanes, droughts, or extreme temperatures. Thunderstorms and storms are weather phenomena that most people have experienced, regardless of their geographical location. I have chosen not to focus on major natural disasters, such as Hurricane Katrina (see Kverndokk 2014) or global climate change, which naturally have cultural significance and should therefore be given due attention (see Crate and Nuttall 2016). The reason for this delimitation is that I want to focus on dramatic but frequently occurring weather experiences. My approach is methodological, with a focus on the elements of "almost" stories drawn from responses to a questionnaire entitled "Nice Weather Today!" ("Vackert väder idag!" in Swedish). My objective is to examine how storm experiences assume a narrative form. I will look at how one talks about their experiences and why they are often dramatic and emotional in character. My hope is also to offer a way to analyze written stories based on experiences. My intention is to deepen the analysis throughout the text, beginning by identifying the elements of the narrative, looking then at narrative form and emotionality, and finally analyzing storm stories as micronarratives, focusing on narrative exaggeration as a way of understanding the dramatization of existential fear.

"Nice Weather Today!" Questionnaire

The questionnaire, which explores the human relationship with and experience of weather, was developed by me in co-operation with the Society of Swedish Literature in Finland (SLS) in the winter of 2015–2016. The SLS has an active network of contributors, and its conventional questionnaire is supplemented by an online questionnaire. The questionnaire consisted of eight main questions dealing with: weather as a topic of conversation; perfect weather; thunderstorms, storms, and heat waves; predicting weather; weather forecasts; gathering information on weather; weather as news; and climate change and its impacts. This chapter only discusses the responses given to the question dealing with the experience of thunderstorms and storms. The fact that weather can give rise to frightening and dramatic experiences was one of the premises for the questionnaire. I know from experience that storm stories do exist, but I also know that people have different attitudes toward thunderstorms and storms. The question on the questionnaire was introduced in the following way: "Many of us have both positive and negative experiences related to the weather. . . .[A]s well as absolutely terrifying experiences with storms and thunderstorms" (Marander-Eklund 2015, 28, all translations by author). The actual question was formulated as follows: "Tell about your experiences with thunderstorms, storms, heat waves or torrential downpours. Please be detailed and also tell how you felt" (Marander-Eklund 2015, 28). The reason that I chose not to analyze the stories about heat waves and torrential downpours that were told in response to the questionnaire was that there were very few given. By contrast, there were numerous stories told about thunderstorms and storms.

The questionnaire was published in the SLS member magazine *Källan* (Marander-Eklund 2015)—only in Swedish—and on the SLS website in December of 2015. The questionnaire was sent out to roughly 450 people in January 2016. A total of seventy-five responses were received, which is a relatively average response rate for questionnaires sent out by SLS (SLS Arkiv 2016). Of the respondents, seventeen were men and fifty-eight were women. A majority of the respondents were from the regions of Ostrobothnia and Uusimaa. Although many of the respondents were born between 1931 and 1955, all age groups were represented. The respondents' occupations ranged fairly widely, but few executives or high-level civil servants responded to the questionnaire. The most detailed and reflective responses were given by contributors who generally respond to nearly all the questionnaires sent out by SLS (see Hagström and Marander-Eklund 2005, 21).

A clarification of the concepts "climate" and "weather" is needed to gain an understanding of storm stories. "Climate" refers to temperature, precipitation, and wind, which are statistically measured over a longer period of time.

"Weather," in contrast, refers to events occurring within a shorter time span, i.e., daily, weekly, or monthly weather (Karttunen et al. 2008, 110). Climate varies from place to place, but the common denominator it shares with weather is that people respond to it by talking or telling stories about it (Strauss and Orlove 2003, 6). Now, I move on, looking at what characterizes storm stories.

Storm Stories

It can be said that stories are an expression of our everyday experiences and give meaning to our lives (Johansson 2005, 23). According to David Herman, stories are accounts of events happening to different people and their interpretation of those events. He contends that, without stories, we are unable to interpret how an event is experienced by the person in question (Herman 2009, 2 and 143). Stories therefore deal with reported events. According to sociolinguist William Labov (1972, 363), at least two events must be reported in a given chronological order. This definition has been questioned because stories can certainly be told with the events portrayed out of chronological order (e.g., Bennett 1984). However, stories always have a temporal aspect, with a beginning, middle, and end. A story also requires that the events in question have a causality that links them together, a cause and effect (Johansson 2005, 124; Young 1987). Without this meaningful context, this causality, a story will not be a narrative, but rather little more than an explanation or description. Among the responses to the questionnaire, there are sixty-three accounts relating first-person experiences with thunderstorms and storms in a causal context.

In addition to temporality and causality, a story consists of narrative components, i.e., events, actors, time, and location (Bal 2009, 3–13). A story contains an event (the plane flying into the storm, Sven), someone experiencing the event (my husband and I), a time (the beginning of December 2013), and a temporality (a beginning, middle, and end, as well as a chronology). There is a causality (it was stormy, and the plane flew into the dark storm clouds, which led to the loud bang) and an evaluation (I was frightened by the bang). After this event occurs, the events of the story return to normal (the flight lands safely at its destination). All these elements can be found in the type of story I call a "storm story."

At this point, I should provide an explanation as to why the storm was dubbed "Sven." Storms have been named since the beginning of the 1950s as a way to simplify working in meteorology, in part by the general public and by the media. The headline "Sven orsakade skador för miljoner" ("Storm Sven Causes Millions in Damage") appeared in the Swedish newspaper *Sydsvenska dagbladet* on December 13, 2013 (Thomasson 2013). The World Meteorological

Organization (WMO) decides which storms and hurricanes are to be named and the names to be used (Ashmore 2008, 40). The practice of naming storms is relatively new in the Nordic countries, where storms are named according to the "name day" on which the storm occurs (SMHI 2014). The personification of extreme weather dates back to the time when weather gods and wind demons were thought to be the cause of thunder, rain, and storms (Tillhagen 1991, 195). The difference between then and now is that the names of gods are no longer used, with names such as Sven, Dagmar, or Gudrun replacing them. I will now analyze an example of a storm story written by a man in his seventies with respect to the components that constitute a narrative.

> In July 2011, we could hear thunder rumbling off in the distance at around eight in the morning. An hour later, we were hit by a big, intense storm, blowing in from the south with heavy rain. A large birch fell on the lawn, and another fell on the garage roof, where it split. On a nearby island, a large birch fell on the roof of a new house. Our heavy wood patio furniture was blown onto the lawn. The storm front was so narrow—roughly a hundred meters wide—that the plastic patio furniture of our daughter's family was untouched on the hill right next to it. By 10:00 that morning, the storm had passed, and the sun was shining. Some neighbors who lived a little ways off had no idea about the storm other than that they had heard some thunder. (SLS Arkiv 2016)

The story presents a temporal framework, both in that it has a beginning, middle, and end and that the events are portrayed in chronological order. It contains reported and experienced events involving a storm that occurred in the summer of 2011. The fact that the storm is dated quite precisely—in July at eight o'clock in the morning—provides a sort of proof for the authenticity of the story as well as for the fact that it is something which actually occurred (see Bennett 1986, 428; Marander-Eklund 2000, 130). The location of the events is the narrator's summer cottage, and the actors involved are the "we," which refers to the people, including the narrator, at the summer cottage on that day. The effects of the storm, its causality, are the trees that fell on the buildings and the patio furniture that was blown away. In the story, the actors express more a sense of surprise than fear regarding the phenomenon. The narrator expresses his wonder at how quickly the storm passed and how localized it was. Everything happened within a two-hour period, from the approaching thunder to the full-blown storm and ending with the sun coming out again. It appears that the neighbors were unaffected by the thunderstorm, which suggests that the narrator felt it was important to mention their involvement in the events.

Folklorist Amy Shuman identifies retrospective accounts, which deal with prior experiences as opposed to simultaneous narratives that are told in the present. Stories that recount events that occurred in the more distant past often deal with something remarkable. The fact that a story is remarkable, interesting, or dramatic justifies its repetitive nature (Shuman 1986, 72; see also Jansson 2017). In the example above, the speed of the thunderstorm is remarkable enough for him to want to write about it as an answer to a questionnaire about the topic. It was a tellable story (see Goldstein 2009). At the same time, it should be kept in mind that what is interesting to one person will not necessarily be interesting to another (Robinson 1981). Weather is a common, benign topic of conversation: an icebreaker. At the same time, there are those, including the questionnaire's respondents, who feel that talking about the weather is tedious and a bit dull (see Melin and Melin 1996, 116). The comment about the neighbors can be interpreted as outside information that reinforces the sense of surprise experienced in the event. In this section I have looked at what components a storm story consists of and especially viewed it in terms of temporality and causality and what is remarkable enough in order to be a tellable story.

Narrative Form and Emotionality

In this section, I will analyze the form and emotionality of storm stories. One way of analyzing the form-related aspects of a story is to determine whether it correlates with the outline for the basic structure of a narrative developed by William Labov (1972) in his classic study *Language in the Inner City*. Prior to this study, Labov did fieldwork among young Black men in Harlem. His task was to study the linguistic skills of young men. In other words, Labov was interested in the social variations of language use. He conducted interviews in "focus groups." Labov found that when the young men talked about their experiences, they did so according to a certain basic structure. According to him, a complete narrative consists of the following parts. The *abstract* contains an encapsulation of the narrative and an exhortation to listen. This is followed by a background *orientation*, during which the people involved in the story are introduced and time, place, and other essential details are presented. The *complicating action* introduces the essential events or problem. These events are subjected to an *evaluation*. Labov draws a distinction between internal and external evaluation. An internal evaluation involves the narrator, for example, citing himself and using gestures, expressive sounds, and repetitions. An external evaluation involves the narrator turning to the listener outside the narrative and the outsider commenting on it. The complicating action creates a tension that is then resolved via an explanation of what happened or how the problem was solved (*result*). The *coda* can be seen as a signal used to indicate that the narrative is finished, helping to bring the listener back to the present from the reported events.

According to Labov, not all narrative parts need to be included in a complete narrative. However, the core of a narrative is the complicating action and its subsequent resolution, i.e., the problem and how it was solved (Labov 1972, 362–66; see also Johansson 2005, 190–92). This is therefore a question of recounted events about which the narrator wants to convey a point regarding that which they feel is surprising, unusual, or significant—in other words, something worth telling. This is also how we learn to tell stories when making the transition from childhood to adulthood, that is to express our experiences in a suitable way (Adelswärd 1996, 22–24). Labov's model has been used in, among others, analyses of childbirth stories, moose-hunting accounts, the experience of childlessness among Indian women, and how pensioners experience music (see, respectively, Marander-Eklund 2000; Adelswärd 1996; Riessman 1997; Hyltén-Cavallius 2005). When Labov's model is applied to one of the storm stories in my written material from the questionnaire, one can see that the model works well.

> [Introductory summary:] One memory that I'll never forget [background orientation:] is from a fall day when I was about nine or ten years old and in third or fourth grade at elementary school. We lived about five kilometers from the school, and I biked there and back. [Complicating action + evaluation:] One day, a big storm blew in, working itself up into a proper gale during the day. When school let out, I had to bike home. It turned out to be a pretty terrifying ride. I was completely alone. The storm roared all around me, and I could hear trees splitting and falling in the woods on both sides of the road. My heart was pounding, but I had to keep going to get myself home. A little past the halfway mark, an enormous pine tree fell right across the road in front of me. I couldn't move—I was virtually paralyzed—but I managed to get myself moving, dragging my bike into the woods and around the fallen pine. [Resolution:] And what a relief it was when I saw that my mother had come to meet me. I wasn't alone anymore, and I knew I could count on her to be there for me! And she was surely just as scared as I was. [Coda:] As we made our way back, we found a number of large trees lying across the road, but we made it home safely. (SLS Arkiv 2016, 49)

The narrator, a roughly seventy-year-old woman, presents her experience with the storm as something still very clear in her memory even though the storm had happened over sixty years before. One of the reasons that her story still feels fresh and recent is most likely that her childhood experience was so dramatic and frightening. These elements make the story worth telling and give the story its point (Polanyi 1985, 12). According to psychologists Peter Salovay and Jefferson A. Singer (1993), certain memories remain so clear due to a strong emotional connection. The narrator evaluates the events, thus revealing that she was terrified. In the story, the narrator's fear is embodied by her apparent paralysis. She was also alone on her way home from school, which emphasizes the feeling of vulnerability and terror: a small child encountering the enormous power of nature—in this case, an intense storm. The resolution of the story comes when the girl is met by her mother, at which time she no longer feels alone and afraid even though the storm is still raging. All the parts of the Labov model can be found in this story, despite the fact that the coda is not very clearly defined because it does not signal a transition back to the present. What is clear, however, is that the story ends here—the girl and her mother return home safely. This demonstrates that it is possible to use Labov's complete narrative model for storm stories. Even though many have used Labov's model, it has also been criticized for its rigid chronological order and for its claim that the subject of the story has to be unique. Katharine Galloway Young (1987, 202–3), among others, points out that the value of a narrative often lies in the realm of the "ordinary." The experience of weather is, in and of itself, something quite ordinary and mundane in nature. However, a terrifying bike ride through a storm is a more dramatic story.

The reader can understand the child's terrifying ride home from school by taking part in the story. But we must then explore what helps us to understand that the subject of the narrative was a dramatic event reported in the form of a story. Philosopher Sara Ahmed has studied emotions and affects and argues that we should focus on what emotions do rather that looking at what they are and that the emotions are created in contact with the object rather than by the object itself. The kind of emotion that emerges is due to cultural ideas about a certain object. Ahmed believes that we can study the emotionality of texts, how emotional words circulate and generate effects in the reader in a close reading of texts, with a focus on figures of speech (Ahmed 2004, 7–13). These can be such emotional phrases as "I am afraid," which directly expresses an emotion. Second, it can be a question of metaphors, like using something to represent or serve as a symbol for something else, such as in the expression "the autumn of life" (see Lakoff and Johnson 1980, 5–15). Third, it can be a question of words with emotive meaning. This involves the use of value-loaded words that express a certain feeling or attitude and are expressive in nature, have a negative or positive connotation, and serve as a counterpoint to more straightforward words. Examples of emotive words are "brilliant" (positive) and "pathetic" (negative)

(Melin and Lange 1995, 23; see also Marander-Eklund 2009). The fact that the above story is an emotional experience is expressed directly by using emotional words, such as "it turned out to be a pretty terrifying ride" and that the speaker felt "relieved." She indirectly states that she was afraid by saying her mother was probably just as scared as she was. Her fear is also revealed in the use of emotive expressions, such as "I was completely alone," "the storm roared," "I was virtually paralyzed," and "my heart was pounding," which can be interpreted as a metaphor for fear or an embodiment of it. The dramatic narrative about a terrifying bike ride has an emotional mood.

Fear can be interpreted culturally and/or psychologically. According to folklorist Jochum Stattin, fears are situational and often involve a culturally learned pattern. These include imaginary creatures, such as the "*brunnsgubben*" (a goblin-like creature who lives in a water well), and other devices for helping people learn how to identify hazards (Stattin 1990, 10–13). Certain nature-related threats were illustrated by the Church in the 1600s and 1700s. In an "*böndagsplakat*" (direct translation: "prayer day plaque"), which issued instructions for special prayer days, God was said to talk about nature's powers. In the proclamation, God used nature as a punitive instrument and was said to afflict a country with disasters, such as storms, crop failure, and famine (Malmstedt 1994, 190–200). Stattin shows how things people were previously afraid of have changed. In earlier times, nature was seen as dangerous (something to be feared), whereas today, nature is more commonly seen as being threatened (something to be concerned about) (Stattin 1990, 157–65). One psychological contention is that fears are genetic and something that keeps us alert. Loud noises and a sudden flash of bright light startle everyone. The human fear of thunderstorms is, by this explanation, universal. Other fears are learned and depend on one's experiences (Kalat and Shiota 2007, 100–123). I suggest that it does not matter which explanation you think is crucial; all the storm stories include figures of speech.

In this section, I have analyzed the form of the stories by using Labov's narrative model. I have also looked at emotions in the stories by using Ahmed's concept of figures of speech; how people narrate about their feelings by using emotional words, metaphors, and words with emotive meaning. These figures of speech stick with the reader and convince us of the fearfulness of storms and thunder.

Micronarratives

Stories or the expression of recorded accounts can also be studied based on the story's stylistic or aesthetic dimensions (Hymes 1975, 11–15). In this section,

I focus on shorter narratives—micronarratives—as structurally cohesive accounts with clearly defined beginnings and endings, dramatic climaxes, and with past events presented as they were experienced. It is thus a question of distinguishing narrative units with a discernible plot from chronological and descriptive sequences (Arvidsson 1998, 25). Micronarratives are comprised of the sequences found in Labov's basic structure, but they involve a more concise narrative. Another way to draw this distinction is to separate narratives in which the narrator wants to convey a message or a point from reports that are more an account of the events, often at the request of the listener (Polanyi 1985). Micronarratives are a concise way of telling a story that are used to draw attention to key themes, exemplify something, and dramatize individual experiences. In certain cases, humor is used to make an experience comprehensible or play down what has transpired (Arvidsson 1998, 25; see also Marander-Eklund 2000, 145).

A narrative is thus recognized, for example, by means of so-called introductory and concluding markers as well as stylistic devices, such as tone, changes of tempo and pitch, and the use of direct speech and speech with emphasis. These markers are often very apparent in oral narratives but can also be observed in a written context (Kaivola-Bregenhøj 2013, 230–33; Marander-Eklund 2002, 113). An example of an introductory marker found in the responses to the questionnaire is the following: "When I was a child, I experienced an intense thunderstorm that passed over us" (SLS Arkiv 2016, 57). Another example is: "I was 15–16 years old and alone one night in our summer cottage when a big thunderstorm rolled in" (SLS Arkiv 2016, 58). The clearly discernible narratives indicate an intention to convey a message or point. In ethnologist Eerika Koskinen-Koivisto's study on the life narrative of a woman, she uses the concept of micronarratives as a way to differentiate narrative units. She also notes that certain narratives are more significant than others. She refers to these as "key narratives" (see also Koskinen-Koivisto 2014, 48). Koskinen-Koivisto's study is based on repeated oral interviews, which clearly reveal the key narratives. In my case, I cannot know with any degree of certainty whether the storm story above is one of the informant's own key narratives. The fact that it is told in a relatively detailed manner despite its description of events that transpired sixty years before would suggest that it is.

> I'll never forget the thunderstorm of 1960, I think it was. It started at around 7 in the evening. I lived on the second floor of a wood house with my mother, and we waited for the thunder and lightning to stop. But the clock kept on ticking, and the storm simply would not let up. The lightning flashed, and thunder cracked—we sat counting the time between the flash and the rumbling that followed, which my mother

explained was the trick to figuring out how far away the lightning was. We sat side by side on the bed, wearing our rubber boots because my mother said that lightning can't hit you when you're wearing rubber. But I'm pretty sure that we were both just as scared [as one another] that the lightning would strike. The thunder and lightning kept up all night long. When we looked out the window, we could see how the entire field was lit up over and over again. I lay down next to my mother on the bed and asked: Mama isn't there a God? (SLS Arkiv 2016, 59)

Even this story about a storm begins with a comment on the importance of the memory in the form of "I'll never forget." The timeframe given is a bit vague, but it still indicates that the event in question took place long ago. By contrast, the narrator, who is a roughly sixty-year-old woman, remembers the time of day when the storm began despite the fact that she was a child when the events transpired. The story has no real climax but is rather more repetitive in nature: the lightning struck again and again. This story is not so much characterized by the lightning striking as by the sense of the vulnerability and fear experienced before it does. It is here that the emotionality of the text is revealed through the use of the emotional words "she was afraid." Fear was evidently not something explicitly expressed at that moment, as the narrator says that she was "pretty sure that we [the narrator and her mother] were both just as scared [as one another]." The mother was there to provide companionship and comfort to her young daughter.

The mother took steps to prepare for the storm by ensuring that she and her daughter were wearing rubber boots. The interesting thing here is not that they were wearing rubber boots to protect them against lightning strikes, but rather that people would believe such a thing in the first place. This action—putting on rubber boots—assumes an almost ritualistic character as a means of protection against danger. Because thunderstorms have always commanded respect throughout human history, many beliefs are linked to them. Such beliefs include the idea that lightning bolts are cast down by storm gods as a warning or to punish people, that lightning will strike if a given god's name is invoked, and that an åskvigga, a type of stone or an axe, provides protection against lightning strikes (Bertell 2003, 207; Tillhagen 1991, 238). Other, older beliefs include the notion that lightning is a power used to scare away or slay a troll (Stattin 1990, 269).

The story above does not end with a transition from the dramatic events back to normalcy, i.e., the passing of the storm. Instead, the story ends with direct speech, with the daughter asking her mother, "Isn't there a God?" By means of direct speech, the author uses the historical present as way to dramatize the events (Wolfson 1982, 29). We also never find out what the mother's answer was to her daughter's question. The question can be interpreted as an expression of existential fear. Turning to God in such cases can be interpreted in terms of the psychology of religion. Pehr Granqvist and Lee A. Kirkpatrick discuss situations in which people turn to an attachment figure that represents a sort of "safe haven." Such situations are: (1) alarming or frightening environmental events; (2) illness, injury, or fatigue; and (3) separation or threat of separation from an attachment figure (Granqvist and Kirkpatrick 2016, 919–920). This case is foremost an alarming or frightening environmental event. The author even expresses a certain suspicion that God has forsaken her. The story also never specifically mentions whether lightning struck, but it probably did not as it would have otherwise been included. The story has a dramatic and existential ending, while leaving open the answer to the question posed.

The responses to the questionnaire also include several stories about times when lightning did strike and the effects of those strikes.

> I will now tell the story about a heavy thunderstorm we experienced in [place omitted for anonymity]. It was a dark night. We were staying in a small waterfront cottage, a few kilometers from the church where the lightning had struck. My husband was not the least bit scared, but I was positively shaking with terror at the intense thunderstorm that seemed to never let up! I have never experienced such an intense storm. My husband convinced me to go outside and look, where we saw that a big fire had broken out in the old wooden church. It was frightening to be outside with all the thunder and lightning. It was the most terrifying thing I had ever experienced. Our cottage was completely lit up by the constant lightning! (SLS Arkiv 2016, 12)

In this story, an older woman tells what happened when a bolt of lightning struck the village church, which caught fire and burned down. The story opens with the dramatic introductory phrase "It was a dark night," which brings to mind the traditional opening of ghost stories. Here, the narrator expresses her fear very clearly with such emotional words as "I was shaking with terror" and emotive expressions as "It was the most terrifying thing I had ever experienced," reflecting Ahmed's (2004, 7) concepts for analyzing emotions through figures of speech. The fact that a church catches fire can be interpreted as powerfully symbolic. A church represents the Christian faith, but it is also a building of cultural-historic importance. The story does not have a clearly defined ending. There is no mention made of the thunderstorm passing. On the contrary, the story is followed by an additional three comments, which the author separated by blank lines. In one, she expresses her opinions of the new church, in another, she expresses her desire to tell this story, and in the third, she explains that she had had a different, less fearful attitude toward thunderstorms as a child. The stories reveal the various attitudes that the narrators have toward thunderstorms and that these attitudes have changed over the years depending on their personal experiences with them. This is readily apparent in the following story, which begins with an account of how the narrator, a woman in her seventies, had previously not been afraid of thunderstorms, but rather actually enjoyed sitting and watching the lightning:

> But now, a massive thunderstorm will definitely make me feel a bit uneasy. Some years ago, I was sitting alone out in the countryside, and a dreadful thunderstorm rolled in. The power went out and everything went dark. Suddenly, there was a deafening crack and a flame shot out of the mains outlet where our answering machine was plugged in. Something shot out across the floor several meters from the wall. It turned out to be the answering machine's power cord and its power adapter (a big, heavy block) that were blasted out of the wall. What had happened was that lightning had struck the telephone pole closest to us and evidently surged straight through the cable all the way into the house. Since then, I've become extremely wary of thunderstorms. (SLS Arkiv 2016, 18)

Here, the experience with lightning told in a narrative form is used as an explanation of why the woman is now uneasy around thunderstorms. In this case, the story is not just about the thunderstorm but also a bolt of lightning that struck. Although the lightning did not hit the house where the narrator was, there was a certain consequence to the strike.

It should be pointed out that some of the responses to the questionnaire contain opinions that express a predilection for thunderstorms. In one case, the narrator explains that it is unusual for people to enjoy thunderstorms: "I love stormy weather, especially thunderstorms (which is very wrong). I've even experienced lightning strikes and had cardiac arrhythmia as a result, but no—I haven't learned my lesson. Thunderstorms always get me to go out on the balcony to enjoy the show" (SLS Arkiv 2016, 16). The following is another example, in which the narrator's attitude toward thunderstorms is ambivalent: "I love intense thunderstorms. It's especially exciting whenever I'm at my summer cottage. At the same time, you do feel a certain amount of fear, because the risk for getting struck by lightning is pretty high when the storm is right over you" (SLS Arkiv 2016, 15). One of the responses discusses the storm "Dagmar" (called "Tapani" in Finland), which occurred during Christmas in 2011, and its impacts: "About four or five years ago, the storm that came on Boxing Day hit hard in Finland, knocking down lots of trees in certain parts of the country. Many homes out in the countryside were left without power for several days.

The storm and the destruction it caused were big news in the papers" (SLS Arkiv 2016, 55). Another response describes the experience with thunderstorms in an almost romantic light: "Thunderstorms awaken feelings. I remember huddling under a bridge in Sweden on a motorcycle with my (future) husband, while lightning struck all around us" (SLS Arkiv 2016, 51). The common denominator for these examples of a positive or neutral attitude toward thunderstorms is that they are not told in a narrative form. For a variety of reasons, there was no need to provide complete narratives (see Bamberg and Georgekapoulou 2008). The experience of a lightning strike and when the couple huddled under a bridge possess all the ingredients required for telling a story as they are based on relatively unusual experiences. The two other examples are more neutral descriptions or opinions that do not have a dramatic climax. The experiences that are considered dangerous and extreme lend themselves more easily to being told as complete micronarratives (see Tangerlini 1998, xxii). In this section, I have analyzed storm stories as micronarratives with a focus on tellable and emotionally charged content. The stories are of vast importance for those who have experienced storms and thunder and are something "they will never forget."

"Almost" Stories

There is only one story found in the responses to the questionnaire that deals with lightning physically affecting the person/narrator, aside from the story above in which the narrator experienced cardiac arrhythmia. It is as follows: "The lightning strike happened in the 1970s. The lightning hit a metal plate on [place omitted for anonymity], and a flame shot out of the mains outlet. I sat at the piano with my back to the outlet and got three burns just above my ankles. The thunderclap was absolutely deafening. Mama was in the kitchen ironing, so the first thing I asked was: 'Are you still alive?'" (SLS Arkiv 2016, 16).

The micronarratives in which danger is lurking around the corner but everything turns out fine comprise a majority of the responses to the questionnaire. One example of this is a story from a long time ago about a long boat trip out to sea. The trip to a remote island is described as a summery idyll: "Yes, the weather during the trip was perfect—light breeze, wispy summer clouds, and around 24 degrees [75° F]. My husband was at the helm of the open motorboat, and I sat comfortably, just enjoying the whole thing" (SLS Arkiv 2016, 20). On the way home the next day, the couple was caught in an unexpected thunderstorm: "We both knew that we were at the highest point on the sea and could be struck by lightning. So, we sought shelter on a beach, which had changing huts. My teeth were chattering, and we didn't talk. But the storm soon passed. We came back home completely drenched" (SLS Arkiv 2016, 20). This is what I refer to as an "almost" story. An almost story is a dramatic experience that is recounted with increasing drama and in which the fear caused by a threatening natural experience is included as an element: "The wind battered us, and the waves mounted higher and higher. The boat lurched forward to the point that my back began to ache. Fear began to take over my whole body and I was sort of half-lying down, frozen stock-still" (SLS Arkiv 2016, 20). Despite the threatening events, everything went well, and the couple made it home cold and wet but otherwise unscathed. The disaster that could have been never happened. It is my contention that, in almost all cases, when an experience is related in a narrative form, it is a question of an "almost" story (with some exceptions). It is a case in which the storm almost knocked trees down, and if they were indeed knocked down, they almost fell on the person in question. With regard to thunderstorms, an "almost" story will relate how the lightning almost struck, and if it did indeed strike, the story will be about how the narrator was almost hit, and if the person was indeed hit, the story will be about how they almost died.

This element of "almost" can also be found in the childbirth narratives I analyzed in a previous project. These narratives deal with how the pregnant mothers "almost" fainted in the shower, "almost" gave birth while they were sitting on the toilet, and how they "almost" needed a caesarean section, vacuum-assisted delivery, or some other more involved procedure. Even in this study, one can see that it is the serious, difficult, and dramatic aspect that lends itself to the shaping of micronarratives (Marander-Eklund 2000, 145). This "almost" can be seen as a narrative strategy, as a way to slightly exaggerate by introducing dramatic elements and a bit of humor. In her study on arguing, Amy Shuman (1986, 52–53) writes about something that "almost" happens: "The events described were often illusory; the stories were about what almost happened; they were often told to make mountains out of molehills." Here, it is a case of using hyperbole as a way of dramatizing the events and making a good story. However, I do not mean to say that this only has to do with creating an entertaining story, but rather that it is a question of existential fear. This is readily apparent in the question the narrator asks her mother in the story above: "Are you still alive?" (SLS Arkiv 2016, 16). Although thunderstorms can indeed be dangerous, they rarely are.[1]

These kinds of near-miss stories have also been studied by Diane Goldstein (2009, 235–55). In her analysis of 9/11 stories, she explores the issue of self-censorship and "untellability" in rumors and legends that circulated in the United States, Canada, and the United Kingdom as a reaction to the events of the eleventh of September. She argues that folklorists should focus both on what is narrated and on what is not. The self-censored and silenced narratives were about people that miraculously got away from the events. When so many people died, it was not suitable to tell stories about narrow escapes. Stories about narrow escapes, preventing people from being trapped in the crisis, have also been told about other disasters, such as the Titanic but also about earthquakes, tornados, and floods. She calls this kind of story a "foreknowledge story" (Goldstein 2009, 237). Stories about storms and thunder are not in the same way untellable, but they also have the element of narrow escape or "almost"—the term I prefer.

When Labov did his fieldwork among the young men in Harlem, his objective was to get them to talk because he was interested in linguistic variation. The question he posed was, "Were you ever in a situation where you were in serious danger of being killed, where you said to yourself—This is it?" (Labov 1972, 354). The question resulted in stories about the times when the men were exposed to violence or in fear of losing their lives. In a later study, Labov elaborates on this idea, explaining that when we report dramatic events, we are telling what happened in relation to what *could* have happened in quite a similar way as Goldstein (2009) does. He calls this a comparison with a parallel universe. Near-death experiences reveal an awareness of our own mortality and trigger an empathetic response in the listener/reader. At the same time, the stories can be seen as a triumph over the dangers that the people faced (Labov 2013, 226–28).

In summary, it can be said that stories about storms—thunder, lightning, and stormy weather—largely deal with a situation in which a person's control is set aside. There are few things as unforeseeable and impossible to influence as when nature takes us to task. At the same time, the weather (even extreme weather) is a way for us to experience nature and assimilate ourselves in the landscape (see Löfgren 2014, 259). Thunderstorms and storms can be seen as nature's drama, which is something powerful, beautiful, and exciting—something one even strives to experience. But thunderstorms and storms can also be seen as frightening and dangerous. They are experiences that lend themselves to dramatic narratives, with fear as the focus.

One interesting observation is that stories often deal with events that happened during childhood. Perhaps we recall our first experience with stormy weather with great clarity. Perhaps we were more frightened, or perhaps it is easier to admit being afraid when we were children. We can also ask ourselves whether it is more acceptable to admit being afraid of thunderstorms when a child is involved. This might be a source of embarrassment for an adult. This is something that could be studied in greater depth. Another observation is that many of the storm stories take place at or around summer cottages; thunderstorms and storms seem strongest in a rural milieu, when spending time in nature away from the safety of lightning rods in the city. The fact that these stories take place in the summer is a matter of course: thunderstorms occur primarily in the summer months.

These dramatic, even terrifying, incidents lend themselves to concludable narratives with a clear element of cause and effect, of causality, and clear markers of internal and external evaluations in the form of emotional words and emotive expressions. The various approaches to the analysis of storm stories produce different but coherent results. The primary goal of this chapter is to draw attention to the characteristics of "almost" stories. As with other crisis situations, the existential fear and awareness of our own mortality are, to a high degree, kept quiet (Labov 2013; see also Sandberg 2016). The attitude toward fears related to stormy weather and our desire to verbalize them evidently evolve throughout our lives. The dramatic thing that "almost" happened can be seen as an exaggeration, a way of recounting something exciting, but it can also be interpreted as a powerful expression of emotion: a fear of the forces of nature. Analyzing storm stories by focusing on their narrative exaggeration thus becomes a way of understanding the dramatization of existential fear.

Note

1. At present, lightning kills one person every other year in Finland. Approximately one hundred years ago, roughly a dozen people were killed by lightning each year because there was a lack of knowledge at that time regarding the best way to take shelter during thunderstorms (Tuomi and Mäkelä 2009).

References

Adelswärd, Viveka. 1996. *Att förstå en berättelse: Eller historien om* älgen. Stockholm: Bromberg.
Ahmed, Sara. 2004. *Cultural Politics of Emotion*. Edinburgh: University of Edinburgh Press.
Arvidsson, Alf. 1998. *Livet som berättelse: Studier i levnadshistoriska intervjuer*. Etnologiska skrifter 10. Lund: Studentlitteratur.
Ashmore, Richard. 2008. "Tropical Cyclone Development, Structure, and Impact on the Louisiana-Texas Coast." In *Geological Society of America Field Guide 14*, edited by G. Moore, 39–47. Boulder, CO: Geological Society of America.
Bal, Mieke. 2009. *Narratology: Introduction to the Theory of Narrative*. 3rd ed. Toronto: University of Toronto Press.
Bamberg, Michel, and Alexandra Georgekapoulou. 2008. "Small Stories as a New Perspective in Narrative and Identity Analysis." *Text & Talk* 28, no. 3: 1–18.
Bennett, Gillian. 1984. "Women's Personal Experience Stories of Encounter with the Supernatural." *Arv Nordic Yearbook of Folklore* 40: 79–87.
Bennett, Gillian. 1986. "Narrative as Expository Discourse." *Journal of American Folklore* 99, no. 394: 415–34.
Bertell, Maths. 2003. *Tor och den nordiska åskan: Föreställningar kring världsaxeln*. Stockholm: Stockholms Universitet.

Crate, Susan A., and Mark Nuttall, eds. 2016. *Anthropology and Climate Change: From Actions to Transformations*. 2nd ed. New York and London: Routledge.

Goldstein, Diane E. 2009. "The Sounds of Silence: Foreknowledge, Suppressed Narratives, and Terrorism—What Not Telling Might Tell Us." *Western Folklore* 68, no. 2–3: 235–55.

Golinski, Jan. 2003. "Time, Talk and the Weather in Eighteenth-Century Britain." In *Weather, Climate, Culture*, edited by Sarah Strauss and Benjamin S. Orlove, 17–38. Oxford and New York: Berg.

Granqvist, Pehr, and Lee A. Kirkpatrick. 2016. "Attachment and Religious Representations and Behavior." In *Handbook of Attachment Theory and Research*, edited by Jude Cassidy and Phillip R. Shaver, 917–40. 2nd ed. New York: Guilford.

Hagström, Charlotte, and Lena Marander-Eklund. 2005. "Att arbeta med frågelistor: En introduktion." In *Frågelistan som källa och material*, edited by Charlotte Hagström and Lena Marander-Eklund, 9–29. Lund: Studentlitteratur.

Herman, David. 2009. *Basic Elements of Narrative*. Chichester, UK: Wiley-Blackwell.

Hyltén-Cavallius, Sverker. 2005. *Minnets spelrum: Om musik och pensionärskap*. Hedemora: Gidlund.

Hymes, Dell. 1975. "Breakthrough into Performance." In *Folklore, Performance, and Communication*, edited by Dan Ben-Amos and Kenneth Goldstein, 11–75. The Hague: Mouton.

Jansson, Hanna. 2017. *Drömmen om äventyret: Långfärdsseglares reseberättelser på internet*. Stockholm: Stockholms Universitet.

Johansson, Anna. 2005. *Narrativ teori och metod*. Lund: Studentlitteratur.

Kaivola-Bregenhøj, Annikki. 2013. "Skrivna berättelser: Textanalytiska aspekter på upptecknade drömmar." In *Folkloristikens aktuella utmaningar: Vänbok till Ulf Palmenfelt*, edited by Owe Ronström, Georg Drakos, Jonas Engman, and Högskolan på Gotland, 229–59. Visby: Högskolan på Gotland.

Kalat, James W., and Michelle N. Shiota. 2007. *Emotion*. Belmont, AU: Wadsworth.

Karttunen, Hannu, Jarmo Koistinen, Elena Saltikoff, and Olli Manne. 2008. *Ilmakehä, sää ja ilmasto*. Helsinki: Tähtitieteellinen yhdistys Ursa.

Koskinen-Koivisto, Eerika. 2014. *Her Own Worth: Negotiations of Subjectivity in the Life Narrative of a Female Labourer*. Helsinki: Finnish Literature Society.

Kverndokk, Kyrre. 2014. "Mediating the Morals of Disasters: Hurricane Katrina in Norwegian News Media." *Nordic Journal of Science and Technology Studies* 2, no. 1: 78–87. https://doi.org/10.5324/njsts.v2i1.2140.

Labov, William. 1972. *Language in the Inner City: Studies in Black English Vernacular*. Philadelphia: University of Pennsylvania.

Labov, William. 2013. *The Language of Life and Death: The Transformation of Experience in Oral Narrative*. Cambridge: Cambridge University Press.

Lakoff, George, and Mark Johnson. 1980. *Metaphors We Live By*. Chicago: University of Chicago.

Löfgren, Orvar. 2014. "Väderbitna landskap." In *Naturen för mig: Nutida röster och kulturella perspektiv*, edited by Lina Midholm and Katarina Saltzman, 257–60. Göteborg: Institutet för språk och folkminnen.

Malinowski, Bronislaw. [1923] 1949. "The Problem of Meaning in Primitive Languages." In *The Meaning of Meaning: A Study of Influence of Language upon Thought and of the*

Science of Symbolism, edited by Charles K. Ogden and Ivor A. Richards, 296–336. 10th ed. London: Routledge & Kegan Paul.

Malmstedt, Göran. 1994. *Helgdagsreduktionen: Övergången från ett medeltida till ett modernt år i Sverige 1500–1800*. Göteborg: Göteborgs Universitet.

Marander-Eklund, Lena. 2000. *Berättelser om barnafödande: Form, innehåll och betydelse i kvinnors muntliga skildring av födsel*. Turku: Åbo Akademi.

Marander-Eklund, Lena. 2002. "Narrative Style: How to Dramatize a Story." *Arv* 58, no. 6: 113–24.

Marander-Eklund, Lena. 2009. "'Kanske den bästa tiden i mitt liv': Analys av emotioner i ett frågelistmaterial om livet i efterkrigstidens Finland." *Tidskrift for kulturforskning* 8, no. 4: 22–38.

Marander-Eklund, Lena. 2015. "Frågelista: Vackert väder idag!" *Källan* 2: 28–29.

Melin, Lars, and Sven Lange. 1995. *Att analysera text: Stilanalys med exempel*. 3rd ed. Lund: Studentlitteratur.

Melin, Lars, and Martin Melin. 1996. *Dösnack och kallprat*. Stockholm: Nordstedt.

Polanyi, Livia. 1985. *Telling the American Story: A Structural and Cultural Analysis of Conversational Storytelling*. Norwood: Praeger.

Riessman, Catherine. 1997. "Berätta, transkribera, analysera: En metodologisk diskussion om personliga berättelser i samhällsvetenskaperna." In *Att studera berättelser: Samhällsvetenskapliga och medicinska perspektiv*, edited by Lars-Christer Hydén and Margareta Hydén, 30–62. Stockholm: Liber.

Robinson, John. 1981. "Personal Narratives Reconsidered." *Journal of American Folklore* 94, no. 371: 58–85.

Salovay, Peter, and Jefferson A. Singer. 1993. *Remembered Self: Emotion and Memory in Personality*. New York: Macmillan.

Sandberg, Christina. 2016. *Med döden som protagonist: En diskursanalytisk fallstudie om dödens makt*. Turku: Åbo Akademi.

Shuman, Amy. 1986. *Storytelling Rights: The Use of Oral and Written Text by Urban Adolescents*. London: Cambridge University Press.

SLS Arkiv. 2016. "Vackert väder idag!" Svenska litteratursällskapet i Finland, Helsinki, Finland, SLS 2303, 1–75.

SMHI. 2014. "Vem namnger stormar?" January 15, 2014. http://www.smhi.se/kunskapsbanken/meteorologi/vem-namnger-stormar-1.18957.

Stattin, Jochum. 1990. *Från gastkramning till gatuvåld: En etnologisk studie av svenska rädslor*. Stockholm: Carlsson.

Strauss, Sarah, and Benjamin Orlove. 2003. "Up in the Air: The Anthropology of Weather and Climate." In *Weather, Climate, Culture*, edited by Sarah Strauss and Benjamin S. Orlove, 3–16. Oxford and New York: Berg.

Tangerlini, Timothy. 1998. *Talking Trauma: A Candid Look at Paramedics through Their Tradition of Tale-Telling*. Jackson: University Press of Mississippi.

Thomasson, Therese. 2013. "Sven orsakade skador för miljoner." *Sydsvenska*, December 13, 2013. https://www.sydsvenskan.se/2013-12-13/sven-orsakade-skador-for-miljoner.

Tillhagen, Carl-Herman. 1991. *Himlens stjärnor och vädrets makter*. Stockholm: Natur och kultur.

Tuomi, Tapio, and Antti Mäkelä. 2009. *Ukkosta ilmassa*. Helsinki: Ursa.

Wolfson, Tessa. 1982. *CHP: The Conversational Historical Present Tense in American English Narratives*. Dordrecht and Cinnaminson: Foris Publications.

Young, Katharine Galloway. 1987. *Taleworlds and Storyrealms: The Phenomenology of Narrative*. Dordrecht: Martinus Nijhoff Publishers.

Section III

TRADITION

Introduction

FEEDING THE STORM

The weather report predicted snow. I hadn't been paying attention, and we needed milk and a couple of other things. The store was packed. There was no milk. There was a storm coming. What food do you buy when you know there is a storm on the way? You run to the store on the way home from work. You have not checked the fridge or the cabinets or pantry. What are the things you will need to get you through the weather?

It depends a little on the storm you are preparing for. Snowstorms require the French-toast list—bread, milk, and eggs. Other items might include hot chocolate supplies and marshmallows and maybe the fixings for a big, heavy, warming stew or pot of soup. Before the snow, Americans seem to crave carbs and dairy. Nova Scotians crave chips (Woodbury 2017). Hurricanes require different planning. A hurricane could last a while. Power could go out for days. There are the hurricane snacks to be eaten while the wind blows and the rain flies—cookies, chips, and crackers. And there are the supplies for later—rice, pasta, canned chili and soup, sausages, peanut butter, and apples. Toilet paper and pet food are bought for all kinds of storms. Everyone buys booze.

Nutritionally, storm shopping shows clear patterns of atavistic survival. Storm food invokes the necessity of human energy production: the turning of calories into cellular power but with the veneer of modern food-science technology. Storm shoppers seek easy proteins in processed forms—peanut butter, dried meat jerky, and salami sticks. They look for heavy carbohydrates to keep them warm and keep them from blowing away—the rice and pasta and bread and pancakes. They savor sugary treats, fast, high-burning energy—chocolate and Oreos and candy—because, after all, who knows how this storm will end. We'll need treats to keep up spirits, to see us through, and food that is slow to perish and filling. Storm food should last both outside and within the body.

There is something temporal about storm food, not only in the sense that these are the foods people might only buy before a storm or that they have an unusual shelf life, but also in the sense that they are consumed at special times within the time lapse of the storm itself. Before the storm, the focus is on what will keep, what the buyer will need to see them through an unknown amount of time between the first precipitation and the return to normality, hours or days later. Foods bought before the storm anticipate the fears of what the storm might bring. They encode "what ifs": What if the power goes out? What if the roads become impassable? What if the storm stalls over us? "Before" foods help ease these fears. They let the buyer know that while they may not be able to change the path of nature, they can do something, this one thing, to prepare.

During the storm, people move from shopping to cooking and consuming. The danger of the first flush of the weather has passed, and the shape of the storm has become clear on the satellite images. There is a sense of what is to come, even if it has not yet fully occurred. This moment, mid-storm and in the storm's first aftermath, suggests the sublime—the sense, in Kantian terms, that while nature may be mighty and produce fear, people have the capacity of reason that ultimately triumphs over it. Storms, as experienced by many in the wealthy, developed world, suggest, in Kant's words, "a power that has no dominion over us" (quoted in Ginsborg 2013, n.p.). The cooking and eating of storm food offers a tangible representation of human survivorship.

During the storm, food becomes a focus of the specialness of the moment, but "during" foods also reflect their specific storm. Diane Tye (2014) analyzes foods cooked on snow days as being a form of play that fills the time on a day when time seems to slow. With these foods, the storm-wracked focus is on their production rather than presentation. The selection of foods suggests a nostalgia for a romanticized past that is believed to be more in touch with the elements (Tye 2014). On snow days, the foods relished are the slow-cooking, rarely made foods of time off. Snow days offer a sense of time out of time, a rare break from the normal schedules of work and school and fitting meals around them. As Tye (2014, n.p.) says, "the day belongs to you rather than those who usually buy your time." Bread is baked, cookies are chilled and rolled and cut, and stews are simmered. There seems almost a celebration in the richness and the warmth—we met the storm, we survived; let's eat. Of course, the emergency itself may not be over, but there is a sense of having weathered the worst.

The habits of storm food provide a familiar comfort, something to do regardless of the specific crisis at hand. One of my earliest memories is of baking cookies as my family rode out Hurricane Frederik (maybe the habits of a family from a snowy place die hard even in the tropics). Such food practices, honed through storm experiences, can be translated to other emergencies of indeterminate length. In the early weeks of COVID-19 lockdowns, the United States saw a rash of bread baking, so much that stores ran out of flour and yeast became hard to come by. King Arthur Flour had to change their production business model and contract with smaller mills to meet soaring demand (Sosland 2020). People baked sourdough loaves by the dozen, despite this type of bread's multi-day requirements. Stuck at home, many turned to the food traditions they knew.

Hurricane snacks are different. The nutritional content is similar to snowy-day food, and the advice handed out by news and food sites alike is nearly identical—*Food and Wine* suggests you "pair crackers with your peanut butter" (Kramer 2019, n.p.), though they do not say which wine to reach for. The vernacular focus, however, is less on cooking, which might not be possible, and more on ease of preparation. And the temporal element centers less on the slowness of the day than on the potential for foods to spoil—eat the most perishable first. Food following hurricanes is strategic, aimed at quantity and clearing things out. Along the Gulf Coast, that often means gumbo (Davis and Fontenot 2005). Pragmatically, gumbo uses the meat from the freezer if the electricity is out, and it feeds a crowd.

Storm foods pose an interesting taxonomic problem though. Thomas Adler (1981) and Warren Belasco (2008) both divide food according to how we go about consuming it, which incorporates how we feel about the food itself, the time taken to eat, who we eat it with, and what its purpose is. For Adler, ferial food is the stuff that keeps us going, the food we eat because organisms must eat to live. Belasco calls the consumption of ferial food feeding. We feed because we have to. Adler contrasts the ferial with festal food, the food we sit down to enjoy, often with others, to savor and celebrate. For Belasco, festal food is consumed through eating, and it implies a commensality with the food itself, with other people, and with our cultural and individual identities.

The two theories seem to line up with each other: eating the festal; feeding on the ferial. Sunday breakfast is festal—a family gathers to eat pancakes specially cooked for the occasion. American Thanksgiving dinner is festal—bring on the turkey and pie. Both imply an enjoyment of the food itself, even if cooked for one alone. There is something special happening around the food, and it is eaten with attention. By contrast, Thursday lunch, in the standard five-day work week of the middle and working class, is likely ferial feeding. Thursday lunch is a quick sandwich or reheated leftovers or vending-machine snacks added to a banana and a cup of coffee. It matters that it doesn't taste bad, but it need not taste good. Storm food breaks the festal/ferial binary.

Food bought in preparation for the storm is meant for feeding, to keep the buyer fed and give them the energy they will need to survive the storm and clean up after it. The intention shows in the focus on all those ready calories and high levels of carbohydrates and protein and sugar. Yet "before" foods are also festal. They do not form a typical shopping list and feature foods not eaten regularly, like cookies and candy and alcohol. Further, before the storm, there is a sense of the communal, as a region prepares at the same time. There is chat in the grocery line about the weather to come. Even if the foods are ultimately eaten by each alone in their own house, the anxieties of being able to feed produced by the coming of the storm are shared anxieties. By contrast, foods eaten during and immediately after the storm suggest the festal in the long-cooking times and focus on meat and sweets usually reserved for holidays, but they are closer to eating in other ways. They incorporate the festal: we have survived; let's enjoy the food. But as hurricane gumbo shows, feeding happens too in the using up of perishables. They are also ferial: let's eat this before it is no longer safe to eat.

Which raises a question: What if the storm passes on and the devastation remains? What happens if there is nothing to celebrate—no assurance that the road will be cleared or that the power will soon be restored? What happens in the wake of the biggest storms? Anecdotally, the first two days are okay. Storm foods are eaten, the pantries and shelves slowly emptied first of fresh produce, then of the tinned foods saved for emergencies. Camp stoves are pulled along with lanterns from basements. For two days, after a summer thunderstorm took out the power for a week or after a tree smashed my roof following an ice storm, I stayed in my house with my family, making tea in a pot on a propane stove, keeping the wood stove lit in the winter. It was almost festal. Almost. Then we decamped, emptying what was left of the food and bringing it with us to relatives' houses. We were fortunate, even though we would not sleep again in our house for months.

But what happens after two days? What of those bigger disasters? What do people eat in the wake of the biggest storms, Hurricane Katrina, the blizzard that hit the northeast United States in 1978, or Snowmageddon of 2010? After the hurricane food was eaten, the grilling and the gumbo that followed the storm emptied freezers and created shared meals with neighbors (Davis and Fontenot 2005). Then came MREs (meals ready to eat): MREs that contained tiny bottles of hot sauce (Horigan 2021). Except when the MREs did not appear, and locals less affected by the storm stepped in where they could (Donlan and Donlan 2008). For some, it is the first hot meal after MREs that stands out, something that didn't just feel ferial for a change (Ingram 2021). And then came volunteers from around the United States. As volunteers and local residents tried to navigate their relationships to each other, food became a site of negotiation. What began as a way for locals to share their culture and offer thanks by making local specialties for volunteers became, as more and more volunteers arrived, an obligation and then a source of discomfort and unease as locals and volunteers, largely divided by culture, race, and class, valued those local foods differently (Harvey 2017, 498–99).

Those biggest storms also produce ripples in vernacular foodways. Houston food was different than New Orleans food (Horigan 2021). After being told for years by a friend from Mississippi about snowballs, I finally tasted one in St.

Louis in the summer of 2006, a year after Katrina. The owner of the snowball stand had been displaced by the storm. It is possible to chart the diaspora of Katrina through the spread of snowballs up the Mississippi Valley. The timelines of the biggest storms seem to stretch out the foodways of storms; it turns eating to feeding to eating again in cycles of survival and relief. It is festal and ferial by turns and at the same time. Storm food is special food. It is survival and celebration all at once.

What will you buy as you anxiously listen to the weather report?

The chapters in this section describe other vernacular responses to the weather. These responses, which incorporate personal, communal, and class traditions, reveal a fraught relationship with the weather. The weather can be prepared for, borne, weathered, but it never seems to quite do the expected. The first two chapters describe how communities make sense of weather events through creative storytelling. This section begins with Jordan Lovejoy's chapter "Framing the Flood: Strategic Environmental Storytelling in Appalachia." Lovejoy takes us through the vernacular art of people in Appalachia as they balance a sharp criticism of the environmental harm being done to their communities with a hope that they and their land will survive. Situating their storytelling within the context of floods links their past and present survival, both figuratively and literally, to the land. In "Weathering the Storm: Folk Ideas about Character," John Laudun examines the ideas that folks have about weather, personal property, and identity through a collection of memes and oral legends that feature trucks abandoned to the weather. The two genres speak both to each other and to larger community concerns about the weather and what it means to be under water both literally through storm induced flooding and financially as a result of those storms.

The last three chapters look at the more personal impacts of the weather and climate change as they are woven with larger local and cultural anxieties. Kristen Bradley's chapter "It Always Rains on a Picnic: Weatherlore and Community Narrative at St. Patrick's Irish Picnic and Homecoming" builds on this discussion of the uses of tradition. Bradley looks closely at the claim that "it never rains" on this long-standing Tennessee event. By engaging a combination of archival research and first-person accounts, Bradley traces this claim to contextualize it within traditional weatherlore while also showing how it speaks to this community's faith, values, and identity. Todd Richardson takes on the rituals of suburban lawn upkeep in his chapter "The Folk Wisdom of Lawns." Weaving together multiple histories—personal, cultural, environmental—Richardson's life writing illustrates just how personal the weather, in all its permutations, can be. We end this section and this collection with Claire Schmidt's chapter "Canning for the Apocalypse: Climate Change, Zombies, and the Early Twenty-First-Century Canning Renaissance." Here, she describes how pop-cultural representations of the zombie apocalypse, often themselves a cipher for climate change and its impacts on food systems, can lead to an intimate response of personal food preservation. Seemingly, historic foodway traditions have been reframed by new generations as survival mechanisms in response to the breakdown of a consistent climate and the failures of vernacular expectations of the weather.

References

Adler, Thomas A. 1981. "Making Pancakes on Sunday: The Male Cook in Family Tradition." *Western Folklore* 40, no. 1: 45–54. https://doi.org/10.2307/1499848.

Belasco, Warren. 2008. *Food: The Key Concepts*. Oxford: Berg.

Davis, Mike, and Anthony Fontenot. 2005. "Hurricane Gumbo." *The Nation*, November 7, 2005.

Donlan, Jocelyn H., and Jon G. Donlan. 2008. "Government Gives Tradition the Go-Ahead: The Atchafalaya Welcome Center's Role in Hurricane Katrina Recovery." *Louisiana Folklore Miscellany* 16–17: 36–44.

Ginsborg, Hannah. 2013. "Kant's Aesthetics and Teleology." Stanford Encyclopedia of Philosophy. Accessed January 12, 2021. https://plato.stanford.edu/entries/kant-aesthetics/#2.7.

Harvey, Daina Cheyenne. 2017. "'Gimme a Pigfoot and a Bottle of Beer': Food as Cultural Performance in the Aftermath of Hurricane Katrina." *Symbolic Interaction* 40, no. 4: 498–522. https://doi.org/10.1002/symb.318.

Horigan, Kate Parker. 2021. Personal Correspondence. January 6, 2021.

Ingram, Shelley. 2021. Personal Correspondence. January 4, 2021.

Kramer, Jillian. 2019. "Keep These Hurricane Essentials on Hand to Ride Out the Storm." *Food and Wine*, August 30, 2019. https://www.foodandwine.com/how/hurricane-food.

Sosland, Sam. 2020. "King Arthur Flour Changes Business Model in Response to Coronavirus Pandemic." *Bake Magazine*, April 17, 2020. https://www.bakemag.com/articles/13303-king-arthur-flour-changes-business-model-in-response-to-coronavirus-pandemic.

Tye, Diane. 2014. "Storm Days: Playing with Food and Time." *Digest* 2, no. 2 (Spring). https://scholarworks.iu.edu/journals/index.php/digest/article/view/34050/37514.

Woodbury, Richard. 2017. "Why Eating #stormchips Says More about Your Identity than Your Appetite." CBC News, February 16, 2017. https://www.cbc.ca/news/canada/nova-scotia/storm-chips-atlantic-canada-snowstorm-hidden-meanings-1.3985805.

Chapter 11

FRAMING THE FLOOD

Strategic Environmental Storytelling in Appalachia

Jordan Lovejoy

In 2003, the Appalachian Program at Southeast Kentucky Community and Technical College formed Higher Ground, "a community arts organization" that creates "photography exhibits, tile mosaics, and plays" that rely "heavily on oral histories collected by Harlan Countians and about Harlan Countians." The first play the community group wrote and performed was a musical, also called *Higher Ground*, that "followed community members as they struggled with a flood of water and a flood of drugs" (Higher Ground in Harlan County n.d.). To create the foundational story for the play, members interviewed around one hundred folks in the Harlan County, Kentucky, area about their lives. The musical was billed as "a new play about celebration and survival" with a tag of "together we can stay afloat" (Higher Ground in Harlan County n.d.).

In 2019, Appalachian author and educator Robert Gipe, who is also the executive producer for Higher Ground, gave a TED talk about his experience working with the group and using art to help the community deal with the opioid crisis. Regarding the interviewing and oral history process, Gipe (2019, n.p.) says:

> We went out and started talking to people. And one of the things that came up quick was stuff about shame. People were ashamed. They'd never had a narcotics addict in their family. So, they were dealing with it inside, inside the house, inside the family. And so, the other thing, because we were interviewing people about their whole lives and not just about their problems, we heard a lot of stories about floods. Because it floods in Eastern Kentucky. Narrow valleys. Water come up quick. And so, one of the things we heard was that in a flood, people will put aside their differences and pull each other out of the mud. So, that became

the metaphor for what we were trying to do with our art project: is that we should deal with this drug-abuse crisis like a flood. Come together.

As Gipe notes, the community—and, therefore, the play—connects two different traumatic events and their lingering effects on the people and place of Appalachia through storytelling; the opioid epidemic and ongoing flooding in the region become tied together in both memory and performance. This intertwining of social challenges and environmental disasters emphasizes the significance of floods to Appalachian life and how other traumatic events become filtered through or linked to ongoing environmental disaster in the region.

Why might Appalachians be linking social struggles to environmental ones? In this chapter, I look specifically at how several Appalachian creators and residents employ flood imagery or use the metaphor of a flood to explore social life and environmental harm through vernacular artistic creations. One explanation for this connection through aesthetic representation, I argue, is to emphasize that floods are not only disastrous and traumatic moments Appalachians historically encounter, but also moments in which creative community survival strategies emerge. Floods, though harmful and destructive, are also reminders of endurance through altered circumstances. The storytelling about and through them strategically paints Appalachians as made up of the very stuff of survival, a depiction that supports a cautious hope, serves as a solemn reminder, and expresses a desire for life and livability despite social, economic, and environmental precarity.

Appalachia, Climate Change, and Writer Activism

The Appalachian region follows the Appalachian mountain range along the eastern United States from Mississippi to New York. As an official region, Appalachia is home to twenty-five million residents from 420 counties in thirteen states (Appalachian Regional Commission n.d.a). In 1965, Congress passed the Appalachian Regional Development Act (ARDA) that formed the Appalachian Regional Commission (ARC) "as a unique federal-state partnership committed to strengthening Appalachia's economy and helping the Region's thirteen states achieve economic parity with the rest of the Nation" (Appalachian Regional Commission n.d.b). According to Congress, the region, "while abundant in natural resources and rich in potential, lags behind the rest of the Nation in its economic growth," and Appalachian "people have not shared properly in the Nation's prosperity" (US Congress 1964). Though heavily focused on economic development, the ARDA also emphasized the need for transportation infrastructure and public resources related to education and health. Importantly, the 1965 act also acknowledged the risks, dislocations, and short- and long-term effects of ongoing changes "in national energy requirements and production" (US Congress 1964). Because of these energy industry changes, Congress argued the region should "take advantage of eco-industrial development, which promotes both employment and economic growth and the preservation of natural resources" and turn to "opportunities for expanded energy production . . . to maximize the social and economic benefits and minimize the social and environmental costs to the region and its people" (US Congress 1964).

Over fifty years after the ARDA and the establishment of the ARC, the Appalachian region, though better off in some regards and increasingly diverse in experience, still elicits stereotypes and is commonly associated with extractive industries, like coal mining and timbering, poor public health, lack of transportation and infrastructure, struggling education facilities, and poverty. Ongoing social challenges, like systemic racism and classism, have been exacerbated by contemporary health emergencies, like the opioid epidemic and the COVID-19 pandemic, further straining the region economically and socially. Environmental damage related to the extractive industry, like mountaintop-removal strip mining, and increased deregulation policies for clean water and air further exacerbate these social challenges, in turn *maximizing* "the social and environmental costs to the region and its people" that Congress vowed to minimize (US Congress 1964). In the Anthropocene, climate change, of course, folds these ongoing social and environmental harms further into one another.

The Anthropocene is our proposed geologic epoch in which human activity has altered the planet on a geologic scale (Crutzen 2002). The rise of the Anthropocene is generally associated with the rise of fossil-fuel-related and industrial activity. The increased emissions of greenhouse gases (like carbon dioxide and methane) and atmospheric aerosols from such human activity contribute to global warming and a host of other interconnecting direct and indirect effects of anthropogenic climate change. Some of these effects include increased land and ocean surface temperatures, altered storm and precipitation weather patterns from poleward displacements of wind belts, and a decline in land ice (glaciers and ice sheets) that coincides with rising sea levels, which contribute to marine-biodiversity and species loss, ocean acidification, and coastal degradation (Letcher 2016). Climate change is also connected to an increased frequency of extreme weather events (IPCC 2014). In Appalachia, climate change provides an additional layer of precarity to communities who have long suffered from the effects of ongoing extractive practices, and as environmental engineer Daniel A. Vallero (2016, 570) notes, these environmental effects are multifactorial and interlock with social ones: "Droughts call for agricultural, potable water and food-related actions. Sea level rise calls for plans to address migrations. . . . Changes in micro- and macro-meteorological conditions, e.g. increased or shifting centers for hurricanes, typhoons, flooding, winter storms and tornadoes, call for both better land use planning and enhanced emergency response actions."

Hans A. Baer and Merrill Singer (2018, 16), in *The Anthropology of Climate Change: An Integrated Critical Perspective*, argue that "many of the most threatening changes [of climate] are a product of anthropogenic climatic interactions with other human-wrought ecocrises such as deforestation, coral reef loss, mangrove loss, wetlands loss, air pollution, and water pollution." Baer and Singer (2018, 16) point us toward "pluralea interactions," "climate/environment interfaces" that develop "an understanding of the pathways and mechanisms through which two or more ecocrises interact to produce synergistic, magnified environmental and human impacts." Heavy extraction like mountaintop-removal coal mining requires few workers and creates steep slopes without forest cover to hold water, producing mudslides and flooding near those sites. This process is compounded by industry deregulation that has led to pollution in nearby waterways and a lack of economic diversification and infrastructure investment in the region. Together, these practices produce "complex sets of interactions between global warming and various anthropogenic impairments" (Baer and Singer 2018, 17). Such pluralea interactions suggest that climate change is a scientific problem that reveals and intersects with social challenges. How climate change becomes meaningful to groups and how its meaningfulness is articulated and experienced within and among those groups are just as important for understanding, addressing, and imagining life in the Anthropocene as the scientific study of the problem (Callison 2014).

In Appalachia, where floods, especially flash floods, are already an issue, increased flooding due to climate change can lead "to increased exposures to water-borne pathogens, in addition to the immediate threats to health and safety, e.g. drowning" (Vallero 2016, 571). Floods are already a historic and ongoing issue in the region, but climate change exacerbates both their frequency and effects. Because floods cause both immediate and delayed problems, they are exemplary of what postcolonial and environmental studies scholar Rob Nixon calls "slow violence," the often ignored or invisible effects of environmental destruction spread through time and space. Nixon (2011, 2) argues that "violence is customarily conceived as an event or action that is immediate in time, explosive and spectacular in space, and as erupting into instant sensational visibility." A flood, especially a flash flood, is that kind of immediate disaster, but its aftereffects linger in both time and space, creating additional infrastructure problems, economic decline, out-migration, institutional abdication, and reinforcement of structural racism and classism. Because attention often wanes after the immediate event, Nixon focuses on the long violences that occur gradually and out of sight, those delayed destructions and their relative invisibility that continue to have violent effects throughout time and space. Because these violences are slow, however, they also present a "representational, narrative, and strategic challenge" to addressing them (Nixon 2011, 2).

One way that artists and activists attempt to address those challenges is through writer activism. According to Nixon (2011, 5), writer activists creatively strategize against the slow, drawn-out violences that harm already struggling communities, such as how corporate and environmental deregulation and petro-imperialism disproportionately affect the world's poor. Writer activists emphasize creative "strategies that emerge from those who bear the brunt of the planet's ecological crises" by "drawing to the surface—and infusing with emotional force—submerged stories of injustice" (Nixon 2011, 23 and 280). In Appalachia, artists, creators, and community members are more frequently turning toward this type of writer activism—or more broadly, this type of strategic environmental storytelling and framing—through creative vernacular representations and memory markers of floods that emphasize their drawn-out, ongoing social and environmental effects. I argue that these forms of strategic environmental storytelling or framing force audiences to witness both ongoing traumatic environmental disasters and the community's ability to endure and survive in spite of industry-fueled destruction and precarity. In the following sections, I explore some of these creative imaginings that speak to the critical role of floods in Appalachia, the continuous toll they take, and the messages these writer activists provide as they imagine how to address, endure, and find life despite environmental disaster, ecological destruction, and socioeconomic struggle.

FRACK!! and Building Scientific Awareness

One of these pieces of Appalachian writer activism that addresses the social and environmental effects of anthropogenic climate change through the context of a flood is *FRACK!! Mountain Musical Comedy Spectacular*, written by J. C. Lacek with music and lyrics by J. Scarborough.[1] *FRACK!!* is a piece of musical theater written for a local group of creatives to perform in southern West Virginia's Raleigh County. Lacek, a local of West Virginia, wrote the play as his MA thesis, but the intended audience for the musical comedy is southern West Virginians deeply familiar with the fossil-fuel economy and its absurd history of exploitation and extraction in the region. The goal of the musical is "to create an event which transcends the average local theater experience, becoming something closer to a community building project designed specifically for the people of the area" (FRACK!! Mountain Musical Comedy Spectacular n.d.). The story is about a family that falls on hard times when a single dad loses his job. The dad, Jim, is approached by Falcone, an executive representative of an overseas energy conglomerate, Grendel Gas Corporation, that wants to put a hydraulic fracturing gas well on Jim's property. The gas company actually places the well inside Jim's home, which leaks gas into the house, causes a lot of hallucinations, and ends up being financially unsuccessful for both Grendel, which files for bankruptcy, and Jim, who loses everything.

The musical tells a local West Virginia story about altered life through hydraulic fracturing or fracking, but it also serves as a representation of the United States's recent fracking boom despite uncertainties surrounding its environmental and health impacts and "mounting evidence raising serious red flags about the impact on drinking water, air pollution, and our climate" (Denchak 2019, n.p.). Fracking is a technique of extracting hard-to-reach oil and natural gas found in shale and tight rock; to release the natural gas, the fracking technique cracks the rock by blasting a high-pressure mixture of water, sand, and various chemicals into the formations. The fracking process requires and subsequently contaminates millions of gallons of water—often fresh groundwater—per well, and "fracking operations can strain resources in areas where freshwater supplies for drinking, irrigation, and aquatic ecosystems are scarce (and often becoming scarcer thanks to climate change)" (Denchak 2019, n.p.). The Appalachian region is home to one of the most productive shale formations for fracking, Marcellus Shale. Leaks, spills, faulty well construction, and inadequate wastewater management have contaminated water sources, and controlled burn tests and gas venting during processing and distribution of the oil and gas can lead to air and methane pollution (Denchak 2019).

Although Jim allows Grendel to place a gas well directly inside his living room, the play is careful to avoid stereotypes of the ignorant hillbilly; Jim states: "I might look like a hillbilly, but I ain't stupid, fella. I heard all about what these fracking gas companies do coming into communities, pollutin', driving property values down, not to mention about that global warming talk" (Lacek and Scarborough n.d., 35). Falcone responds to these concerns by presenting Grendel as invested in helping, making Jim an equal partner and attempting to mend the image of the fossil-fuel industry's historic exploitation of Appalachians. Jim's financial strain from overdue bills, lack of job opportunities, and anxiety over economic insecurity alongside Falcone's deception combine to force him into a hesitant partnership with the gas company. The international energy conglomerate places the gas well in the living room of Jim's double-wide trailer, further highlighting how interconnected local people are with global practices on social, economic, and environmental levels. As shown in figures 11.1 and 11.2, the strategic decision to place the hydraulic fracturing gas well in the center of Jim's home brings into dramatic focus the effects of the fossil-fuel industry on life and livelihood and makes visible the harms wrought by a typically invisible agent. The setting of the play emphasizes the interconnections among invisible toxins, bodies, and exploitative industrial practices that exacerbate environmental and social harm on both global and local scales. The gas well, continuously leaking toxins into the home and bodies of the family throughout the play, directs immediate and constant attention to what would normally be considered an invisible slow violence.

The musical also does a lot of work to warn the audience about the dangers of fracking and builds scientific awareness and understanding about what fracking means for socioenvironmental life. In the sixth song of the musical, "FRACK!! Theme," the performers break down a long list of chemicals involved in the extractive process:

> Before you sign on the line you should prolly know the risk
> They're injecting chemicals, I'll run you down a grocery list
>
> There's sodium carbonate methanol ethylene
> with glycol I know they pump it into coal seams
> Lutrol # 9 potassium chloride
> Naphthalene know what I mean magnesium peroxide
>
> Hydrochloric acid fractures the rocks while
> Phosphonium sulfate, that's a micro-biocide
> There's surfacants [sic] alcohols silica and sulphates
> Butanol acetate, stop now before it's too late
>
> Dezomet [sic] acetylene hear what I'm telling you
> Glycerol Isopropanol, just to name a few
> Ammonium enzyme and limonene
> Ethyl alcohol and acetates and ethylbenzene.
>   (Lacek and Scarborough n.d., 84–85)

The upbeat, poppy song—complete with a dance number performed by the entire cast—is a strategic form of local storytelling that attempts to build scientific understanding among the audience by emphasizing the range of chemicals and gases released in the process that can cause bodily, social, and environmental harm. The lyrics also serve as a warning of how quickly the extractive process can get out of control, forcing people to leave their homes due to contamination and pollution. The song ends with a "talking part," words spoken directly to the audience instead of through singing:

> It doesn't exactly take a brain scientist to know that that stuff doesn't belong down there.

> It also don't take an accountant to know peoples got to pay their dang bills. . . .
> We just wanna say that our hope is . . .
> as we all inch towards the future together,
> we do it with our eyes open.
> Hopefully, we can all take a second and consider how the choices we make today will affect things further on down the road. (Lacek and Scarborough n.d., 85)

This ending of the song evokes a more serious attempt to stress how those already socially and economically marginalized and environmentally at-risk people are often forced into even riskier situations due to a local wealth of fossil-fuel resources and a lack of social ones. The play provides a glimpse into the choices many are forced to make just to try to survive and points to how large fossil-fuel corporations exploit and further harm the poor. It also begins suggesting a move toward a more collective imagining of future life over an individualistic approach to survival.

The final scene of the play opens and closes with an impending flood, the familiar sounds of heavy rain, and the emergency alert system warning for a flash flood, a warning sound very familiar to many Appalachians. Throughout the scene, the sounds of the flood exacerbate the disorientation and confusion of the characters and often cause them to miss important information. Jim learns that the gas company has gone under and his attempt to save his family has ended up bankrupting and poisoning them instead. Jim's final song, "Ain' Nothin' the Same," tells us his mortgage is late, and he feels like giving up. This final scene also begins and ends with his friend Larry attempting to reach Jim to warn him about the flood and help him find a job, but Larry can't get in touch with Jim because of the storm. Finally, Jim gives everything to his kids as they leave the house and resigns himself to the coming waters. The last sound we hear is heavy rain as the lights fade on Jim sitting alone. We don't know whether Jim survives the flood, but we do see the toll these events can take on the individual when singled out for exploitation and isolated from community or group support and problem solving.

The play ties three important and significant threads—and threats—of Appalachian life together: the historical and ongoing practice of the irresponsible and corrupt extractive industry, the historical and ongoing risk of floods, and how climate change is connected to both. *FRACK!!* illuminates the slow violences of historic extraction and shows one version of life in which the loss of community, stability, and support leads to ruin when the individual (represented by Jim) or a marginalized place (represented by Appalachia) is isolated by "capitalism's innate tendency to abstract in order to extract" (Nixon 2011, 41). Doubt about climate change and socio-environmental damage are strategically sewn by the fossil-fuel industry (represented by Falcone) to divert attention from the slow violences these practices directly and indirectly provoke throughout time and space. As our world and our environmental problems become increasingly global, strategic forms of local storytelling, like *FRACK!!*, that force immediate witness to the negative effects of the fossil-fuel industry and climate-change-exacerbated disasters, like floods, aim to remind audience members of such connections and violences.

Folk Music and the Thousand-Year Flood

In June 2016, a series of thunderstorms over southern West Virginia brought heavy rainfall totaling up to ten inches in some parts of the state (Austin et al. 2018). Climate scientists consider such a massive rainfall within a twenty-four-hour period to be a thousand-year event because "there is only a 0.1% chance of an event of this magnitude happening in any given year" (Di Liberto 2016, n.p.). The historic downpour led to deadly flash flooding, killing more than twenty people and making it one of the state's deadliest floods (Di Liberto 2016). Climate scientist Tom Di Liberto (2016) of the National Oceanic and Atmospheric Administration (NOAA) notes that ongoing flooding in the region will likely worsen: "warming temperatures due to climate change are projected to increase extreme precipitation for all U.S. regions. For mountainous regions like West Virginia, increases in extreme precipitation also mean a greater flood risk, especially in valleys where water collects." Yet climate scientists are not the only ones noting the serious toll floods take on the region. Local singers and songwriters have also taken notice of the seriousness of increasingly common and intense floods and are using their creative platforms to both tell the story of the flood's devastation and share a message of how best to endure and find life despite it.

Fletcher's Grove is an Appalachian folk-rock band out of Morgantown, West Virginia, that often tells folk and environmental stories through music. The band has a devoted local audience in and around northern West Virginia, but they are known throughout the state. Fletcher's Grove often tours and performs at festivals across the country, and they recently created their own state festival, Groovin' with the Grove, to showcase local talent. In 2017, they released the album *Fletcher's Grove* and debuted the song "Mourning Mountaineer (Creek's Gonna Rise)" as the first single. The song, inspired by 2016's thousand-year flood, walks through the process of a flash flood and connects the flood event to accompanying mudslides:

> Good Lord! Creek's gonna rise. Better save somebody you love.
> It's coming down heavy. It's coming down quick.
> It ain't gonna stop. It ain't gonna quit.
> Good Lord! Creek's gonna rise. Better save somebody you love.
>
> Well, the rain came down in a thick haze (*'bout a foot of rain in about a day*).
> And the water can rise (*flash floods, mudslides*)
> When the rain came down heavy on the hard of living.
> Mourning mountaineer. (Fletcher's Grove 2017, n.p.)

The lyrics also note how environmental disasters further strain those already socially and economically marginalized—"the hard of living"—found certainly within Appalachia but also across the globe. Although the song notes how difficult and painful flooding is for residents of the mountain state, the chorus also urges listeners to turn to the community and "save somebody you love." Despite the swiftness of a flash flood, the lyrics also suggest that residents only have more of these historic flood events coming and emphasize the long-term effects and slow violences they bring: "It ain't gonna stop. It ain't gonna quit" (Fletcher's Grove 2017, n.p.).

In composing the song, band member Ryan Krofcheck wanted to create "a hopeful feeling and message from the start." He continues, "I was really influenced by the New Orleans style of music that celebrates tragedy and death with a marching lively backbeat" (Krofcheck 2020, n.p.). The desire for hopefulness despite tragedy and a strategic emphasis on endurance through destruction and precarity is also why Krofcheck chose to include a locally familiar John F. Kennedy quote[2] near the end of the song after expressing how community support and love can turn traumatic memories to mud:

> Like the rain, love started to pour (*out of every mountain galore*).
> Memories turned to mud. (*Shoveled away. Turning gray.*)
> Well, the sun doesn't always shine in West Virginia.
> Oh, but the people do.
> Shine gold and blue
> On every mourning mountaineer.
> Every mourning mountaineer. (Fletcher's Grove 2017, n.p.)

This ending of the song not only moves against the stereotypical fatalism often associated with Appalachians but also weaves a form of strategic localism into listeners to create pride in endurance despite repeated disaster and trauma.

Eastern Tennessee's EmiSunshine is a teenage singer, songwriter, and multi-instrumentalist who, in 2016, also wrote, with her mother, a song about West Virginia's thousand-year flood in order to support disaster relief. The young singer sparked national attention through a 2014 YouTube video of her singing at a flea market, and she has since performed at the Grand Ole Opry and frequently tours around the country (Cash 2020). Her musical style encompasses Americana, bluegrass, and country or what she calls "old-time music turned upside down" (quoted in Shaffer 2018, n.p.). The singer's 2016 song, "As the Waters Rise," tells a story of watching the flood ravage West Virginia, take lives, and cause devastation:

> Water poured out of the creeks
> And from the rivers flowed.
> The mountain flood grew higher
> as the night was coming on.
> In darkness, we held on
> for morning to be seen. (EmiSunshine 2016, n.p.)

EmiSunshine (2016, n.p.) links the flood event to darkness and uncertainty, and the song is especially focused on how environmental disasters, like flooding, have the power to destroy imagining: "as we watched the waters wash away our dreams." Yet, much like the Fletcher's Grove song, "As the Waters Rise" turns to the power and endurance of a group as it comes together to support one another as both the waters and the darkness recede:

> The flood may take our houses
> and wash away our land.
> But it will never take our strength,
> and together we will stand
> as the waters rise. (EmiSunshine 2016, n.p.)

Although the 2016 flooding has concluded within the song, EmiSunshine's (2016, n.p.) lyrics also hint at the likelihood of the waters rising once again; the song ends with a final reminder that we will continue to watch "the rain raging on." In reminding listeners that only more flooding will come, the song brings back into focus environmental memory of flooding that transcends time and space to reach into the future. It also strategically paints the region's residents as capable of enduring floods and increasing disasters because they are able to continue imagining despite darkness.

What lessons might these local musicians be imparting about our lives within a world facing a climate crisis, turmoil, and change? Speaking on the band's aim to reflect on our shared humanity, Fletcher's Grove drummer Tommy

Bailey states, "great music teaches us a little something about the world we live in and maybe a little something about ourselves" (Human 2019, n.p.). Despite institutional abdication and industrial exploitation within the Appalachian region that make facing climate crisis more difficult, both songs suggest a will to find livability through mutual aid and local community building. Mountaineers survive, mourning and changed, because they know how to shovel through the mud together. These songs are both forms of strategic storytelling that employ the environmental memory of flooding to uncover the historic and ongoing slow violences that climate crisis will only continue to exacerbate, and they work to counter the image of stereotypical hopelessness and encourage imagining as a way to continue finding livability despite precarity.

Vernacular Memorialization in Wyoming County, West Virginia

Another significant and historic flood event hit the southern West Virginia area in July 2001. That time, the water levels of the flash flooding "surpassed those of 500-year floods" and was "at the time termed the worst, and most expensive, in state history" (Brooks 2019, n.p.). Twenty years later, residents of southern West Virginia's Wyoming County (the hardest hit county) still remember and tell stories about the flood and their ongoing attempts to overcome and endure the insidious social and environmental effects that linger. In the heart of rural Appalachia's coal country, Wyoming County residents have related the flood to increased out-migration, a rise in opioid use and overdoses, further economic strain alongside the coal industry's decline, failing or lacking infrastructure, and even government and institutional abdication of America's rural communities. As a local of Wyoming County with vivid observations of the lingering socioeconomic and environmental aftereffects, I grew up understanding the flood, which took one life and directly affected "nearly one-fifth, or 5,000, of Wyoming County's 25,000 residents," as a central event and defining traumatic moment for the area (Brooks 2019, n.p.). In 2019, I returned home to Wyoming County to conduct ethnographic interviews with flood survivors and attempt to understand why a twenty-year-old environmental disaster remains so prominent in our local memory and storytelling.[3]

The year 2001 also brought other traumas, like the opioid epidemic and the 9/11 terrorist attacks, and storytellers often relate the local flood to these larger disasters. The folks I interviewed often carefully navigated both their understanding of and frustration with the loss of institutional aid and media attention after the 9/11 attacks occurred. Some interviewees felt the loss of attention and aid slowed recovery efforts that are still needed as county residents continue to reckon with the slow violences wrought by the flood. Several lawsuits claimed that mountaintop-removal coal mining and timbering practices increased water runoff and worsened the 2001 flood; one mountaintop-removal operation settled its lawsuit, and two timbering companies were found liable for flood damages (Burns 2007). Yet coal and timbering operations continue in the area in the familiar boom-and-bust cycles; in my interviews, most residents acknowledged the need for economic diversification, and many were careful to not directly criticize the fossil-fuel and extractive industries' role in the flood—in part due to their economic and cultural legacies in the region. Climate change, whether the scientific consensus of its existence was agreed with or not, was also often acknowledged when discussing environmental memory of the flood, its aftereffects, and what the future of the area might look like.

Often during my fieldwork, people pointed me toward aesthetic expressions and reminders, like unofficial memorials and flood markers, that residents continue to publicly display, especially in local businesses damaged by the 2001 flood. Some of these flood memorializations are unintentional, such as the lingering flood line marked by replacement drywall in an insurance office (figure 11.3) and the watermarks left on a framed public accountant certification (figure 11.4). Other memorializations of the flood were more intentional though, such as the framed flood photograph (figure 11.5) and flood-line sticker (figure 11.6) at Charlie's Pharmacy and Castle Rock Restaurant's framed photo of the restaurant surrounded by the floodwaters (figure 11.7). Joe Hill, a Mullens, West Virginia, barber, also filled his shop with now yellowing images taken during and after the flood and dozens of fading newspaper articles chronicling the devastation and recovery attempts (figure 11.8). If the 2001 flood is generally viewed as a major traumatic moment for the county that residents also connect to other public traumas and struggles, why frame visual reminders of the event? What do these framings suggest about how residents view themselves, their resilience, and their ability to endure despite ongoing flooding or socio-environmental destruction and disaster in the face of climate crisis, precarity, and resistance within a marginalized place?

Folklorists like Dorothy Noyes and Kate Parker Horigan have critiqued the concept of "resilience" as a neoliberal abstraction that "indexes the decline of institutional willingness to assume responsibility for the collective wellbeing" (Noyes 2016, 420) and seeks to paint the individual as solely responsible for their own survival or recovery—"and if they do not, it is a personal failure, not a public one" (Horigan 2018, 103). As such, these vernacular memorializations of the 2001 flood are less indicators of resilience than strategic reminders of local group endurance through long and slow violences despite historic destruction, marginalization, and abdication. In calling attention to the perpetuity of the flood's effects on social and environmental life, residents provide one way to express their shared trauma that continues to inform life and livability in the area.

As the local economy becomes increasingly dependent on ATV tourism and municipalities focus on providing enjoyable (temporary) experiences for tourists while locals continue to struggle, the vernacular memorials force recognition of how the past flood still pervades daily life: you cannot visit, eat out, get your hair cut, or pick up your medication here without being reminded of the insidious and often invisible socioenvironmental effects of economies of extraction. The vernacular memorializations also continually bring to the forefront a particular moment of the community's past that others might suggest they just move on from already. When presented with the question of why you would display and remind others of a decades-old traumatic event, these unofficial memorializers respond by pointing out that while the immediate event is over, its effects violently linger. As Noyes (2016, 392) notes, historically, "displacement and image repair [have] take[n] precedence over regulation and land repair" after environmental disasters in Appalachia. Although not overtly critical of such practices, these vernacular memorializations provide space to criticize notions of successful recovery and redemptive suffering often touted by government agencies and fossil-fuel-industry representatives after intense environmental and natural disasters with lasting, often unaddressed damages left for the already marginalized to deal with themselves (Horigan 2018; Lawrence and Lawless 2018). In framing the flood, these locals are also framing themselves as resistant to powerful attempts to further marginalize them and ignore the traumatic experiences they have long been forced to endure for capitalist gain.

Conclusion: Environmental Memory and Imagining Life

In the absence of official support for (re)building livability after destruction, residents force continuous witnessing to both their trauma and their endurance of flooding. Through these strategic forms of environmental storytelling in local theater, folk music, and vernacular memorialization, Appalachians employ aesthetic tools to organize shared experiences among a range of time-spaces, build scientific understanding, address traumatic history and environmental memory, and demand witness to the continuous, collective trauma of floods throughout time and space. As a region, Appalachia and its people have long been stigmatized as backwards, uneducated, and unmodern hillbillies who deserve the harms they endure because they vote against their own interests or refuse to leave a place damaged beyond repair. More generous onlookers might simply ignore this exploitation that benefits industrial modernity and energy consumption. As with other marginalized groups in stigmatized places around the globe, Appalachians have long carried the violent weight of industrialization, and they are expected to bear the brunt of a climate crisis that will only intensify their ongoing social struggles. As Baer and Singer (2018, 24) note, "The vulnerability felt by marginalized populations in a time of climate turmoil is directly tied to elite practices regarding disposable underserving bodies, affirming the very sociopolitical nature of human environment/climate relationship." This dynamic is perhaps one reason why some Appalachian creatives have turned to strategically local environmental storytelling as a way to force witnessing and memory of what traumas have occurred and continue to affect life in the region.

In his TED talk, Gipe notes that many folks whose families were dealing with the opioid crisis felt shame, and that shame created isolation and attempts to deal with trauma and pain on an individual level instead of through a group level—reflecting our neoliberal approach to many large social problems that continue to plague groups around the globe. Popular ideas for addressing the social and environmental effects of climate change are also often presented as methods an individual should take (like reusable shopping bags and water bottles) despite the incredible role fossil fuels and corporations play in worsening the climate crisis. The Higher Ground organization chose to approach the opioid crisis like a flood because residents noted that a flood—an environmental disaster—forces people to come together. In using the metaphor of the flood, these storytellers call upon environmental memory to suggest that while a disaster can affect the individual, a group can respond to and endure the disaster and make sure the slow violences are always brought into dramatic focus.

Such slow violences, according to Nixon (2011, 9), of course present representational challenges: "In the long arc between the emergence of slow violence and its delayed effects, both the causes and the memory of catastrophe readily fade from view as the casualties incurred typically pass untallied and unremembered." Yet the use of strategic environmental storytelling brings to the forefront these slow violences by "devising iconic symbols that embody amorphous calamities as well as narrative [and aesthetic] forms that infuse those symbols with dramatic urgency" and render "them apprehensible to the senses through the work of scientific and imaginative testimony" (Nixon 2011, 10 and 14). The Appalachian examples above strategically and dramatically confront these slow violences through an environmental storytelling that attempts to build scientific understanding and directly or indirectly connects familiar and increasing flooding to climate change. They urge local witnesses or audience members, already deeply familiar with the traumatic effects of flooding, to consider and imagine a future collective livability through the context of an environmental event they have already endured and will endure more often due to the ongoing extraction and burning of fossil fuels. The flood is coming at the end of the play, has come in the musicians' stories, and returns through the vernacular memorializations; the individual who is isolated for corporate gain and neoliberal exploitation does not fare as well as the community that bands together to imagine life beyond such patterns of exploitation and despite ruin. To endure, the storytellers suggest, we turn to environmental memory and imagine altered life in a livable future with that memory always in mind.

Figure 11.1: Gas well used in the 2017 production of *FRACK!!* at the Raleigh Playhouse. Beckley, West Virginia. Image by Brad Davis for the *Register-Herald* and reprinted with permission.

Figure 11.2: Gas well used in the 2019 production of *FRACK!!* at the Raleigh Playhouse. Beckley, West Virginia. Image by author.

Framing the Flood: Environmental Storytelling in Appalachia

Figure 11.3: Drywall marker of the 2001 flood line in Butch McNeely's State Farm Insurance office. Mullens, West Virginia. Image by author.

Figure 11.4: Watermark left by the 2001 flood on Robert Toler's framed public accountant certification. Mullens, West Virginia. Image by author.

Figure 11.5: Framed photograph of Charlie's Pharmacy during the 2001 flood that hangs in the pharmacy. Mullens, West Virginia. Image by author.

Figure 11.6: Flood-line sticker at Charlie's Pharmacy indicating the height of the 2001 flood. Mullens, West Virginia. Image by author.

Figure 11.7: Framed photograph of Castle Rock Restaurant during the 2001 flood. Jesse, West Virginia. Image by author.

Figure 11.8: Barber Joe Hill points at photographs and newspaper clippings of the 2001 flood hanging in his barbershop. Mullens, West Virginia. Image by author.

Notes

1. A copy of the unpublished script was generously provided to me by the musical's writer, Lacek.

2. On June 21, 1963, the one-hundredth anniversary of West Virginia's statehood, President John F. Kennedy gave an outdoor speech in the rain to a crowd of citizens in Charleston, West Virginia. Noting the rain, Kennedy opened his speech with a now-famous line in the state: "The sun does not always shine in West Virginia, but the people always do, and I'm delighted to be here" (quoted in Stafford 1963, n.p.).

3. The collection of interviews, titled "Lovejoy, Jordan_Beyond the Flood," is available in both Columbus, Ohio, and Oceana, West Virginia. The Wyoming County Historical Museum in Oceana is a 501(c)(3) organization focused on the cultural heritage of Wyoming County, West Virginia. Learn more about the museum by visiting https://wyomingcountymuseum.webs.com or emailing wyomingcomuseum@gmail.com. The Center for Folklore Studies Archives at Ohio State University is a research repository that houses a variety of folklore and ethnographic collections. Learn more about the folklore archives by visiting https://cfs.osu.edu/archives or emailing cfs@osu.edu.

References

Appalachian Regional Commission. n.d.a. "About the Appalachian Region." Accessed January 1, 2021. https://www.arc.gov/about-the-appalachian-region/.

Appalachian Regional Commission. n.d.b. "ARC's History and Work in Appalachia." Accessed January 1, 2021. https://www.arc.gov/arcs-history-and-work-in-appalachia/.

Austin, Samuel H., Kara M. Watson, R. Russell Lotspeich, Stephen J. Cauller, Jeremy S. White, and Shaun Wicklein. 2018. *Characteristics of Peak Streamflows and Extent of Inundation in Areas of West Virginia and Southwestern Virginia Affected by Flooding, June 2016.* Ver. 1.1 (September). Richmond: US Geological Survey. https://doi.org/10.3133/ofr20171140.

Baer, Hans A., and Merrill Singer. 2018. *The Anthropology of Climate Change: An Integrated Critical Perspective.* 2nd ed. London: Routledge.

Brooks, Mary Catherine. 2019. "2001 Floods Revisited—No One Imagined What Was to Come after May Floods." *Beckley (WV) Register Herald*, July 29, 2019. https://www.register-herald.com/news/lifestyles/2001-floods-revisited-no-one-imagined-what-was-to-come-after-may/article_10fba659-cc4b-5caf-8f55-a8d7f335fbbf.html.

Burns, Shirley Stewart. 2007. *Bringing down the Mountains: The Impact of Mountaintop Removal Surface Coal Mining on Southern West Virginia Communities, 1970–2004.* Morgantown: West Virginia University Press.

Callison, Candis. 2014. *How Climate Change Comes to Matter: The Communal Life of Facts.* Durham: Duke University Press.

Cash, Reilly. 2020. "Music and Manatees: EmiSunshine Performing at Crystal River Manatee Festival." *Citrus County (FL) Chronicle*, January 10, 2020. https://www.chronicleonline.com/news/local/music-and-manatees-emisunshine-performing-at-crystal-river-manatee-festival/article_7af68eca-2d8a-11ea-ac98-e73585d082f2.html.

Crutzen, Paul. 2002. "Geology of Mankind." *Nature* 415, no. 6867: 23.

Denchak, Melissa. 2019. "Fracking 101." Natural Resources Defense Council, April 19, 2019. https://www.nrdc.org/stories/fracking-101#whatis.

Di Liberto, Tom. 2016. "'Thousand-Year' Downpour Led to Deadly West Virginia Floods." National Oceanic and Atmospheric Administration, July 8, 2016. https://www.climate.gov/news-features/event-tracker/thousand-year-downpour-led-deadly-west-virginia-floods.

EmiSunshine. 2016. "As the Waters Rise." Track 14 on *American Dream*. Self-published, CD Baby audio.

Fletcher's Grove. 2017. "Mourning Mountaineer (Creek's Gonna Rise)." Track 3 on *Fletcher's Grove*. Self-published, CD Baby audio.

FRACK!! Mountain Musical Comedy Spectacular. n.d. "About FRACK!!" Accessed January 1, 2021. http://supportfrack.com/?fbclid=IwAR2jANcQPP55X9pMFopWUgDu4UiLdqYLMz6Nd5SxMmvYOw4fFRj5re1q5FE#intro.

Gipe, Robert. 2019. "Finding Higher Ground through Art." Filmed October 15, 2019, at TEDxCorbin, Corbin, KY. Video, 9:47. https://www.ted.com/talks/robert_gipe_finding_higher_ground_through_art.

Higher Ground in Harlan County. n.d. "Higher Ground in Harlan." Accessed January 1, 2021. https://www.highergroundinharlan.com/.

Horigan, Kate Parker. 2018. *Consuming Katrina: Public Disaster and Personal Narrative*. Jackson: University Press of Mississippi.

Human, Tiny. 2019. "Fletcher's Grove *Waiting Out the Storm* Out 05.17." Grateful Web, April 27, 2019. https://www.gratefulweb.com/articles/fletchers-grove-waiting-out-storm-out-0517.

IPCC. 2014. *Climate Change 2014: Synthesis Report. Contribution of Working Groups I, II and III to the Fifth Assessment Report of the Intergovernmental Panel on Climate Change*, edited by R. K. Pachauri and L. A. Meyer. Geneva: IPCC.

Krofcheck, Ryan. 2020. Facebook message to author. December 29, 2020.

Lacek, J. C., and J. Scarborough. n.d. "FRACK!! Mountain Musical Comedy Spectacular." Unpublished manuscript, typescript.

Lawrence, David Todd, and Elaine J. Lawless. 2018. *When They Blew the Levee: Race, Politics, and Community in Pinhook, Missouri*. Jackson: University Press of Mississippi.

Letcher, Trevor M. 2016. *Climate Change: Observed Impacts on Planet Earth*. 2nd ed. Amsterdam, Netherlands: Elsevier.

Nixon, Rob. 2011. *Slow Violence and the Environmentalism of the Poor*. Cambridge, MA: Harvard University Press.

Noyes, Dorothy. 2016. *Humble Theory: Folklore's Grasp on Social Life*. Bloomington: Indiana University Press.

Shaffer, Victoria. 2018. "EmiSunshine at CMA Fest: Old School Music Turned Upside Down." *Guitar Girl Magazine*, July 8, 2018. https://guitargirlmag.com/interviews/emisunshine-cma-fest-old-school-music-turned-upside-down/.

Stafford, Thomas F. 1963. "It Was a Rainy Day, but a Glorious One! Bright Future Seen by Kennedy." *Charleston (WV) Gazette*. June 21, 1963. http://www.wvculture.org/history/government/jfkcentennial.html.

US Congress. 1964. *United States Code: Appalachian Regional Development Act of, 40a U.S.C. §§ 1–405 Suppl. 1 1964*. https://www.loc.gov/item/uscode1964-015040a001/.

Vallero, Daniel A. 2016. "Societal Adaptation to Climate Change." In *Climate Change: Observed Impacts on Planet Earth*, edited by Trevor M. Letcher, 569–83. 2nd ed. Amsterdam, Netherlands: Elsevier.

Chapter 12

WEATHERING THE STORM

Folk Ideas about Character

John Laudun

On September 22, 2017, a cavalcade of musical celebrities gathered in the Frank Erwin Center in Austin, Texas, to perform for the Harvey Can't Mess With Texas benefit concert. Two of the night's stars, James Taylor and Bonnie Raitt, added to the event by performing a somewhat more gospel version of a folksong that has come to be known as "Wasn't that a Mighty Storm?" whose origins, at least in recordings, lie with Sin-Killer Griffin. At the time of John Lomax's recording of Griffin in 1934, Griffin was a prisoner in Texas's Darrington State Farm (Lomax, Lomax, and Griffin 1934). The recording captures Griffin leading his small congregation of prisoners who accompanied him as he sang "Wasn't that a Mighty Storm?"

For those not familiar with the song, it represents the events occurring during a hurricane in Galveston in which people lose their lives either because they failed to heed the warnings or, perhaps through no fault of their own, they became trapped in the middle of a natural catastrophe and they had no choice but to die; the role of fate is highlighted in the oft-repeated line "When death calls you got to go."[1] Thus, like other moments in oral tradition, the song is a meditation on human nature in a time of crisis when, according to one strand of American tradition, one's true character is revealed out of necessity.

Indeed, success in the face of adversity is to "weather the storm." And so, this chapter's contribution to the current volume is, in many ways, to explore the verb side of "weather." That is, the weather is rarely simply the weather. It is, as "wait five minutes" suggests, a commentary on what we can know and the kinds of claims we make about knowing. To suggest to someone "wait five minutes" is to suggest that the future is unknowable or at least what we think about something as seemingly knowable as nature—we live, after all, in an age of science and reason where the natural sciences are the model for all other forms of science—is to rebuke those epistemological structures for their inability to do something as simple as predicting the weather.

At the heart of this chapter lies two instances of the idea of weathering a storm: a collection of memes that tend toward the comic and an oral legend, indexical of a small collection of similar texts, that trends toward the tragic. Both the memes and the legend focus on historical weather events, and both feature abandoned trucks. As students of the two forms are well aware, both allow for a wide variety of form and meaning, and yet, when it comes to trucks weathering the storm, there is a striking continuity, a conservatism, in the ideas, both latent and manifest. Situated at the heart of American car culture and heightened in dramatic form when manifested as a truck, we have vernacular considerations of the status of the American dream. In addition to topical matters, what follows is also an exploration of folk ideas as they relate to forms like the meme and the legend and how viewing the forms through the lens of ideas might offer us a way to reimagine the relationship between forms.

Weathering Storms of Emotions

A range of expressions, from proverbs to sayings to metaphors, all reveal the role of stormy weather in our imagination. The premonitory forms tell us to be wary of "the quiet [or calm] before the storm" or warn us of the "clouds on the horizon." Admonitory forms caution us "not to rain on my parade" or resign us to the fact that "into every life, a little rain must fall." Disembodied and distinct from us, storms represent misfortune. Embodied, however, they reveal ill temper, like when someone "storms" into or out of a particular situation or their anger is revealed by their face being "clouded." When bodies accumulate, we worry when there are "rumbles" of discontent, and, of course, should those crowds cross the barricades, it is quite probable they will *storm* a building.

Such freighted metaphorical usage is not limited to folk speech, of course. Its role in literature is quite extensive, and we need only recall, in the context of the US South where our memes and legends are also situated, the storms that lie at the heart of Zora Neale Hurston's *Their Eyes Were Watching God* (1937) and William Faulkner's *The Old Man* (1948). In both, the protagonists are tested by events that unfold as a storm rages: in the former, Janie faces having to kill the man she loves, and in the latter, the unnamed protagonist saves a life, helps bring a new life into the world, and ultimately sacrifices himself. As Izabela Żołnowska notes in her discussion of weather as a source domain for metaphorical expressions, storm events, like rain, wind, thunder, lightning, and fog, are all associated with difficult times or problems. While each can also signify other dimensions of human existence (e.g., fog stands in for confusion), together they reveal a tendency to imagine bad weather simply as bad. The result is, as Żołnowska (2011, 176, emphasis original) concludes, a metaphorical system within which "WEATHER CONDITIONS ARE PROBLEM INDICATORS." Hence, weather can be not only a problem in its own right but also a metaphorical realm into which we project out understanding of problems. And thus, in some fashion, weather shapes how we understand and act upon those problems. It is, for example, quite different to spit into the wind or wade through a difficult situation than to be drowning in your problems.

So, storms in stories are more than reports about the weather: external events are closely tied to internal events and not just any kind of internal events but those that challenge us and force us to reveal something of our "true nature." In the process, such external events make manifest that which is normally internal. To be sure, this notion is as commonplace as the ideas about storms just mentioned. The saying "adversity does not build character, it reveals it," often attributed to James Lane Allen, is sometimes described as a bromide, so commonplace are the pairings of crisis and character.[2] Similar sorts of statements have been attributed to Charles de Gaulle and Arthur Schopenhauer as well as a host of lesser luminaries.[3] And such a list would be incomplete without including perhaps the most famous of them all and which so bears on the topic at hand that it would serve just as well as the title of this chapter: "There is a tide in the affairs of Men which taken at the flood leads on to fortune" (Shakespeare 1998, 218–19).

The point is that memes and legends draw upon similar reserves, reserves which we might describe as folk ideas, but enact them differently. While some might argue that all folklore forms are ideological in nature, in keeping with Roman Jakobson's notion of the dominant, I would like to suggest that some forms are more ideological in nature than others: they foreground that ideas are in play. (And "play" is an operative word here.) As Jay Mechling (2004) notes, the ties between certain kinds of authorized forms, like proverbs and myths, are well established. And yet no legend scholar would dispute the moral implications of any number of legends. This does not require a hierarchy of any kind: different forms provide users with different levels of commitment to the ideas they understand to be latent within any particular event that is either under discussion or being created.

Setting aside this larger framework, let us return to our discussion of the nature of character, especially American character, as found in folklore. Legend is one form where character comes into play. Adjacent to legend in many discourse communities that I have studied, and often adjacent in form and in topic, is the anecdote. The two together offer a kind of communal and individual testimony on the nature of what it means to be a moral actor. Where proverbs end and legends begin, or where myths end and memes begin, is not really the question. All participate within a larger complex that some have called "worldview."

Within the realm of possible responses to adversity as presented by storms (and other weather events), there seem to be four possibilities: those who do the right thing and are heroes; those who do the wrong thing and are villains; those who do things of uncertain moral nature and are tricksters; and those who have things of uncertain moral nature happen to them revealing them to be numbskulls. Of these four character types, the two that dominate the memes and legends discussed here are tricksters and numbskulls. Why the characters of uncertain moral valuation predominate will be taken up after we have a fuller account of them in action.

Numbskulls

The numbskull has long been a favorite in a variety of American traditions. In many communities, numbskulls are commonly featured in jokes. In Louisiana, like elsewhere in the United States, numbskulls appear regularly as a pair: dumb and dumber. Boudreaux & Thibodeaux jokes feature the pair in a variety of activities, some represented as generally American and some as particularly part of the area—those jokes that hinge on some interplay of French and English or on game peculiar to the region are often examples of more widely available localizations. While in a small number of instances the native wit of Boudreaux and Thibodeaux is the focus, in most cases it is their misapprehension of simple things—throwing away nails with the heads on the wrong end, for example—that makes the joke work. When their families are featured, most often one or both of their wives, they are there largely as props for the behavior of the two comedic protagonists. In the case of the following joke, however, it is the wives who dominate the joke, which might underline that what's at stake is human nature set against mother nature:

> One night, a torrential rain soaked South Louisiana. The next morning, the resulting floodwaters came up about 6 feet into most of the homes there. Mrs. Boudreaux was sitting on her roof with her neighbor, Mrs. Thibodeaux, waiting for help to come. Mrs. Thibodeaux noticed a lone baseball cap floating near the house. Then she saw it float far out into the front yard, then float all the way back to the house. It kept floating away from the house, then back in. Her curiosity got the best of her, so she asked Mrs. Boudreaux, "Do you see that baseball cap floating away from the house, then back again?" Mrs. Boudreaux said, "Oh yes, that's my husband; I told him he was going to cut the grass today come Hell or high water!"[4]

The saying "come Hell or high water" appears to have a mid-nineteenth-century American origin, with its first appearance in print coming in 1882 in the Iowa newspaper the *Burlington Weekly Hawk Eye*. It is not unlike the common southern saying "Lord willing and the creek don't rise." In both cases, a (literally) supernatural event and natural event are conjoined, marking the importance of proper conduct in human affairs and the possibility for failures of character.

Those familiar with Hurston will immediately be reminded of the story about the Johnstown flood that features early in *Mules and Men* (1935). In it, a man who dies in the flood goes to Heaven and proceeds to tell everyone he sees about the terrible flood that killed him until he tells one man whose only response is "'Shucks! You ain't seen no water!" When he asks Peter who that man was, Peter, after asking a few questions to clarify the man's identity, responds that it was Noah. Peter and the joke finish with "You can't tell him nothin' 'bout no flood" (Hurston 1935, 13). The joke, in other words, is on the seeming protagonist of the story, who thinks, having been drowned by a historical weather event, he has experienced something extraordinary. He appears the fool in comparison to the biblical event.

Fools and floods seem to go hand in hand. The numbskull is such a popular trait that KRON TV 4 in San Francisco has a series called *People Behaving Badly* that features, in addition to the usual assortment of people being rude to each other, people being too adventurous in regard to waves or flooded roads. One episode even features the reporter, Stanley Roberts, interviewing drivers as they queue to plunge into an obviously flooded stretch of road. Such adventurousness not only has a place in news but also in law, with a number of states enacting "stupid-motorist" laws. In Arizona, the statute states: "a driver of a vehicle who drives the vehicle on a public street or highway that is temporarily covered by a rise in water level, including groundwater or overflow of water, and that is barricaded because of flooding is liable for the expenses of any emergency response."[5]

One such foolish driver captured national attention when his red Jeep Grand Cherokee appeared to have been abandoned on Myrtle Beach as Hurricane Dorian made its way to South Carolina. As with all hurricanes, the combination of winds pushing in water in the form of higher tides and water coming down in the form of torrential rain led local officials to order coastal towns evacuated. Sitting in the center of sixty miles of beach known as the Grand Strand, Myrtle Beach has a history shaped by hurricanes—one of its national landmarks was historic for having survived Hurricane Hazel in 1954. With a population of twenty-seven thousand, Myrtle Beach annually hosts an estimated fourteen million tourists who come to the Grand Strand throughout the year, staying in high rises that are strung along its beaches and possibly playing on one of the hundred-odd golf courses. And yet, with all this revenue, the average take-home pay for someone living in Myrtle Beach is $23,000.

As the surf rose and the winds began to blow, some who had stayed behind looked out on one of the beaches and saw something odd: a late model, ruby-red Jeep Grand Cherokee parked on the beach. While it is hard to know who first noticed the Jeep, the first moment it appears to have been brought to a broader public's attention is in a story posted at 10:21 a.m. by local NBC affiliate WMBF. A little over an hour later, at 11:45, Kathleen Serie, a reporter for the local Fox affiliate WZTV, was on the scene and captured not only video of the waves breaking over and the water swirling round the Jeep, but also a small cluster of ten or more onlookers who appeared simply to be standing there, waiting to see what would happen next. Later that day, WZTV included the video as part of a story that ran under the headline: "Jeep Abandoned on Myrtle Beach as Hurricane Dorian Rages; Onlookers Take Selfies." Their report includes the following:

> This Jeep owner clearly left things in Mother Nature's hands. As Hurricane Dorian raged off the coast of the Carolinas on Thursday, a unique sight appeared on the shores of Myrtle Beach, S.C. A Jeep SUV was seen stuck on the beach as fierce waves from the hurricane could be seen battering the vehicle. . . . As time progressed on Thursday fierce waves engulfed the Jeep, knocking its bumper off. Onlookers could be seen taking photos of the Jeep and selfies before police cleared the beach area. One man was seen posing for a photo on top of the vehicle as it was engulfed by water. WZTV reporter Kathleen Serie said that as the waves got more violent, the Jeep's tires became loose.[6] (Fedschun 2019)

That the phrase "fierce waves" occurs twice appears to be part of a larger trope that grants nature considerable agency: the storm rages, knocks off the Jeep's bumper, and then both engulfs it and becomes more violent. The storm itself is part of a larger agency, Mother Nature, into whose hands the Jeep's owner has passed its fate.

While arguably the prose of the local news post is a bit overblown, if readers will forgive the pun, it does capture something of the Jeep as not simply another object dotting the beach but as a kind of objet trouvé, a found object that refocuses our attention and our understanding of events in a particular way. How else to explain the explosion of interest that followed, except that the Jeep is not simply an inert thing but an abandoned thing, a thing that asks questions about what it means to buy and pay for and own a reasonably expensive object like a Jeep Grand Cherokee only to strand it in a storm?

The first image in the emergent vernacular we have are the selfies, which happened also to be documented by the local news. A steady stream of individuals acting ostensively, coming out either to see what's happening to/at the Jeep or to be part of what's happening, followed. The simplest form of ostension was the selfie; the more complex forms involved other kinds of performances, the most notable of which was a man playing the bagpipes while circling the Jeep as the water came up around it. The series of events was captured by David Williams (2019) of CNN:

> The owner of the Jeep abandoned on Myrtle Beach during Hurricane Dorian will probably think twice before giving someone his car keys. The red SUV became a social media sensation Thursday as it was bashed by strong surf whipped up as the hurricane moved past South Carolina and up the Atlantic coast. People posed for selfies with the Jeep and some even climbed on top of it. One man dressed in black walked solemnly around the vehicle in his flip-flops while playing "Amazing Grace" on the bagpipes.

The Jeep survives, and yet it is also dead, the subject of a spectacular funeral procession, as CNN and the world watched. The scene underlines the long relationship news outlets—first newspapers and later websites—have had with legends: as much as news outlets wanted to be outside the zone of ostension themselves, they were an integral part of the legend's circulation, with a number of stations actually establishing video feeds of the Jeep.[7]

What followed was a number of selfies posted various places, the establishment of at least two Twitter accounts, and even a parody produced by a grade-school child for the amusement of her friends and parents. Two minutes after Serie tweeted her video of not only the Jeep but also of people taking selfies of the Jeep, @daily_staley posted a video of someone sitting on the Jeep cross-legged, as if meditating. He tagged local weatherman Ed Piotrowski, who would himself become a part of the growing body of lore and ostension surrounding the Jeep. While people continued to flock to the Jeep and at least one local news station set up a video feed, Twitter was momentarily quiet. Then, an hour later, @the_Jameson asked, "does the Myrtle Beach Surf Jeep have a twitter account yet[?]" (Jameson 2019, n.p.).

Within two hours of that question being posed, two Twitter accounts for the Myrtle Beach Jeep sprung to life. The first, @DorianJeep, was created shortly after 2 p.m., and its first post was at 2:13 p.m. and was, as was only fitting, a photo of the Jeep firmly ensconced in water with the hashtag #selfie (DorianJeep 2019, n.p.). Twenty minutes later a second account was created, @MyrtleBeachJ33p, and its first post was also an image of the Jeep with waves crashing around it.

It too only had a hashtag for text: #NewProfilePic (Myrtle Beach Jeep 2019, n.p.). In this way, both accounts reinforced the notion that there would be an account from the Jeep's point of view. By the time the storm and interest in the Jeep had run their parallel courses, @DorianJeep had tweeted thirty times, with the last coming sometime later on the same Thursday afternoon as the account's creation. By contrast, @MyrtleBeachJ33p posted 496 tweets, principally on Thursday and Friday with limited activity on Saturday and then sporadic activity through September 17.[8]

Of the almost five hundred tweets from @MyrtleBeachJ33p, 136 were outbound tweets originating with the account, sixty-eight were retweets, and 292 were replies.[9] Of the media created and circulated in the course of the day, there were sixty photos, ten videos, and eight GIFs.[10] It will surprise no one that the texts of tweets often anticipated or echoed media incarnations. The first photos and videos are mostly concerned with the Jeep itself, with images of the water and of people interacting with the Jeep amid the incoming waves, but later in the afternoon, Twitter users had found ways to make the inanity of an abandoned Jeep and people wandering during a storm taking selfies with it into something more. It began with a tweet from @alwaysjathis with a screenshot from a smart phone of a listing for the Jeep on the Facebook Marketplace. While the other vehicles have expected price tags, the Jeep is listed for one dollar. @alwaysjathis's comment is: "@MyrtleBeachJ33p wtf. I believed in you" (Shelter 2019).

Tricksters

Insofar as the Jeep is the main character of the memes, then "numbskullery" is the feature, though it is interesting to note that often the local media stories, which depend more on the anecdote for structure, tend to feature a somewhat larger repertory of character types: heroes, villains, and occasional tricksters. Perhaps no better example can be had than the piper circling the Jeep as the storm waded onto shore. Of course, the action is ludicrous: it foregrounds play and makes judgement of the actions as good, bad, wise, or foolish impossible. The same holds true for a handful of legends that emerged after a rare but disastrous weather event that struck south Louisiana in 2016. Like the preceding set of memes and other kinds of texts, the legends focused on property, and in the case of a particular legend that was performed for me by an acquaintance, they featured a truck.

The August 2016 storm itself was eventually described by the National Oceanic and Atmospheric Administration as an "inland sheared tropical depression" (Dolce 2016). For most residents of the areas affected, it looked a lot like a typical Gulf Coast thunderstorm with the accompanying possibility of flash flooding. Such rains can last for a couple of days, usually coming in waves that offer some relief to infrastructure focused on moving the water out of harm's way. Residents of Louisiana are used to this kind of storm and are also used to flash flooding, and they keep track of which streets are impassable—this task is simplified by the fact that some streets and/or areas are known to flood. People plan their day according to an emergent mental map they develop based on local reports culled from news outlets as well family and friends.[11]

While the idea of threading one's way through flash floods might strike some as foolish, it is a relatively quotidian experience for area residents. Almost everyone in the area has lived through a named storm that did not completely flood an area or otherwise cause a clear and present danger. It is the experience of most residents, and thus a kind of folk wisdom, that even the most punishing of storms does not pummel a region evenly: most take it as a given that it is better to try to turn back then not to try at all if you need to go to the grocery store or get to work. In fact, many businesses remained open throughout the weather event simply because they, and enough employees, were not affected by the unevenly occurring storm. Thus, it was a fairly common trope of the stories that emerged after the storm to feature individuals driving around and encountering something usual or unusual, depending upon the focus of the text. In nonnarrative texts, this was a moment to describe or explain something that had happened. In narrative texts, the usual or unusual would signal, dramatized within the text itself, that either the protagonist did not yet know what was to come or that things were, in fact, coming.

A month after the "inland tropical depression," September 17, on what was a warm Saturday afternoon, I sat on a set of small bleachers with a fellow dad as our daughters practiced nearby. The two of us were catching up on events in each other's lives, our conversation followed the general conventions of most such speech events in North America, with an especial focus on the recent dramatic event, the flooding of the area, acting as a lens through which other topics passed. In the middle of what had become the usual course of conversations about the flooding as described above, my friend told me about vehicles being abandoned in the rising water, seemingly for the insurance money. Used to the current conventions of "storm stories," I did not notice the shift in genres at first, but as the telling unfolded and people and places went unspecified, I realized I was in the middle of a legend performance. My attention shifted and as soon as the larger conversation was done, I transcribed what I had heard as closely as I could. The text as I recorded it that day went like this:

> So, my buddy was out. And he went to cross the bridge, and he'd been across it not long before and it was okay, but now he could see it was

kind of deep. So, he got part way in, and then he decided "nah ah" he didn't want to risk it. But, you know, some trucks were. . . . My buddy said he saw some nice trucks. Some nice trucks got flooded out. Some guys just drove their trucks in the water. I guess they were already under water with their payments, so they thought why not, you know? Anyway, my buddy says he saw some trucks and their windows were rolled up. You know if you got stuck, you'd roll your windows down to climb out. But their windows were up. So, they were pushed. People got to the edge of the water and then pushed their trucks in. He said he'd seen a bunch of nice trucks with their windows up. You know, I guess people were just doing what they felt they had to do. I'm not saying it's right. But I can understand it.

As noted above, it was not the case that information about whether and where the flooding was happening was well established or evenly distributed. Stories about places flooding that do not normally flood were part of the larger set of genres, and they are often framed in terms of driving and coming upon such a place rather suddenly or unexpectedly. So, when the story began with an account of a friend, who lives in a nearby town along one of the major waterways in the area, it was not remarkable that the text's protagonist was out driving.

Flood stories, as they came to be called, had a predictable sequence in which streets, and even bridges, which were normally passable were discovered to be impassable. In most textual interchanges, this opened up the conversation to a discussion about water actually flowing over the bridge and even how deep it was. It was not unusual for the anecdotes and reports to feature a decision to turn around to emphasize the impassable nature of the route. So, when this particular text has the protagonist look around him to see other trucks like his that have attempted to pass and have not made it, it does not seem all that remarkable. The narrative wobble—where the text has to back up to repeat the point of view, of "my buddy," and changes point of view from the friend to trucks to unnamed individuals—marks the shift from anecdote to legend.

Because the trucks are a defining feature of the legend, their presence is worth consideration. Southern Louisiana has been dominated first by agriculture and then by the oil industry, both sectors that require a fair amount of heavy equipment that needs to be hauled from one job to another. Pickup trucks are ubiquitous, and while most will think of trucks the size of a Ford F-150, it is not uncommon to see the likes of F-250s and F-350s or their competitive counterparts parked in driveways. Extended and crew cabs add to the overall size of these large trucks, which are also often elevated, because both farmers and oil workers tend to work in a variety of environments. Large pickup trucks are the coin of the realm, and where other regions might focus on fancy cars, foreign or domestic, as status symbols, in south Louisiana it is the truck.[12]

Just as importantly, in terms of these legends, trucks like these are expensive, some approaching the cost of a small home. Because car loans are shorter than most mortgages, many face monthly truck payments exceeding their mortgages. But farmers have to have them and fold them into the cost of their farm operations, and many workers in the oil field and adjacent industries just regard them as part of what you own and who you are. When times are good in the oil industry, men can make six-figure salaries that make a $700-a-month note look like a pittance. In the two years leading up to the 2016 flood, gas prices had dropped, and the oil industry had shed twelve thousand jobs in the region. Area newspapers slowly switched from hopeful stories about what would happen when oil returned to a price above fifty dollars a barrel to stories of how workers and their families were adjusting to what was coming to be considered the new normal (see, e.g., Truong 2016). It was becoming increasingly clear that a good number of the jobs would never be coming back.

The legend addresses the current situation for many head-on: many people before the flood occurred were already "under water" financially. The flood was simply a physical manifestation of a less visible economic and emotional reality. The rising waters were a concretization of a landscape that was already flooded with despair and possible destitution. The mud-filled waters that creeped and then drowned people's homes and left behind the danger of black mold were simply confirmation that the current situation could not hold. For many more than have been reported, it is said, it was simply the last straw, and they left their homes behind, much like what we were told happened in Las Vegas when the housing bubble burst in 2007.

After encountering the legend above, I asked others if they have heard anything about vehicles being abandoned during the flood. That was the extent of the prompt—"vehicles abandoned during the flood"—with no mention of insurance, though sometimes I substituted "car or trucks" for "vehicles." I received a number of positive responses, many of which cited as evidence for their claim a report seen on a local television station. Inquiries to local news outlets turned up only stories that were basically requests from authorities either urging avoidance of or caution when using local roads due to debris or abandoned vehicles. For example: "The following areas are still reporting problems with high water in the roadway. The public is asked to avoid these areas until the water subsides and the streets are cleared of abandoned vehicles" (KPEL 96.5 FM 2016, n.p.).

Finding no evidence for actually abandoned cars in the historical record, it made sense to seek out possible verification from the foil in the legend, the insurance industry.[13] A representative of one of the big three insurance companies, who notes that his company probably insured one car in six in the region, responded that, while such cases were not unknown, there were no such cases pending before his company in the months following the flood. He notes that it was important to understand that the industry regularly distinguishes between clear-cut cases of fraud and insured individuals simply being stupid—the determining the latter, he emphasizes, is simply one function of insurance, catching us when we are not at our best. While there had been a few cases of people having driven their vehicles through a flooded area and then claiming that the vehicles had been flooded in place, there were no cases, of which he knew, of people driving cars or trucks into flooded areas and leaving them there.

Moreover, the insurance representative notes in those confirmed cases of insurance fraud of this kind that it is rarely the case that the insurance disbursement will pay off what is owed completely. The nature of being upside down, he observes, is that your vehicle is already worth less than what you owe. Even receiving full compensation will leave you with money owed on a loan. In most instances—and it was especially his experience with the flood—people plead to have their vehicles not totaled. In these cases, he said, cars and trucks are usually paid for, and the person is not able to afford, or would rather not take on, a monthly loan payment. Unfortunately for the insured, insurance companies will not pay more to repair a car or truck than it is worth in terms of resale value. In terms of veracity, then, not only is there little evidence to support our legend, but the stories from the other side reveal that far from pushing trucks into the onslaught of flood waters, most individuals were really left "high and dry" when their insurance company declared their vehicle a total loss and simply issued them a check for its current value, which is often far less than the value the vehicle has for them or what it would take to replace it.

The fit of the legend here is not simply the topical context of flood discourse but also the larger ideological network: of people already drowning metaphorically, such that abandoning an artifact that not only has high utility, but also high social status, achieves such effective narrative closure. Another legend very similar to the current one repeats the idea of insurance fraud, but in the case of the versions I heard, none of them are as long or as structured as the one about trucks. The texts tell of a local business owner, sometimes owning a store and sometimes another kind of business, who, already on the brink of financial difficulty, decides to open the door to the flood, causing the contents of his business to be lost to the rising water. All the versions of this variant I heard located the business in the eastern flood zone, an area east of Baton Rouge in which a number of small towns were flooded.

In the case of at least one town, Walker, its mayor, Rich Ramsey, protested that the flooding was the result of the refurbishing of nearby Interstate 10 that runs east-west through Louisiana, and, with its newly raised central divider, actually acted as a weir, preventing the rising waters from flowing, as they normally would have, southwards and toward the wetlands that border the western and northern edges of Lake Pontchartrain. Ramsey argued in the weeks that followed the flooding that "what they did was effectively create the largest retention pond in the world" (Jones 2016, n.p.), maintaining that he had requested additional drainage options, especially after similar flooding occurred in 1983.

The abstract "they" is significant here. Similar complaints about large construction projects have been made by Louisiana residents elsewhere, and it is part of a larger network of ideas that bureaucracies, both governmental and corporate, are at the very least indifferent if not hostile to local inhabitants and dismissive of their knowledge.[14] There are a variety of narratives that circulate that focus on the revelation that if an official of one type or another had simply listened, he or she would have learned that: that land has always flooded, that no one has ever grown anything there, and that you can't use that tool that way (even though the manufacturer says you can). In some ways, it is the quintessential American folk idea of "book learning" versus "common sense," with the common here being knowledge hard-won by years of living and/or working in a particular place. Barry Ancelet (2008) has documented similar kinds of narratives that followed the 2005 hurricanes, concluding that those from outside the community ignore local knowledge at their own peril, and that those who respect such knowledge, almost always represented as officials from inside the community, succeed.

Nonplayer Characters[15]

In his consideration of legends that circulated within communities affected by Hurricanes Katrina and Rita, Carl Lindahl (2012, 143) notes that "Disaster legends may not report the facts, but they are an essential vernacular tool for expressing how the tellers feel about the prevailing social order and for helping their communities seek explanations that square with their convictions." Lindahl describes these kinds of stories as "the right to be wrong": they should not be judged by referential veracity but by their sensitivity to larger systems that for most are more often felt than understood. Legends allow their tellers to articulate, through decisive dramatic action, the dynamics at work within their world and to bring those dynamics into some kind of dramatic enactment with their worldview. Thus, while the legend here does not condone insurance fraud, it asks its audience to understand what lies behind the desperate actions that led to the trucks being under water.

While the legend foregrounds the existential ambiguities that a storm highlights, the memes that surrounded the Myrtle Beach Jeep offer additional vernacular dramatizations. All these texts present, by offering a version of the things as they should be (if things felt were made manifest), an alternative world. As Hilary Dannenberg (2008) notes in her discussion of multiple temporal dimensions in novels, the effect of creating an alternative possible world is to intensify, or sometimes frustrate, the logical force of texts by offering more than one possible version of events. I think it is fairly safe to say that the force of these texts is that they not only inherit the general traits of the actual world, as most mimetic narratives do, but that they also inherit traits of a particular set of folk ideas—folk ideas that run counter to the ideas presented in official press releases and in journalistic outlets.

In the wake of almost any severe weather event, news outlets feature stories about local heroes, some of whom are emergency workers and some of whom are merely ordinary citizens. Part of the larger genre of "human-interest" reporting, these mediated heroic tales have all the features that might make them available for vernacular reproduction: they have a clear theme, usually stated in brief in the opening voiceover; they feature dialogue with either the hero or those whom he saved giving brief, plain-spoken accounts of the events that transpired; and, perhaps more than anything, they feature a clear emotional core. Human-interest stories and legends have either overlapped or fit into the same space in local newspapers, as a number of folklorists have pointed out over the years, and yet we have never really examined how some stories pass mediated to vernacular circulation or, at least, which texts do. In the case of weathering the storms, it would appear to be the case that, as intertwined with media as memes and legends have long been, there were no circulating accounts of heroes nor villains. No doubt there were such stories performed by one person for others, but those stories do not seem to have gained much distribution. And this lack of distribution occurs in spite of the overwhelming flood of such stories, especially those focused on heroes, that flood local and national news outlets after storms.[16] It is, in fact, the relative absence of the expected character types from the vernacular texts that marks them.

The vernacular texts by and large do not explore notions of heroism or villainy. Instead, they are drawn to characters of more ambiguous nature, the numbskull and the trickster, characters whose motivations or intentions are opaque to us, perhaps as opaque as our own actions might be to ourselves should we face a similar plight: How might we fail in the face of disaster? How might we do the stupid thing or the wrong thing for perfectly reasonable reasons? Certainly, one dimension of the ambiguity in these texts, as Lindahl might note, is in the very "there-but-for-the-grace-of-God-go-I" dimension of them. This ambivalence is captured in the legend itself with the following closing lines: "You know, I guess people were just doing what they felt they had to do. I'm not saying it's right. But I can understand it." This closing series of assertions, which acts as a kind of coda in the Labovian model, opens with a reference to the audience in order to distribute the load pragmatically. The rest of the sentence establishes a clear triangulation: adding *I* and *they* to the initial *you*. The action of abandoning their trucks for the purposes of claiming the insurance money is further distanced by reducing everything to *it*. The ending couplet reinforces this moral distance: something seen need not be understood.

The Folk-Idea Complex

One element of these vernacular responses that we have not yet considered is the role of the truck. While cars might be desirable for a number of reasons—luxury, performance, value—trucks are almost marketed on the basis of their ruggedness. They are, after all, according to the three major American makers: "built Ford tough" so that they are (one supposes) "like a rock" (Chevrolet), which they would have to be if you were going to "grab life by the horns" (RAM). The association between trucks and masculinity in America is fairly well established.[17] Thus, these texts do provide a gender, either directly or indirectly. The oral legends, both the one focused on the truck and the ones featuring businesses, were all masculine in nature, and so, it is tempting to consider these vernacular texts as part of an ongoing, diffuse exploration of masculinity in America. And, even within the sphere of trucks, there is further gendering: in the oral legend, "some guys" were *driven* to push their trucks into the water, while in the meme, the SUV seems to offer a slightly less gendered enactment, featuring as it does a ruby-red—like Dorothy's slippers?—Jeep Grand Cherokee. Does an SUV and not a truck incline the meme to be a numbskull performance rather than a trickster one?

So far in this discussion, I have used "trickster" over other kinds of designations. I have largely done so out of the felt nature of the tale in which the performer clearly saw the owners of the abandoned trucks as amoral characters: moral men forced into immoral choices, but because they were forced, they could no longer be classified within the accepted moral arena and, thus, had escaped it and become amoral actors. Another possible designation for this particular character type is that of bandit hero, a widely popular type in American folklore in general and in Louisiana folklore in particular.[18] The actual designation is less important than the fundamental ambiguity that underlies these texts. As Lindahl (1986, 6; emphasis original) notes in his consideration of modern legends: "the central symbols of many of these no longer behave according to obvious moral patterns. Many of the newer figures are deeply and thoroughly ambiguous in identity, in behavior, in their implications for *human* behavior."

Lindahl's emphasis on the human is a product of his argument about modern legendry—which might now be expanded to include vernacular forms like memes, which often achieve their effects, he suggests, through a kind of doubling. In the case of the UFO sightings of the seventies that Lindahl examines, the aliens represent both technology taken to an extreme—turned back on us, as it were—as well as a larger encroachment on what had once been spiritual matters. Where once gods dwelled in the heavens, now come alien forms to probe, abduct, or scare us. Questions about alien existence, he notes, create "a symbolic format against which to work out our hopes and fears concerning the value of the technological world and its ability to displace successfully the spiritual world to which people formerly turned to salvation" (Lindahl 1986, 8).

In the current complex composed of memes and legends with trucks abandoned in the face of storms, we have a similar doubling. If we begin with the idea from which our complex descends, we have something simple: *person leaves object*. Leaving can be, of course, both transitive and intransitive; we regularly observe "people leaving." Closer consideration reveals however that people leave places or groups; that is, the expression is still transitive, but the object, what is left, is often simply left off or out. The act of leaving, as many narratives reveal, is often simply a prelude to a return. We use "abandon" to emphasize the lack of the possibility of return, and that is certainly the idea that both meme and legend present.

Such an interpretation of the verb is debatable, since it occurs nowhere in the text, but the object is not: what is abandoned is a particular kind of object, one that has several dimensions. First, the object itself *moves*, or, rather, it moves us. We get in it: it carries us. Thus, its second dimension is that it *holds* us: we get in it, trusting it to keep us safe as we move through the world inside of it, and so *trust* is a corollary of the second dimension. Its third dimension is extrinsic to its objective qualities and ties it more closely to the institutional dimensions upon which the complex depends: it is expensive. A car or truck is, for most people the second most expensive thing they will purchase in their life, and for those who never purchase a home, it will be the single most expensive item they own. And, as the legend itself intimates, it is like a home for most of us, something we pay for but perhaps never fully own. It is thus a fraught object, one that tests our sense of what is proprietary as well as what is property: what belongs to us and what does not. The other images that often follow storms are those of houses emptied of people but still full of the things that make up their lives. The very notion of abandoned homes runs counter to the fundamentals of the American dream, and images of blocks of such houses after hurricanes or the 2008 financial crisis, which we conventionally imagine as a storm, are one of the principal forms of enacting devastation, both in its literal and in its figurative senses.

This doubling of devastation, in its focus on things that are both objects and places, in the American context, is made clear by the focus on trucks, which are themselves scenes of rugged individualism. What the folk-idea complex makes clear is that the storm is an enactment of both powerful forces already at work: the natural storm reveals what has already happened, that an institutional storm has already eroded our place in the world, washed away our perceived and/or desired independence. We have to wonder, then, given the emphasis in American folk culture(s) of braving one's way through a storm, of facing storms of emotions or events, why these memes and legends focus on people, men in particular, who simply walked away. As Lindahl (1986, 10) notes, "Modern legend tellers, then, find themselves asking the same question earlier posed in the classic legends: of course, we want help in our lives, but at what cost?"

At the end of his survey of America in Legend, Richard Dorson (1973, 310) observes, "In every period bodies of folklore have reflected the currents of change, the new values, the new ethics, the new road to happiness." Legends have long astonished us with their seemingly limitless ability to enact our concerns and have remained one of the more vibrant bodies of folklore. Other vernacular forms have arisen as well in order to be able to articulate our concerns in a way that is separate from any direct statement about belief or truth with a simple "this happened." Quite often, it should be noted *what has happened*, either within the text or in the world without, is extraordinary and not so simple. In a world where the sheer complexity of it all threatens our very ability to process *what's happened*, these texts offer us a way to process, or begin to process, a large "what" into smaller, cognizable chunks or facets of "whats."

The relationship of the "whats," of ideas to their expression in various forms, remains an ongoing part of the work of folklore studies. Tim Tangherlini's (2017) recent explorations of online discourse focus on antivaccination efforts, the so-called Bridgegate, and a number of other rumors and legends have begun to articulate a methodology to map the relationships between ideas and the forms in which they are embedded. A lot of work still remains, and as is suggested here, the parallel work on conceptual metaphors is worth our consideration. How the rich and dense networks of ideas—which as encountered in the world are far too inconsistent and nonhierarchical to be considered ideologies—and vernacular expressions interact, each shaping the other, is a domain well suited to the interests and abilities of folklorists.

Notes

1. Griffin claimed to have composed the song himself but spirituals like it circulated among African American churches in the early 1900s. Because the origins of the song are not clear, there has been some debate about which of the Galveston hurricanes is chronicled in it: the famous 1900 storm which claimed between six thousand and twelve thousand lives or the later 1915 storm which claimed 275.

2. Allen was a prolific novelist from Kentucky who published twenty-odd books between 1891 and 1926, many of them considered to be regional in nature. The quotation itself appears nowhere in Allen's work. Rather, it seems to be a compression of the following passage from the 1892 novelette "John Gray: A Kentucky Tale of the Olden Time":

> All this feeling has its origin in my contemplation of the character of the President. You know that when a heavy sleet falls upon the Kentucky forest, the great trees crack and split, or groan and stagger, with branches snapped off or trailing. In adversity it is often so with men. But he is a vast mountain-peak, always calm, always lofty, always resting upon a base that nothing can shake; never higher, never lower, never changing; from every quarter of the earth storms have rushed in and beaten upon him; but they have passed; he is as he was. The heavens have emptied their sleets and snows on his head—these have made him look only purer, only the more sublime. (Allen 1892, 703)

3. A search for proverbs on the topic turned up one by a Robert McKee (1997, 101, emphasis original): "*True character* is revealed in the choices a human being makes under pressure—the greater the pressure, the deeper the revelation, the truer the choice to the character's essential nature." Another one, by a Dr. Paul TP Wong, is as follows: "A person's true character is often revealed in time of crisis or temptation. Make sure that you have what it takes to be your best in such times" (quoted in Quantum Leap Capital n.d., n.p.).

4. While I have heard this joke a number of times in conversation in south Louisiana, I do not have a recorded version. This particular text is from a website, the Boudreaux & Thibodeaux Cajun Humor Page, that has been around so long that it is at least venerable, if not traditional (CajunGuy n.d.). As proof of the website's reputation, I offer that I have even had students plagiarize from the website in the past when given fieldwork assignments.

5. Arizona Revised Statute 28-910: Liability for Emergency Responses in Flood Areas. https://www.azleg.gov/ars/28/00910.htm. Accessed August 25, 2022.

6. Please note that the text has been edited for readability with removal of the extensive paragraphing used by news websites. This was done to all the news texts.

7. From later reporting, in an updated version of the original WMBF post, we learn that the Jeep had come to be on the beach due to some carelessness followed by confusion:

> The owner of the Jeep, who does not want to be identified, reached out to WMBF News and explained what happened and why it was abandoned while a hurricane hit the coast. "My cousin has been around, he rides a motorcycle so I thought I'd let him borrow my jeep because the weather has been so bad. This morning he thought it would be cool to go on the beach and take a quick video of the sunrise

before the storm came," the Jeep's owner said. But the ride on the beach took a turn for the worst. "So he got on the beach and started driving it. I guess there's that runoff there and he didn't realize it was in front of him, he was looking out the window when he went off and got stuck, which you can see he actually banged up the bumper a bit," the owner explained. (Rambaran 2019, n.p.)

With the mystery of the Jeep revealed, there was only its hauling away to come, which the city promptly accomplished with a backhoe and some chain, removing, from their point of view, a public hazard, but one which had been a welcome point of diversion for those waiting out the storm. (This story appears to have had an original posting date and time of September 5 at 10:21 a.m. How much local activity, and legendry, may have sprung up and/or been enacted before such documentation like this is currently unknown. This particular story indicates it was updated the next day, Friday, September 6 at 12:57 p.m.)

8. It's interesting to note that the @DorianJeep account was largely accessed through a web browser and that @MyrtleBeachJ33p through the Twitter app for iPhone. How the difference in devices involved was affected by power outages or mobility during use is something worth possible exploration.

9. Online activity, as principally evidenced by the more active of the two Twitter accounts, @MyrtleBeachJ33p, roared relentlessly for the first twenty-four hours, but by Friday afternoon, around 4 p.m. local time, appears to have wound down: the attention of pranksters and punsters had moved on.

10. The videos on the Myrtle Beach Jeep (@MyrtleBeachJ33p, emphasis original) Twitter feed during that timeframe include a time-lapse recording of one of the television live feeds of the Jeep with the caption "IGHT // IMMA HEAD OUT" super-imposed, the aforementioned bagpiper, which was enormously popular, and a young girl impersonating a reporter and recreating the entire event with a toy car. Hers was not the only toy reference of the day; another photo posted revealed a toy red Jeep ensconced in a zip plastic bag of rice with the caption "24 hours and it'll be good as new" (Myrtle Beach Jeep 2019).

11. This kind of "folk cartography," for lack of a better term, has been explored in a variety of ways by folklorists and anthropologists. Gerald Pocius's (2001) treatment in *A Place to Belong* is one of the more comprehensive, and Michael Jackson's (1995) *At Home in the World* is one of the more hermeneutical. In my own work, I have examined the role of unseen topographies and histories of tracts of land by Louisiana farmers (Laudun 2016).

12. With the rise of the medical industry in the region and the seeming preference of doctors for more traditional forms of status vehicles, this trend seems to be changing somewhat, and Lafayette now has its fair share of very expensive SUVs and German sports cars.

13. The insurance industry has, in fact, a term for this particular kind of fraud—opportunistic fraud, and they are acutely attuned to the opportunities offered by storms. As one study notes: "We first find that among the insured who encountered a typhoon hit, the insured who purchased automobile theft insurance but did not purchase typhoon/flood insurance had a significantly higher possibility of filing a total theft claim than the insured who purchased both types of coverage" (Pao, Tzeng, and Wang 2014, 93).

14. In keeping with the idea of outsiders not understanding, it should come as no surprise that one genre of anecdotes that circulated after the floods were those that targeted Red Cross meals for derision and disgust, with at least two different Facebook posts

substantiating their claims with photos and with captions that offered first despair: "Well today I'm in Baton Rouge helping a friend clean a house out. Well we stoped [sic] by the Red Cross van on way to [the] house at about the time for lunch. So we stopped and got a few plates. Look at picture and I will say no more." And then the other post offered disgust: "This was served to the people in Springfield La tonight by the Red Cross!! Disgusting! And the govt pays Red Cross $8 per meal they serve!! Really. Why don't our governor run them out of the state." Both accompanying photos featured partitioned Styrofoam meal containers opened to reveal their meager or inappropriate contents. The containers themselves are the kind often used by local lunch houses that specialize in generous portions of the kind of rice and gravy cuisine for which the area is known. The containers are also quite often used by churches and schools for fundraising, when Sunday dinners are bought by the ticket. The containers themselves, then, come with a fair amount of semantic weight, and, while the Red Cross is generally well received, its conflation with government bureaucracies, which here not only are not doing what they are supposed to do (provide a decent meal) but are doing so corruptly (getting more money than was spent on the food).

15. In a video game, you can interact with two different kinds of entities: those animated by other players and those animated by the game's programming. Depending upon the game, these nonplayer characters, or NPCs, as they are called, can occupy a number of functions, including helpers, henchmen, and even villains. In most games, the players are the heroes. Nonplayer characters are, in other words, those provided by the game and, as such, have fairly limited functions.

16. A web search for any given weather event will, inevitably, after the meteorological entries, feature individuals doing (mostly heroic) things. While no quantitative analysis was undertaken, most of the featured heroes are men, with most images and reports featuring a healthy, middle-aged man, often white but sometimes black, helping a woman or an elderly person. In some cases, it is a group of teenaged boys. On the other side of the moral axis, there are news reports, for example, of thieves breaking into cars abandoned during a snowstorm in Iowa (McGee 2020).

17. Further examples are a web search away: simply pairing a name of a truck and the word *slogan* turns up ample official and vernacular texts that abound in masculinities.

18. Orrin Klapp's (1949) catalog of folk heroes includes the conquering hero, the clever hero, the Cinderella, and the martyr—he also goes on to note that there are really only two kinds of villains: traitors or persecutors. To this list, Bruce Rosenberg (1972) adds the martyred hero with his study of legends and folk histories of George Armstrong Custer. For more on bandit heroes, see Graham Seal's (2009) work on "the Robin Hood principle" as well as, more recently, work by Nicholas A. Curott and Alexander Fink (2012) on the sociological and economic roles of bandit heroes. Barry Ancelet has also explored bandits in Louisiana folk culture in a number of papers presented at the annual meetings of the American Folklore Society.

References

Allen, James Lane. 1892. "John Gray: A Kentucky Tale of the Olden Time." *Lippincott's Monthly Magazine*, June 1892, 641–709.

Ancelet, Barry Jean. 2008. "Vernacular Power: The Social and Cultural Implications of Katrina and Rita." *Louisiana Folklore Miscellany* 16–17, no. 1: 28–35.

CajunGuy. n.d. "And More . . . ." Boudreaux & Thibodeaux Cajun Humor Page. Accessed September 22, 2022. https://cajunguy20.tripod.com/jokepage7.html.

Curott, Nicholas A., and Alexander Fink. 2012. "Bandit Heroes: Social, Mythical, or Rational." *American Journal of Economics and Sociology* 71, no. 2: 470–97.

Dannenberg, Hilary P. 2008. *Coincidence and Counterfactuality: Plotting Time and Space in Narrative Fiction*. Frontiers of Narrative. Lincoln: University of Nebraska Press.

Dolce, Chris. 2016. "Why the Louisiana Flood Happened, and 4 Other Things to Know." *The Weather Channel*, August 15, 2016. https://weather.com/storms/severe/news/louisiana-flooding-why-it-happened-things-to-know.

DorianJeep (@DorianJeep). 2019. "#selfie." Twitter, September 5, 2019. https://twitter.com/dorianjeep/status/1169690011870085120.

Dorson, Richard. 1973. *America in Legend: Folklore from the Colonial Period to the Present*. New York: Pantheon Books.

Faulkner, William. 1948. *The Old Man*. New York: Signet.

Fedschun, Travis. 2019. "Jeep Abandoned on Myrtle Beach as Hurricane Dorian Rages; Onlookers Take Selfies." *Fox News*, September 5, 2019. https://www.foxnews.com/us/hurricane-dorian-myrtle-beach-jeep-abandoned-waves.

Jameson (@the_Jameson). 2019. "Does the Myrtle Beach Surf Jeep have a twitter account yet." Twitter, September 5, 2019. https://twitter.com/the_jameson/status/1169667795484889090.

Hurston, Zora Neale. 1935. *Mules and Men*. New York: Harper Perennial.

Hurston, Zora Neale. 1937. *Their Eyes Were Watching God*. Philadelphia: J. B. Lippincott.

Jackson, Michael. 1995. *At Home in the World*. Durham: Duke University Press.

Jones, Terry L. 2016. "Walker Mayor Says DOTD, Barrier Wall Responsible for Town's Flooding, Threatens Lawsuit." *Baton Rouge (LA) Advocate*, August 25, 2016. http://www.theadvocate.com/louisiana_flood_2016/article_59c52500-6a28-11e6-af6e-877659a40ce5.html.

Klapp, Orrin E. 1949. "The Folk Hero." *Journal of American Folklore* 62, no. 243: 17–25.

KPEL 96.5. 2016. "Lafayette Parish Road Closures (Latest Update)." August 15, 2016. http://kpel965.com/lafayette-parish-road-closures-latest-update/.

Laudun, John. 2016. *The Amazing Crawfish Boat*. Jackson: University Press of Mississippi.

Lindahl, Carl. 1986. "Psychic Ambiguity at the Legend Core." *Journal of Folklore Research* 23, no. 1: 1–21.

Lindahl, Carl. 2012. "Legends of Hurricane Katrina: The Right to Be Wrong, Survivor-to-Survivor Storytelling, and Healing." *Journal of American Folklore* 125, no. 496: 139–76. https://doi.org/10.5406/jamerfolk.125.496.0139.

Lomax, John Avery, Alan Lomax, and Sin-Killer Griffin. 1934. "Wasn't That a Mighty Storm." Song at Palm Sunday service, Darrington State Farm, Sandy Point, TX. Audio recording. https://www.loc.gov/item/afc9999005.593/.

McGee, Lakyn. 2020. "Thieves Break into Abandoned Cars Left on Roads after Winter Storm." *We Are Iowa*, January 20, 2020. https://www.weareiowa.com/article/news/local/thieves-break-into-abandoned-cars-left-on-roads-after-winter-storm/524-5c3fc272-bcb0-4356-8c85-21070bc6ecb7.

McKee, Robert. 1997. *Story: Substance, Structure, Style and the Principles of Screenwriting.* New York: ReganBooks.

Mechling, Jay. 2004. "'Cheaters Never Prosper' and Other Lies Adults Tell Kids: Proverbs and the Culture Wars over Character." In *What Goes Around Comes Around*, edited by Kimberly Lau, Peter Tokofsky, and Stephen Winick, 107–26. Logan, UT: University Press of Colorado.

Myrtle Beach Jeep (@MyrtleBeachJ33p). 2019. "@papaspack." Twitter, September 5, 2019. https://twitter.com/myrtlebeachj33p/status/1169700063746113542.

Pao, Tsung-I, Larry Y. Tzeng, and Kili C. Wang. 2014. "Typhoons and Opportunistic Fraud: Claim Patterns of Automobile Theft Insurance in Taiwan." *Journal of Risk and Insurance* 81, no. 1: 91–112. https://doi.org/10.1111/j.1539-6975.2012.01498.x.

Pocius, Gerald. 2001. *A Place to Belong: Community Order and Everyday Space in Calvert, Newfoundland.* Montreal: McGill-Queen's University Press.

Quantum Leap Capital. n.d. "Our True Selves." Accessed September 22, 2022. https://quantumleapcapital.com/blog/our-true-selves.

Rambaran, Vandana. 2019. "Jeep Owner Explains Why Vehicle Was Left on South Carolina Beach During Dorian." *Fox News*, September 16, 2019. https://www.foxnews.com/us/owner-of-abandoned-jeep-explains-why-it-was-left-on-the-beach-during-hurricane-dorian.

Rosenberg, Bruce A. 1972. "Custer: The Legend of the Martyred Hero in America." *Journal of the Folklore Institute* 9, no. 2–3: 110–32.

Seal, Graham. 2009. "The Robin Hood Principle: Folklore, History, and the Social Bandit." *Journal of Folklore Research* 46, no. 1: 67–89.

Shakespeare, William. 1998. *Julius Caesar*. Edited by David Daniell. 3rd ed. The Arden Shakespeare. London: Bloomsbury.

Shelter (@alwaysjathis). 2019. "@MyrtleBeachJ33p wtf. I believed in you." Twitter, September 5, 2019. https://twitter.com/alwaysjathis/status/1169718621196750853.

Tangherlini, Timothy. 2017. "Toward a Generative Model of Legend: Pizzas, Bridges, Vaccines, and Witches." *Humanities* 7, no. 1: 1.

Williams, David. 2019. "Owner of the Jeep Abandoned in the Surf on Myrtle Beach During Hurricane Dorian Explains How It Got There." CNN. Last modified September 6, 2019. https://www.cnn.com/2019/09/06/us/dorian-myrtle-beach-jeep-owner-trnd/index.html.

WWL Staff. 2016. "Low Prices Devastating Louisiana Oil Industry Businesses, Workers." WWL TV 4. Last modified May 14, 2016. http://www.wwltv.com/news/local/lafourche-terrebonne/low-prices-devastating-louisiana-oil-industry-businesses-workers/191042106.

Żołnowska, Izabela. 2011. "Weather as the Source Domain for Metaphorical Expressions." *Avant: The Journal of the Philosophical-Interdisciplinary Vanguard* 2, no. 1: 165–79.

Chapter 13

IT ALWAYS RAINS ON A PICNIC

Weatherlore and Community Narrative at St. Patrick's Irish Picnic and Homecoming

Kristen Bradley

St. Patrick's Irish Picnic and Homecoming is a large barbecue and festival held in McEwen, Tennessee, on the last Friday and Saturday of July. The event, which will hence be referred to as the picnic or the Irish Picnic, has taken place every year since 1854. The event is important in its community, as it acts as the primary fundraiser for St. Patrick's Catholic School, a K–8 institution serving the area's Catholic population. The picnic attracts an estimated ten to fifteen thousand people annually and provides significant tourism to the area, having a large economic impact on the community. Local narratives indicate that, despite being an outdoor event held in the middle of summer, the picnic has never been rained out. Even more peculiarly, flying in the face of a popular belief that it always rains on a picnic, local newspapers, including the *Tennessean*, formerly known as the *Nashville Tennessean*, claim that it has not rained on the picnic for more than one hundred years. These narratives are congruent with weatherlore of religious communities, which often invoke saints as protectors or predictors of weather trends. Furthermore, the narratives associated with the weather at the Irish Picnic are grounded in the community's Irish heritage and their Catholic faith. Though it has, in fact, rained on the picnic, the stories reported are significant in their demonstration of community values and identity.

McEwen is a small town with a population of 1,770 people located approximately sixty miles west of Nashville in Humphreys County, Tennessee (US Census Bureau 2019). The earliest settlers came to the area prior to 1805, when the first school in Humphreys County was opened on the banks of White Oak Creek (Garrett 1963, 152). These early settlers included a large faction of immigrants from Ireland (Bullington 1994). In 1842, five thousand acres of land were purchased at a bankruptcy sale by Dr. Frederick Knapp, a physician, and his brother-in-law James F. Neale, a businessman, from New Orleans in order to start a sheep ranch (Garrett 1963, 26). Knapp chose more than two dozen Irish immigrants from New Orleans, whom historian Thomas Stritch (1987, 121) calls "the poorest of the poor," to bring with him to the area to tend the sheep. The descendants of many of these original Irish settlers still live in the area today, and their Irish heritage plays a large role in community identity.

The Catholic Diocese of Nashville began serving the settlement and eventually built what is now St. Patrick's Church. A mere five years after the community's establishment, the diocese started providing priests to serve the area as a mission ("St. Patrick's" 2013, 14). The first of these priests, Father William Fennelly, served McEwen from Rutherford County, visiting every two months from 1842 to 1844. During this time, Knapp donated one thousand acres of his land to Bishop Richard Pius Miles, the first bishop of Nashville, to begin "a church for the workers" (Anderson 2007, 9). Knapp asked that additional parcels of his land be sold at a rate of twenty-five cents per acre "to attract more Catholic families to settle in the area" (Whalen 2013, 2). Father Ivo Schact became the second priest in the community, and he began construction on McEwen's first church, a log chapel located on the road between McEwen and White Oak (Whitfield 1979, 21). This church was dedicated by Father Schact on Easter Sunday, March 23, 1845, as Saint Dominic's (Whitfield 1979, 57). The *Catholic Almanac* for the following year, however, lists the church as St. Patrick's, and "no explanation was given as to the mission name change" (Whalen 2013, 3). Others, however, attribute the change, in honor of the patron saint of Ireland, to the growing number of Irish immigrants in the area (*Nashville Tennessee Register* 1995, 14).

The first Irish Picnic occurred in 1854 when parishioners held a barbecue picnic to raise money to purchase a new bell for their church. Today, the event is hosted by the parishioners of St. Patrick's Catholic Church, descendants of those original Irish settlers, as a fundraiser for their parochial school. Such church fundraisers are common in the South. Beverly Gordon documents these events in their nearly two-hundred-year history in the United States. They are ubiquitous throughout the region and "almost synonymous with religious institutions" (Gordon 1998, 1). In the state of Tennessee, both the town and the church have become synonymous with the event, an occurrence largely due to the event's age, but also due to its size. The picnic at St. Patrick's is no small affair. Each year, hundreds of volunteers get together to barbecue more than eighteen thousand pounds of pork shoulder and nearly four thousand chicken halves. The event attracts ten to fifteen thousand visitors every year and funds the school's operational costs, allowing the school to have one of the lowest tuition rates in the Nashville Diocese. The picnic is reported to have operated annually since its inception in 1854. As such, the event is the oldest continually running festival in the state, and its continuity is a point of pride for the parishioners and volunteers.

The importance of the event is immeasurable within the community. There is certainly financial importance, but there is more to it than that. The school could not operate without the funds raised by the picnic. The parishioners and volunteers know this; a large percentage of them are alumni of the school. Being a St. Patrick "Fighting Irish," the school's mascot, runs deep in the community, and many volunteers dedicate their lives to serving the school. The largest part of doing so is work on the picnic. Though it is a two-day event, work on the picnic is year-round. There are clean-up days, board meetings, and planning committees. There are meats to be ordered, prizes to be bought, and volunteers to be coordinated. Every family has a role, and their role becomes part of their identity. There are "porkers," men and women who deal in barbecuing, selling, and chopping pork shoulders; there are "pluckers," those who do the same with chickens, and there are families who become associated with specific booths, some of whom have had the same roles for generations. With so much of their lives tied up into the event, the parishioners and volunteers need to be assured of its success.

A good portion of the event's success can be linked to the weather; the event simply cannot be a success if the weather does not hold. The picnic takes place outdoors. There is a small, covered bandstand to house musical performances, but electricity and band equipment require dry weather. The barbecue can be consumed in the Dinner Stand, a large wooden building that houses a serving line and picnic tables, but no more than 150 people can be served in the building at one time. There is hardly enough room for the picnic's usual crowd. Carnival-style games scattered across the dirt picnic grounds contribute to the funds raised by the picnic, but rain would wash away their popularity and leave patrons muddy and with nowhere to go for shelter. With many visitors traveling from miles away, even the forecast of rain could drive away business, and a lack of business would diminish the school's operating budget, perhaps leaving it unable to stay open.

Because of the importance of the event in the community, the Irish Picnic must take place, rain or shine; fortunately, in the parlance of the volunteers, the picnic has always been blessed with good weather. The indicated pattern of weather has been true enough that it has become local lore. Starting in the 1950s, newspapers out of Nashville picked up on the trend. Before the 102nd Irish Picnic, the *Nashville Tennessean* published an article detailing the event, stating: "If precedent means anything there will be no rain here on Saturday. Never since the annual Irish picnic became an annual institution 102 years ago has it rained here on the picnic day and Saturday is the day for this year's festivities" (*Nashville Tennessean* 1959, 23). Here, the newspaper reported an amazing trend, perhaps coincidence, that in more than one hundred years, it never once rained on the day of the Irish Picnic. Throughout the 1960s, the *Nashville Tennessean* continued to report this same occurrence. A 1960 article about the picnic states, "If the sun shines a little brighter today than it did yesterday or the day before, it will be no surprise to the picnickers, for the tradition is that it has never yet rained here on the picnic day" (*Nashville Tennessean* 1960, 3). Again, the newspaper emphasized the amazing coincidence that it had never, in more than one hundred years, rained on the day of the picnic. Both of these news snippets reveal something more as well: not only had it never rained, but the writers began using this fact as a prognostic for what the weather would be like at each year's upcoming picnic, including assertions that "there will be no rain here on Saturday" or that the sun would shine "a little brighter" that day.

Weather forecasting based in this lore became more common as the decade progressed. In July 1964, the *Nashville Tennessean* listed weather as a featured attraction for the picnic, alongside barbecue, dancing, and pretty girls (Thoni 1964, 15A). Staff-writer DeAnna Thoni (1964, 15A) wrote: "In addition to all this, a visitor will receive an added treat—if the traditional luck of McEwen's Irish holds out as expected. The picnic's oldest devotees say they cannot remember an occasion when rain ever started before the last of the barbecue supply was being digested." In this statement, Thoni not only lends credence to the idea that it has never rained on the Irish Picnic by citing the picnic's oldest participants, but she also adds a timeline: the weather is sure to hold out for the entire event, making it an added attraction for any prospective visitors who will not have their festivities ruined by inclement weather.

Thoni's article relies on two important aspects of weatherlore, the first of which is observation. When the St. Patrick's community and, likewise, Thoni invoke the community's oldest members, they are not only adding credibility to their argument, but are also showcasing weatherlore's basis in observation. In an early study of weatherlore, Fanny Bergen and W. W. Newell (1889, 204) quoted a Lieutenant Dunwoody as saying, "many of these sayings express in a crude form, the meteorological conditions likely to follow, and have resulted from the close observation on the part of those whose interests compelled them to be on the alert in the study of all the signs which might enable them to determine approaching weather changes." Per this statement, the lore must first be based on observation, which then indicates what will come; next, these observations must be made by people who are most affected by shifts in the weather. For the Irish Picnic, as the participants are certainly affected by the weather, they have a vested interest in observing it closely.

Thoni is not only basing her remarks on observation, however, which follows a larger trend in weatherlore studies. Bergen and Newell (1889, 204) note that observation-based weatherlore is not the only kind, as they classify weatherlore into two separate categories: "those which are the result of observation, and those which are the expression of superstition." The community of St. Patrick's often invokes a statement of such superstition, one that Thoni denotes in her article: the luck of the Irish. Thus, the weatherlore surrounding St. Patrick's Irish Picnic shows basis in both observation and superstition, the latter invoking the community's identity as people of Irish heritage. If their (Irish) luck holds and precedent remains, it will not rain on the picnic.

For the people of the parish, having observed the trend to be true for many years and having an easily identifiable cause—the luck of the Irish—the lore becomes the predictor of what will happen. Another collector of weatherlore, H. A. Hazen (1900, 191) once said, "In order to be of value, a weather saying should be based on a sufficient number of coincidences between the sign and the supposed resulting weather to make it represent a law." Hazen suggests that though it may merely be coincidental, a particular expression of weatherlore will only become valued as it is observed; the more it is observed to be true, the more it becomes accepted as fact. The reports concerning St. Patrick's Irish Picnic are certainly based on observation, and more than one hundred years of the same occurrence—no rain on the day of the picnic—provides "a sufficient number of coincidences." The lore becomes law, then, and begins to be used as prediction.

When it comes to forecasts, the lore surrounding the Irish Picnic weather serves another purpose. Bergen and Newell illustrate that weatherlore often involves items or events that indicate what the weather will be like; in other words, they tell the future. This is consistent with M. G. Wurtele's categorization of weatherlore. He calls the first category "operational." Operational weatherlore, according to Wurtele (1971, 293), "consists of more or less reliable forecast rules, warnings, advices, etc." Statements of this kind tell listeners what the weather will be like and offer advice for day-to-day activities and behaviors. Wurtele (1971, 293) argues, "People observed this sequence of events; they generalized and they verbalized it in a way which makes it easier for the hearer to confirm the truth of the statement on his own. It is a guide to action, like our daily forecast." Irish Picnic weatherlore fits with Wurtele's observations here as well; people observed the dry weather for more than one hundred years, and this became the easily digestible statement sold to the public in the area's newspaper: it never rains on the picnic. In Wurtele's words, it became a guide to action; the patrons knew they could attend the event, and the parishioners knew the event would be successful because both groups of people knew the weather would be good.

Naturally, such meteorological luck—even the luck of the Irish—could not hold out forever. By 1970, the narrative surrounding the picnic's weather changed. In an article about the picnic, the *Nashville Tennessean* reported: "Many humorous and unusual stories, the type Irish tell so well, have been told and retold over the years. One that is absolutely not blarney is about the weather. It has (on rare occasions) sprinkled, to settle the dust, it is said, but never in 116 years has the picnic been rained out" (Thoni 1970, 37). Again, Thoni asserts the veracity of her tale, but here, we see a shift in narrative. The lore is no longer that it has never rained on the picnic, but that the picnic has never been rained out. The shift speaks to the community's values: it is not important that the picnic never be rained on; it is important that it never be rained out. Due to the event's status as a fundraiser for the school and role in community identity, the picnic simply must go on. Additionally, the new lore—that the picnic has never been rained out—is more easily believable and verifiable, as the older generations pass on.

To follow the formula presented by Wurtele, the fair-weather fact has been boiled down into a phrase often repeated. Today, nearly every parishioner asked can repeat the phrase "rained on, but not rained out" (Bradley 2019, n.p.). The commonality can be attributed to the phrase's use historically. In 1980, the church's residing priest, Father Joseph Brando, stated that the picnic had never been rained out (East 1980, 20). Then, in 1987, Barbara Hooper, the publicity coordinator for the picnic that year, stated, "We have been rained on, but never rained out" (Easley 1987, 4). In 1995, a report in the county newspaper, the *Waverly News Democrat*, stated, "in all years since its inception in 1854 the picnic has never been rained out" (*Waverly News Democrat* 1995, 6). The newspaper interviewed Frank Walsh for the same article. Walsh, who was the picnic chair for many years and the sole guardian of the picnic's secret sauce recipe, stated: "It's rained a lot of times, but has never been rained out. There are just too many people working, year after year, to pull this off" (*Waverly News Democrat* 1995, 6). Walsh's statement shows a different ideal than many others. No matter what happens, the community will work together to make the picnic a success. This could be interpreted as being true come rain or shine, so, to an extent, the weather would not matter. Barring anything too torrential, the weather will not spoil the picnic; the community will not let that happen. As former picnic Chair Tommy Hooper stated, "The only serious threat is the possibility of running out of barbecue sauce" (East 1980, 20). His statement, also printed in the newspaper, shows a dismissal, even if facetious, of weather as a threat to the event.

Though the narrative has changed over the years from never rains to never raining out, the picnic will still continue. Importantly, the *Nashville Tennessean* stated in a report after the 1964 Irish Picnic, "Rumbling thunder and a nearby cloudburst steered away from McEwen's 110th [sic] annual Irish picnic as some 8,000 celebrants stayed dry and happy yesterday. McEwen's oldest residents say they cannot remember a day when the picnic was spoiled by rain and yesterday was no exception" (*Nashville Tennessean* 1964, 3A). Here, the note indicates, in much the same way as above, that, no matter what happens, the picnic will not be spoiled. As Wurtele suggests, the often-repeated phrase becomes a call to action. If the picnic will not be ruined, if it will continue as planned, then festivalgoers can be assured that their weekend plans are safe. They can visit the small town of McEwen without fear of anything, including the weather, spoiling their good time. The same is true for volunteers and parishioners. They can know that the weather will not spoil their fundraiser. As Walsh stated, there are just too many of them working too hard to allow that to happen.

Many have noted that the forecasts presented in weatherlore are not always true. In older studies, it seems to be pretty common practice to try to debunk weatherlore. In 1900, Hazen did exactly that, looking into hundreds of pieces of weatherlore just to prove them untrue. He opines, "such forecasts may be quoted by the hundred, and it is easy to see their worthlessness" (Hazen 1900, 192). However, he also states, "after a saying based on hasty generalization is once started, it may be handed down to later generations" (Hazen 1900, 191). Here we see the importance of weatherlore: it is not, in fact, worthless, as its value lies outside of the realm of truthfulness. As Bergen and Newell (1889) so dutifully noted many years ago, weatherlore is not meteorological but rather anthropological. Instead of asking whether or not the weatherlore is factually correct, we must wonder what it might reveal to us about the community who shares it.

Whether or not the narratives are meteorologically correct, they reveal something more about the community; the lore associated with the Irish Picnic weather is closely tied to the identity of the community. They are Catholics of Irish descent, and the patron saint of their parish is St. Patrick. Many of the parishioners take pride in their Irish American culture, and their heritage and Catholic roots play a large factor in community identity and in the narratives surrounding the Irish Picnic. The community performs its Irish American identity throughout the picnic, with bagpipe performances, homecomings for distant relatives, and signs bedecked in shamrocks and printed with slogans in Irish Gaelic. *Nashville Tennessean* articles from the 1960s latched onto this aspect of the picnic, often using Irish American stereotypes to portray the community. For example, articles would note the Irish's penchant for storytelling, describe the female participants in the picnic as "colleens" but with southern accents (Thoni 1964, 15A), and include references to "blarney," stressing the parishioners' Irish ancestry. Interestingly, the same quip about Irish storytelling and "blarney" was printed in the Nashville newspaper in both 1970 and 1997, confirming the historical continuity of the weatherlore.

The same emphasis on the community and its Irish heritage continued in other reports of the event and its weather. The *Nashville Tennessean* reported:

"The picnic has the distinction of being one of the oldest celebrations in Tennessee and a celebration that has never been rained out. It has rarely been known to even sprinkle on picnic day, a fact which some say testifies to the luck of the Irish" (Thoni 1965, 18F). Here, the report echoes a familiar saying in the community and attributes the good weather associated with the Irish Picnic to the parishioners' Irish heritage and the Irish's fabled good luck. Similarly, in 1970, Thoni wrote for the *Nashville Tennessean*: "Irish people are well known for their loyal and steadfast faith. The Sunday before the great event, the people pray for beautiful weather and never seem surprised that the Sun always shines on their day. Heavy rain that surrounded McEwen last year would have made an atheist wonder, but seems to be just what the Irish expected" (1970, 37 and 52). In this instance, Thoni's statement stresses not only the community members' Irish heritage but also their Catholic faith.

A third category of weatherlore, as recognized by Wurtele, can account for this type of belief. This category includes lore that "invokes the heavenly bodies, birds, animals, and saints' days" (Wurtele 1971, 299). There are a large number of ecclesiastical traditions that fall under this category, not the least of which is weather associated with saints and saints' days. There are traditions linked to the feast days of the saints Swithin, Valentine, Bartholomew, Joseph, and Thomas, among others (see Bergen and Newell 1889; Eden 2008; Reed 1986; Reynolds-Ball 1927). Furthermore, early collectors of weatherlore note, "Particular saints have also been selected as exerting special influence over the weather" (Inwards [1869] 2007, 5). Though it is common to associate particular saints with patterns in weather, St. Patrick is not commonly thought to have such weather-controlling abilities.

Instead, the community's invocation of St. Patrick as a guardian over the weather at the picnic relies on the saint's ties to the community identity. When the church was founded in 1849, it was named for St. Dominic, but as missionaries became aware of the large Irish settlement in the area, the church was renamed for St. Patrick, the patron saint of Ireland. The renaming of the church demonstrates the important ties the community has to St. Patrick by virtue of their Irish heritage. The community often ties their "luck" with weather to both their heritage and to this patron saint of Ireland.

In traditional saints' days weatherlore, the importance of the day and of the church lead the community to lend the saints extraordinary power. Events of these days, per Wurtele (1971, 299), "must have special significance, special potency as signs and portents," meaning these days are significant enough that they hold a certain power over the weather. The same can be said for St. Patrick's Irish Picnic. Despite the fact that the Irish Picnic takes place in July and not on March 17, the feast day of St. Patrick, the date holds large enough significance within the community to hold the same potency as feast days in the other instances of weatherlore. For the parishioners, the date is powerful enough to control the weather. While saint-related weatherlore was common in Europe, Bergen and Newell note fewer instances of the same in America. Phillip Eden (2008, 33) contends that the lore surrounding the saints Swithin and Bartholomew "originated in the early Middle Ages when the Church was intimately involved in people's everyday lives." The declining involvement between the people and the Catholic Church in contemporary American society can help account for the declining importance of this saint-related weatherlore. However, for the parishioners of St. Patrick's, their everyday involvement with the church and, more specifically, the Irish Picnic matches earlier accounts of ecclesiastical weatherlore.

For St. Patrick's parishioners weatherlore reveals not only the importance of the event but also the community's Christian faith. Beliefs about weather in the community do not indicate cause and effect, as noted above, but instead invoke a type of divine protection, with stories of storms suddenly clearing or veering in other directions on the day of the picnic. Longtime picnic Chairman Michael Bradley relates a story of two or three of the Irish Picnics from about 2000–2006, stating: "local emergency management officials and others with weather radar apps were concerned about storm systems coming into Humphreys County from the west. The storm cells either dissipated or split around McEwen and did not impact the Irish Picnic. Parishioners and the Dominican Sisters attributed the continued good weather to Divine protection of the event" (Bradley 2019, n.p.). This has been true for other times of year as well. In October 2019, Humphreys County was ravaged by some of the worst storms middle Tennessee has ever seen, remnants of a Gulf Coast hurricane that passed through the area at higher than usual speeds. The storm knocked out power in most of Humphreys County and McEwen, but St. Patrick's Church and school had minor damage and did not lose power. Staff at the school attributed it to protection of St. Patrick and stated that the worst of the storm "split" to go around the school and church grounds (Bradley 2019, n.p.).

Historically, it is not uncommon for people to devise links between God and weather. In his philosophical study of Aristotle's *Meteorology*, Craig Martin (2010, 261) asserts that Christians "often saw meteorological phenomena as proof of divine providence or God's wrath." Historically, people often attributed weather to God's will rather than to meteorological causes. He continues, "the Catholic Church maintained that destructive weather was the result of divine punishment throughout the sixteenth century" (Martin 2010, 274). If God controls destructive weather, then so too does he control fair weather. Martin (2010, 275) further states that people have also "interpreted 'fruitful weather' as proof of God's love for humanity. Thus, in the sermons on Luke 21, while rare events were ominous or apocalyptic, fair weather might signal

God's protection." The idea that God provides protection for people through weather has been a persistent belief throughout history and remains true today.

The link between fair weather and God's protection is perhaps nowhere more prevalent than it is in the story of Noah and the flood. In the Old Testament, the story of Noah says that God promised to never again destroy the world through flooding. His promise comes in the form of a rainbow. Therefore, today, a rainbow has two causes. There is the physical one, that rain or the condensation in the air and the refraction of light causes a rainbow to appear. The second is teleological; it is a religious sign that "God will not bring another *diluvium* because the rainbow is a sign of the pact between humans and God" (Martin 2010, 278). The latter explanation, that the rainbow is a sign from God, is common in religious communities. When the *Nashville Tennessean* reports that the community prays each year for good weather, they are demonstrating this belief and thus demonstrating the importance of the Christian faith in the St. Patrick's community.

St. Patrick's Church is no exception; they hold their own narrative regarding rain and rainbows. Earlier lore that held that it had never rained on the picnic signifies a type of divine protection for the event, but the same is true for the "rained-on-but-not-rained-out" narrative. Late July in Tennessee is known for being hot and extremely humid, which can sometimes lead to terrific rainstorms. The fact that the picnic has never been rained out is extraordinary, and this trend leaves room for a rainbow narrative to arise. In 1987, the *Nashville Tennessean* reported that it rained, and hard, on the Friday morning of the Irish Picnic. Newspaper staff interviewed Barbara Hooper, Irish Picnic public relations coordinator, who stated: "Friday when the men were loading the pit, we had a storm. While it was pouring rain just over the school from where we were standing under a shelter, we could see a giant rainbow" (Easley 1987, 4). She continued: "I felt then it was a sign of good tidings—and I think it has turned out that. The rain soon stopped and we had wonderful weather" (Easley 1987, 4). Her interpretation of the event emphasizes divine protection. She invokes the second cause of the rainbow; it acts as a sign from God. The rain did stop before the public would come to enjoy the festivities, but even before it stopped, the rainbow provided a sign that the event could (and thus must) continue. This type of community narrative reifies the importance of the event to those who repeat the story.

Hooper related the story of the 1987 picnic to *Tennessee Crossroads* (2012) reporters in 1992. The show's host, Joe Elmore, said, "Every year the luck of the Irish has been with the picnic. Even when storms threaten" (*Tennessee Crossroads* 2012, n.p.). His introduction to Hooper's story again invoked "the luck of the Irish," playing on community identity in the same vein as earlier newspaper articles, but the story continued by again morphing into the biblical reference to Noah and the flood. As in the earlier telling of the story, the rain began early Friday morning, around 4 a.m., when workers were tending to the barbecue chicken pits. Hooper said:

> I can remember one year, it was early morning, and we were doing chickens around four o'clock in the morning, and it was just coming a flood. And I thought, well, this is the year that ... the rain is going to hurt us. And at seven o'clock, it quit raining. And there was a beautiful rainbow right behind the school.... And we have a cross that's right in behind it. And ... the cross was in the middle of that rainbow. (*Tennessee Crossroads* 2012, n.p.)

Hooper's retelling of the story gives religious significance to the event. Her observations highlight not only the continual fear that the picnic will not be a success, that this will be the year that the rain hurts rather than helps, but also highlights the profound hopefulness provided by the rainbow and the parishioners' unfailing faith. Paired with Elmore's preface introducing the luck of the Irish, Hooper's emphasis on religious elements reflects the community's Irish Catholic heritage. The variation in the story featuring the rainbow with the cross in the middle is purposely reminiscent of the Old Testament promise God made to Noah to never again destroy the world through a flood. For St. Patrick's purposes, it can be seen as a promise that the picnic will not be destroyed by rain. As Martin (2010, 272) states, "Only religious faith allows for the understanding that these purposes exist." In any other context, it is just weather.

The narrative surrounding weather at the picnic has changed significantly over the years, perhaps as the overall climate has changed. It is no longer true that it never rains on the picnic. In fact, some younger volunteers and parishioners remember it raining more often than not (McMillan 2019). The truth behind the weatherlore remains the same, however; the weather will not ruin the picnic. It will not be rained out. The stories this community tells about itself reveal a different kind of truth—not meteorological, but cultural (Bergen and Newell 1889). They reveal who this community considers themselves to be; they have a strong Irish heritage and strong Catholic faith. Perhaps more importantly, they have a strong work ethic. For this reason, the story will continue: the picnic has been rained on, but not rained out.

References

Anderson, Win. 2007. "20,000+ Descend on McEwen for 153rd Irish Picnic and Homecoming." *News Democrat*, July 27, 2007, 9.

Bergen, Fanny, and W. W. Newell. 1889. "Weather-Lore." *Journal of American Folklore* 2, no. 6 (July–September): 203–8. JSTOR.

Bradley, Michael. 2019. Interview with author. November 25, 2019.

Bullington, Joyce Enochs. 1994. *Pie Suppers and Cake Walks: A History of Liberty School Located on Little Blue Creek, McEwen, Tennessee.* Self-published, J. E. Bullington.

Easley, Bill. 1987. "St. Patrick Picnic Seen Successful." *Nashville Tennessean*, August 5, 1987, 4. ProQuest.

East, Vickie Kilgore. 1980. "Annual Irish Picnic in McEwen Saturday." *Nashville Tennessean*, July 24, 1980, 20. ProQuest.

Eden, Philip. 2008. "The Weather Week: Country Lore Still Has a Dry Tale to Tell. St. Bartholomew's Day Mystique May Have Its Roots as Much in Meteorological Trends as Rural Hokum." *Sunday Telegraph*, August 24, 2008, 33. Gale OneFile.

Garrett, Jill Knight. 1963. *A History of Humphreys County, Tennessee.* Self-published, Jill Knight Garrett.

Gordon, Beverly. 1998. *Bazaars and Fair Ladies: The History of the American Fundraising Fair.* Knoxville: University of Tennessee Press.

Hazen, H. A. 1900. "The Origin and Value of Weather Lore." *Journal of American Folklore* 13, no. 50 (July–September): 191–98. JSTOR.

Inwards, Richard. [1869] 2007. *Weather Lore: A Collection of Proverbs, Sayings, and Rules Concerning the Weather.* Whitefish, MT: Kessinger Publishing.

Martin, Craig. 2010. "The Ends of Weather: Teleology in Renaissance Meteorology." *Journal of the History of Philosophy* 18, no. 3: 259–82. Project Muse.

Nashville Tennessean. 1959. "102nd Irish Picnic Slated at McEwen." July 24, 1959, 23. ProQuest.

Nashville Tennessean. 1964. "8,000 at Irish Picnic Gay, Well-Fed and Dry." July 26, 1964, 3A. ProQuest.

Nashville Tennessean. 1960. "McEwen Prepares Barbecue for Big Irish Picnic Today." July 30, 1960, 3. ProQuest.

Nashville Tennessee Register. 1995. "St. Patrick's History Tied to Irish Heritage of Its Parishioners." March 1, 1995, 14.

Reed, Mary. 1986. "Weather Talk: Of Soggy Seaweed and Assorted Saints." *Weatherwise* 39, no. 6 (December): 327. Taylor and Francis Online.

Reynolds-Ball, Eustace. 1927. "Piedmontese Weather-Lore." *Folklore* 38, no. 4 (December): 368–71. JSTOR.

Stritch, Thomas. 1987. *The Catholic Church in Tennessee: The Sesquicentennial Story.* Nashville, TN: Catholic Center.

Tennessee Crossroads. 2012. "McEwen Irish Picnic." July 25, 2012. YouTube video, 5:32. http://www.youtube.com/watch?v=qOZytPNX21c.

Thoni, DeAnna. 1964. "Colleens with Drawls to Preside at McEwen." *Nashville Tennessean*, July 19, 1964, 15A. ProQuest.

Thoni, DeAnna. 1965. "McEwen's Irish Picnic Day Nears." *Nashville Tennessean*, July 25, 1965, 18F. ProQuest.

Thoni, DeAnna. 1970. "Irish Clan to Meet at McEwen Picnic." *Nashville Tennessean*, July 23, 1970, 37 and 52. ProQuest.

US Census Bureau. 2019. "Annual Estimates of the Resident Population: April 1, 2010 to July 1, 2018." US Census Bureau, October 9, 2019. https://factfinder.census.gov/faces/tableservices/jsf/pages/productview.xhtml?src=bkmk.

Waverly (TN) News Democrat. 1995. "St. Pat's 141st Picnic Will Be Bigger and Better." July 26, 1995, 6.

Whalen, Carol. 2013. *Diocese of Nashville: 175th Anniversary. St. Patrick's Church.* Nashville, TN: Diocese of Nashville.

Whitfield, Margaret. 1979. *Humphreys County Heritage: A Collection of Historical Sketches and Family Histories.* Waverly, TN: Humphreys County Historical Society.

Wurtele, M. G. 1971. "Some Thoughts on Weather Lore." *Folklore* 82, no. 4 (Winter): 292–303. JSTOR.

Chapter 14

THE FOLK WISDOM OF LAWNS

Todd Richardson

Everything in this chapter was better expressed in the opening sequence of David Lynch's *Blue Velvet* (1986). The film opens on red roses against a white picket fence beneath a blue sky. The image gives way to an immaculate, classic-looking firetruck rolling by in slow motion, a kindly fireman, Dalmatian by his side, waving from the truck's running board. The scene crossfades back to flowers, again against a white picket fence, only they are yellow tulips now, before crossfading again to well-dressed children obediently crossing the street, a gray-haired crossing guard smiling and nudging them along. The sequence eventually settles on a well-maintained, two-story home. Outside, a middle-aged man waters his lawn while, inside, a woman sips coffee as she watches something ominous, a gun being pointed, on the television. The gurgle of the hose, the first thing heard other than Bobby Vinton's "Blue Velvet," which has been playing all along, disrupts the suburban tableau. The man watering his lawn struggles to untangle a knot in his hose when he suddenly collapses, falling to the ground, the hose still in his grip. As he writhes on the ground, a small dog tries to drink from the hose's jet while a curious baby, popsicle in hand, ambles into the scene. At this point, the sequence slows down as the camera creeps closer, first on the dog, whose attempts to drink from the hose become horrific in slow-motion, then down into the lawn itself. The sound of "Blue Velvet" relents as life beneath the grass canopy is revealed, a world where insects gnaw and mash their way through earth and one another, their collective mastication a grotesque thrum.

In sum, beneath all the wholesome associations, lawns are awful to ponder.

The suburban lawn functions as both symbol and site of the American dream. A happy mom watches children cavort through a sprinkler, with a dad in the background grilling burgers and grinning. Yet, as the opening of *Blue Velvet* suggests, beneath the surface of such Edenic scenes lurks a horror show. While this description may seem hyperbolic, my own unsettling and surreal experiences with lawns and lawn care certainly fit the assessment. When I was thirteen, mowing the lawn became my responsibility, the thing I did to keep our household functioning properly, or at least presentable, and I soon after went professional, becoming a lawn boy at Westgate Elementary School. Throughout my indoctrination into lawn culture and its customs, I was made to believe that I was learning important lessons about being an American contributor while earning my place in a decent society, yet looking back, I feel the experience deformed me.

Because this is a collection about weather and folklore, I will start with climate's role in lawns and lawn care, how climate is somehow both foundational and irrelevant when it comes to proper lawn care. Wherever in America a person happens to live does not, *must* not, deny them the luscious green grass emblematic of a proper suburban home. While a location's climate, a desert for instance, can make it difficult to grow and maintain turfgrass, a lawn is realizable anywhere, provided the homeowner wants it enough and is willing to work at it. In this way, the lawn, as an ideal, indulges in the dream logic of the American dream, which insists that success is a product of desire and toil, not of circumstance. And just as it's un-American to acknowledge that the zip code a person is born in predicts their prospects better than their abilities or ambition do, it is un-American to acknowledge the importance of climate. Nay, it is downright patriotic to deny the reality of climate.

In its own way, the lawn, that infinitely reproducible trophy of a successful, middle-class existence, facilitates climate denialism. Lawns foster a thoughtless relationship with one's environment, a sort of mass mediation of nature. Michael Pollan (1989, 22) calls lawns a form of television in that they "lift our gaze from the real places we live in and fix it on unreal places elsewhere." Growing up, I marveled at the Bradys's backyard and its precisely green Astroturf carpet, and while it was clearly fake, I nevertheless envied the reliability of it, along with the pleasant memories the Brady family made on it. It was a far better place than the unpredictable one I lived in. The suburban lawn may have originally aspired to replicate the expansive grounds of English manors, but it has come to invoke Southern California's lack of seasons, with that evergreen everywhere of American fantasies that Lynch describes in daily weather reports he uploads to YouTube. His reports rarely vary. Taken as a whole, they grow deranged in their invariance, Lynch's enthusiasm for the weather seeming more and more performative and desperate with each day's ritualistic invocation of Los Angeles's "blue skies and golden sunshine" (Lynch n.d., passim).

Please keep in mind the lack of weather within the homes that all those lawns buffer. Suburbs consist of a series of personalized ecospheres, each one set to the inhabitants' preferences and tendencies. In a sense, thermostats are enchantments, giving people the once-unthinkable ability to remake nature or, at the very least, to mute it. Or perhaps they're more properly *talismans*, charmed objects that provide protection from evil. Either way, they're unnatural, technological witchcraft that is called upon to keep human bodies relentlessly comfortable. As a whole, the suburban indoors, like the insides of office buildings and shopping malls, stands indifferent to whatever the weather outside happens to be.

None of this is sustainable, but sustainability is not the point. If anything, it's the opposite. As far as I can tell, the point of suburbia and its well-disciplined lawns is the propagation of an ideal of genteel self-sufficiency. Yet as a social reality, suburban life is much more effective at obscuring the reality of human interdependence, and sustainability, as an ideal, is necessarily rooted in interdependence, whether it is between people or between people and the world they live in. In short, sustainability is incompatible with the ideal of suburbia.

When I was sixteen, I drove myself to the Service Center, the maintenance headquarters for Omaha's School District 66. I had been instructed to go there for training prior to reporting to Westgate Elementary, the school I would work at as a lawn boy throughout the summer of 1990. My parents, believing it would build character, made me get the job; they had, in fact, arranged it. My father taught in District 66, and it was common practice there to provide teacher's kids with summer jobs as lawn boys.

Because my parents provided necessities and I wasn't saving for anything big like a car, I did not see the point of this job, but the adults in my life assured me that the job would teach me responsibility. More than once I was told the experience would "make a man of me." Summer jobs certainly are a conventional rite of passage for American youth, a way of easing adolescents into the working world, and lawn mowing stands as one of the most popular, or at least most romanticized, summer jobs. Dripping with honest sweat and nostalgia, lawn mowing registers one notch below a newspaper route on the wholesome scale. Looking back on it, I can appreciate why my parents thought the job would steer me in the direction of good citizenship.

On August 2, 2017, White House Press Secretary Sarah Huckabee Sanders read from a letter Frank Giaccio sent to President Donald Trump: "It would be my honor to mow the White House lawn some weekend for you." Giaccio wrote, "Even though I'm only ten, I would like to show the nation what young people like me are ready for" (Moyer 2017, n.p.). The following month, Giaccio got his wish, mowing the White House's Rose Garden, an event best remembered for the iconic photo it produced. "That's the real future of the country right there," President Trump told reporters at the event. "Maybe he'll be president someday" (Moyer 2017, n.p.).

The rhetoric in and inspired by Giaccio's letter draws heavily from the myth of the American dream. "Myth," as a term, has taken on an almost entirely negative connotation in contemporary culture, suggesting falseness or naïveté, but I use the word in the folkloristic sense, as in a treasured narrative that conveys key values in support of an ideology. In this folkloristic sense, the provability of the American dream is irrelevant. What matters are the values expressed through the story or stories. Put another way, whether or not Giaccio really will end up being president someday is far less significant than the underlying notion that unglamorous toil, such as mowing a lawn, is a gateway to greater things. Perhaps it would be better to take Alan Dundes's (1971) advice and identify this "myth" as a folk idea instead. Dundes (1971, 95), acknowledging the increasingly pejorative connotations of "myth," offered "folk idea" as an alternative term for "traditional notions that a group of people have about the nature of man, of the world, and of man's life in the world." As an example, he offers the widely held American belief that "any object can be measured in monetary terms." Expressed through proverbs like "money talks" or "you get what you paid for," this idea isn't exclusively proverbial. Nor is it a superstition exactly, even though it may function like one. It is, simply, a folk idea. And so is this American insistence that a person's desires are more important than their circumstances. This idea is at the root of all American-dream narratives, the most enduring of which is Horatio Alger's *Ragged Dick* ([1868] 2014), so much so that rags-to-riches stories are sometimes called "Horatio Alger tales." In the original *Ragged Dick* novel, a young bootblack, with the help of some new, more affluent friends, is instructed in his disreputable condition and thereafter commits himself to a project of self-improvement. Already gifted with a strong work ethic, Dick endeavors to acquire an education, eliminate vices, like smoking and gambling, and save his money, all in order to make himself "'spectable." Throughout the novel and by various clear-sighted and well-intentioned characters, he is told his seemingly low toil as a bootblack is nothing to be ashamed of.

    The first time I read the novel, I immediately recognized that rhetoric. It was the same motivational fluff people shared with me when I started as a lawn boy.

    Lawns, or at least a lawn mower, feature prominently in another Lynch movie, *The Straight Story* (1999). The movie is based on the true life of Alvin Straight, a World War II veteran who, in 1994, drove his lawn mower from Laurens, Iowa, to Blue River, Wisconsin, to visit his older brother. Alvin's story received a great deal of national attention when it happened because he seemed to capture something essential about the American spirit, a heroic resourcefulness and determination. After all, the seventy-three-year-old Alvin did not allow the reality of not having a driver's license to get in the way of doing what he needed to do. Lynch's film capitalizes on this idea, but being a Lynch film, a surreality shades the narrative. As inspirational as Alvin's lawn-mower journey seems, it is, considered more deeply, rather pitiable and debased. None of his friends and neighbors offer to drive him to Wisconsin, which would have taken half a day, not the six weeks it took Alvin to get there on a lawn mower. Nor does Alvin accept a ride when it is finally offered, preferring to do things his way, no matter how dangerous and impractical his way may be.

In the movie, as happened in real life, Alvin's mower breaks down at one point, and he spends a few days in Charles City, Iowa. He grabs a drink with a fellow World War II veteran while there, but their conversation does not revel in good old days and the righteousness of the cause. Alvin confesses to his fellow veteran that he mistakenly shot and killed a young man in his own division. His fellow soldiers thought the young man was killed by Germans, but Alvin knew the truth, a truth he shares with someone else for the first time in this bar in Charles City. The scene is excruciating, engaging with PTSD frankly and refusing any nostalgic appeal. Beneath the surface of this plucky American character lurks unspeakable horror.

And then Alvin is off again, defiantly riding his little lawn mower across America to visit his dying brother. This, too, is Americana in that it is typical of America. As much as we want the American dream to be best represented by charmingly self-reliant characters like Alvin or Dick, it is also represented in a sad man riding a lawn mower across Iowa.

Back in the summer of 1990, it wasn't just that I didn't understand why I needed a job. I didn't understand why this particular job, lawn mowing, needed to be done at all. Other landscaping responsibilities popped up from time to time, but the bulk of my working days passed on the back of a riding mower, slowly rolling over Westgate's lawn. Once I'd get the front lawn to a uniform length, the back lawn would no longer match, so I'd have to cut it again, but when I finished there, the front lawn would now be too long. Day in and day out, I mowed the expansive grounds and its relentless grass. The summer was Sisyphean. All that time outdoors made me hate the outdoors. I learned to like the rain more than sunshine because it meant I got to go inside and relax. Being a lawn boy fundamentally altered my relationship to the weather, turning conventionally good weather into bad weather and vice versa, upending the sensible order of human experience.

The continental United States contains more than forty million total acres of lawn (Milesi et al. 2005). Altogether, turf grasses take up as much of the nation's space as Iowa, the twenty-third largest state. And it is all non-native vegetation. Kentucky Bluegrass may ring of Americana, but the grass itself, like all the most popular turf grasses, is an alien in North America, its popularity arising not from its suitability or sustainability but from its verdancy. The verdancy doesn't come easy as maintaining that extreme green requires a stunning amount of water, twenty trillion gallons in the United States annually, making *lawn* the country's most irrigated crop. Add to this ecological tally seventy million pounds of pesticides used annually, along with billions of pounds of CO_2 emitted by gas-powered mowers. In short, the lawn, a tricky fixture of American culture, takes up a grotesque amount of space.

"Our research has convinced us that one of the main purposes of the garden is to assert dominance over nature," E. N. Anderson Jr. (1972, 187) observes in "On the Folk Art of Landscaping," the sole folkloric assessment of suburban lawn customs to date, "or at least to keep nature at a distance, beyond a barrier of carefully controlled and manipulated planting." He points out that vegetation incompatible with the environment is often chosen "so that the grower can feel a sense of accomplishment in making it grow" (Anderson 1972, 187). In other words, lawns, little fits of manifest destiny that they are, express a self-destructive, climatic braggadocio characteristic of American culture. "This begs the question," Anderson continues in his analysis of the folk nature of landscaping practices, "of why American suburbanites feel the need to assert the triumph of artificial order over natural events, and why they perceive the latter as chaos," before backing off, insisting "this takes us beyond communication into individual psychology, and thus beyond the scope of this paper" (Anderson 1972, 187).

It is not, however, beyond the scope of this chapter, as I want to understand how lawns have influenced—warped may be a better verb—my individual psychology. It's why I'm exploring the wisdom of lawns and lawn care and why I'm doing so as a folklorist. It's fitting, after all. Lawns and lawn mowing are dense with both folklore and folksiness. In terms of the former, a traditional, informal, and unwritten rulebook encodes and enforces the proper performance of lawn-care customs, which is the epitome of folklore as a mode of cultural expression and transmission. As for the latter, lawn mowing occupies a nostalgia-drenched space in the American imagination, a space where fathers teach sons how to mow in straight lines and, in the process, prepare them to be good neighbors and employees.

Lawns and lawn care have long been tied to popular notions of good citizenship. In his book *Lawn People*, Paul Robbins (2007) suggests that lawn ecology creates a certain kind of civic subjectivity that, in effect, lawns and lawn care don't just exhibit citizenship, they create citizens. As he puts it, "the lawn is a system that produces a certain kind of person—a turfgrass subject" (Robbins 2007, xvi). In many ways, Robbins's argument is a dense, sophisticated articulation of a more common idea, one many people have but have not thought through. Simply put, a person's lawn conveys that person's social character. Ideally, the message sent is one of solidarity as the rhetoric surrounding lawns invokes democratic ideals of good citizenship. Maybe not citizenship in a legal sense, but it certainly demonstrates the less formal *folk* relationships a person maintains with their neighbors and neighborhood. While a person born in the United States immediately possesses legal citizenship, they must learn proper citizenship through example and experience, which they then exhibit by adhering to common customs, like lawn mowing.

While estates with meticulously tended grounds predate the American iteration, the notion that everyday folk can and should have their own lawn emerged in America in the late nineteenth century, later exploding in the postwar suburban boom of the mid-twentieth century. In *The Lawn: A History of an American Obsession*, Virginia Scott Jenkins (1994) explains that a variety of actors promoted the idea that well-tended turf grass was a status symbol, not the least of which was the golf establishment. Lawns and golf courses grew up together, a coordinated effort to green America in the most literal sense. Strategically placed articles in popular publications regarding the positive qualities of a lawn bolstered their popularity, yet ultimately, no force promoted the lawn more strongly than peer pressure. From my own experience, I would not maintain my lawn were it not for the pressure—or just the thought of such pressure—that I feel from my neighbors and their lawns. In order to be a worthy member of a suburban community, one must maintain a consistently trim and vibrant green lawn. To deviate from either of these qualities is to risk being labeled a neighborhood blight.

Ecologically speaking, lawns are monocultural, but the same can be said of their expressive function. A smooth carpet of green blending in with its suburban environs demonstrates a community's solidarity, making it appear that everyone lives in a single, unbounded park. In his article "Why Mow? The Case Against Lawns," Pollan (1989, 22) calls the suburban ecology "an egalitarian conceit, implying that there is no reason to hide behind fence or hedge since we all occupy the same middle class." Such uniformity does not arise without coercion however, no matter how much folks insist otherwise. Pollan recounts the summer his father opted not to mow and the many ways in which his family became neighborhood pariahs. "No one said anything," Pollan (1989, 22) writes, "but you could hear it all the same: Mow your lawn or get out."

Not long ago, I turned onto my street to see my neighbor's son mowing their lawn. Being the suburbs, the sight would not be surprising were it not for the time and place: Nebraska in February. Granted, the weather was unseasonably warm, but it wasn't *that* warm. Moreover, the uptick in temperatures came during a dry spell. Nevertheless, my neighbor's son paced both front and back yard, running the mower over brown, dormant grass for reasons unclear to me. When my wife got home and I told her what I saw, she asked, not entirely joking, "Does this mean we have to mow our lawn now?" While our neighbors did not announce anything verbally, the message came through, and sure enough, their neighbor to the north was out mowing the next day, leveling the infrathin difference running along their otherwise invisible property line.

My on-the-clock existence as a lawn boy officially started at the District 66 Service Center, a windowless brick building filed away in an anonymous industrial neighborhood. Like administrative offices and other educational facilities without students, the building's indifference unsettled me. One of the three seasoned custodians who trained me that morning actually smoked throughout the orientation. While smoking may have been more normal in 1990, it still didn't feel right. This was, after all, school property, or at least the painted cement aesthetic gave it a scholastic ambiance. I was struggling to resolve the scene's dissonance while the three men went about pretending to explain to me the fundamentals of routine mower maintenance. Mostly, however, they just made it clear to one another that training lawn boys was beneath them, and after forty minutes of instruction, shooed me away to Westgate. The only things I took with me were the sense I did not know why I needed to do the things I would be doing and the knowledge that I would be doing them regardless.

I no longer mow my own lawn. I pay someone to do it. Once a week, Lovely Lawns, a private mowing company operated out of a pickup by a charming fellow named Scott, cuts our lawn, trims the edges, and blows away the remains. He never leaves uncut tufts like I do when I mow, nor does he trust the wind to sweep away clippings. After I mow, the yard looks like an amateur haircut, which is fine with me, but Scott's a crackerjack. He makes the grass look symmetrical and professional, thus giving the impression I care more about the look of the lawn than I really do. In this way, the arrangement resembles my approach to other civic issues. For instance, while I used to speak out against injustice through protest and direct action, I now give money to the ACLU to do that for me. In effect, I outsource my civic responsibilities.

And make no mistake: mowing is very much a civic responsibility, a necessary demonstration of one's earnest investment in collective well-being. Suburbanite slackers of the past were approached informally about their unruly lawns, but many contemporary neighborhoods have raised the stakes, codifying proper landscaping via housing covenants that compound general opprobrium with monetary fines. Where there are no covenants, zoning ordinances and local laws often criminalize "nuisance vegetation," a process Sarah Baker (2015) discussed in a widely shared editorial, "My Township Calls My Lawn 'a Nuisance.' But I Still Refuse to Mow It." Baker, an avid gardener, opted to not mow her lawn one summer, and by June, officials were threatening her with a one-thousand-dollar fine and a visit by a police-escorted landscaping crew. She eventually relented, cutting her lawn to eight inches while she and her husband "figured out their next move" (Baker 2015, n.p.).

Pulling into Westgate, I took in its large, sloping front lawn. Mostly too steep for a riding mower, it wasn't uniformly too steep, which is why I would push the physics of the situation throughout that summer, riding the incline from side to side while bringing the mower within a whisper of it tipping. So much of that summer was spent on that front lawn, probably more time than all the students of Westgate combined spent on it during a regular school year. Under no circumstances were students allowed to step foot on that grass. Like most American front lawns, the grass was ornamental, a platonic ideal to be pondered, not walked upon. Back at my elementary school, I had been on safety patrol, and in addition to helping people cross the street safely, we were responsible for keeping students off the school's lawn. Members of the safety patrol were instructed to "report" any student who had to be told three times to get off the grass. Drunk with this childish authority, our warnings came in quick succession—"off-the-grass-off-the-grass-off-the-grass"—after which we wrote down the troublemaker's name, later transcribing it in large and shaming letters on their classroom's chalkboard, letting their teacher and classmates know they were terrible citizens. We all did it, each and every year. Such is how tradition works.

My new boss, a tiny man with an oversized mullet, bounded out to greet me. Mr. Varley proved to be a righteous dude. He thought it was funny when he found me asleep in unusual places, and he never batted an eye as I gradually, minute by minute, pushed the start of my days until arriving an hour late became somehow acceptable. The assistant custodian, in contrast, was never charmed by my laziness, and when Mr. Varley took a two-week vacation to Hawaii—he returned with a cast after breaking his arm learning to surf—the assistant tore into me with red-faced rage.

Mr. Varley took me straight to the groundskeeping shed, a concrete closet containing all the tools I'd use that summer. He pulled the door open, and a dense odor of cut grass poured out. While people frequently name this smell as a favorite, its appeal does not strike me as inherent. Because smell is the sense most associated with memory, cut grass likely reminds people of childhood summers, but I can't disentangle it from the pungency of small engine exhaust. With a dash of theatricality, as if bestowing some great honor upon me, Mr. Varley handed me a key to the groundskeeping shed. He turned to walk away, but not before nudging me toward a specific task. "First thing you'll want to do," he said over his shoulder, "is sharpen the blade on that push mower. Hasn't been sharpened since last summer." And with that, he was gone.

My parents taught me how to mow the lawn when I was thirteen, which is to say, when I was thirteen, my mother showed me how to start a lawnmower after my father said it was time that I took on more responsibility. I hesitated at first, asking, "Why do we need to cut the grass?" It was a sincere question.

"Because it's getting long," My father answered.

I followed up, "What's wrong about that?"

"It looks bad," he said.

"Why?" Again, it was a sincere question.

"Our grass is the longest on the block!" He said this in a way that let me know that was my last question on the subject.

I mowed the lawn that day, poorly, and I would continue to mow it without enthusiasm or precision for a number of years, treating it with the reverence I felt it deserved. Because no one fussed about the sloppy edges and uncut bits, I quickly figured out that my family mows to avoid reproach, never to set an example. Nevertheless, my parents foisted the chore on me for the same reasons they arranged for my job as lawn boy: because they had been led to believe the responsibility would build character.

Other, more appropriately acculturated lawn boys with sturdier cultural inheritances may have known how to sharpen mower blades, but I did not. Perhaps I should have demanded Mr. Varley stay to show me how, should have stood up for my adolescent right to not know things, but I was too callow, too green to advocate for myself. The training, which was barely an hour in the past, included, I thought I remembered a cursory pantomime of blade sharpening, so I conjured a vague vision of the process and reconstructed the chore from that one poorly observed example.

I squatted on the ground, tipped the mower over, and assessed the situation. I needed a wrench and a file, both of which were on hand, yet how to use the tools eluded me. If I had a cell phone and had it been the future, I could have consulted with the great and all-knowing tradition-bearer YouTube, but it was just me and my memory in that shed. Honestly, I can't recollect how I went about the task. There's a fuzzy image of me trying to hold the rotating blade steady with my foot while sharpening the blade in its place, but I refuse to believe I was that stupid. However it happened, the blade swiped into the back of my right hand, cutting deeply, and the gash, white for a slow moment, flooded suddenly with deep-red blood. I knew immediately I couldn't handle the injury on my own, but I was too ashamed to ask for help. It was my first day, after all, and I had mangled myself in less than an hour.

Life in the suburbs had taught me to keep to myself, that it is better to suffer in silence than rock the boat. It was right there in the idea of a lawn itself, an expansive green space designed to keep people away. The literature on lawns often talks about suburbs being designed to look like endless parks, all those uniform lawns blending together, but anyone who has spent time in the suburbs can tell you there are boundaries everywhere and not just the ones marked by the ubiquitous chain-link fencing we have trained ourselves not to see. Lawns, front lawns in particular, are inviolate spaces. One may tread on another's lawn only accidentally or to avoid danger, and, even then, only momentarily. Lawns are like those living rooms one of your friend's families had, a room that was off-limits, existing only as an idea to be kept clean, abandoned use value at its most indulgent.

I headed to the nearest restroom to tend to my wound. Running cold water over it, I saw how deep I had been cut. Deep. I tried stanching the bleeding with brown paper towels from a dispenser, but their absorbency was laughable, the blood-soaked towels instantly sticking to my skin.

There was no way I could take care of this on my own. I went to Mr. Varley and showed him my hand. His eyes widened, and he exclaimed, "Holy Shit!" I asked him what I should do, and he shrugged. "I don't think there are Band-Aids big enough in the nurse's office, but it doesn't matter because I don't have a key for it anyway." There was a story there—What had he done to be denied a key to the nurse's office?—but I had to stay focused. "Here's what you do," Mr. Varley offered. "Go down to ShopKo and get yourself some Band-Aids. It's a big, bad cut, but it doesn't need a doctor." I liked Mr. Varley, and I don't want to sell him out, but he had no idea what he was talking about. But I knew even less, so I got in the car and headed to ShopKo.

ShopKo was close, less than a mile, so I was there in moments. Mr. Varley had given me some shop rags for the wound, which were an improvement on the worthless paper towels. Wrapped tightly around my balled fist, they kept the bleeding at bay, so I only looked half as damaged as I felt as I pushed my way through the front doors. A large "Health and Beauty" sign beckoned to the left, so I headed in its direction.

The first-aid aisle offered an array of bandages, but I only had a couple of dollars on me—it was, remember, my first day of work ever—so my choices were limited. As much as I needed one, I simply could not afford the Band-Aid brand bandages big enough to cover my sizable wound. A ninety-nine-cent box of child-sized Teenage Mutant Ninja Turtle bandages was in my budget, but all of them combined might not cover the wound and, even if they could, the stress and sweat of the work to come would have them sliding off my hand instantly. I decided to remove a single, large Band-Aid brand bandages from its box and tuck it inside a box of the Ninja Turtle bandages. I then paid for my contraband and exited the store.

Ethics are learned informally, through customary example. I cannot say I ever saw my parents steal anything. They engaged in some ethically questionable behavior—my father cheated on my mother, and my mother will strategically fib to evade blame, but I wouldn't figure out either until I was older—but, for the most part, my parents were decent citizens who modeled appropriate conduct. They may not have taught me how to mow a lawn, but they taught me I *ought* to mow a lawn, and that, ultimately, is the essential lesson in America. Anyway, because I believed myself to be a responsible and ethical person—in my pocket, the key to the groundskeeping shed provided proof of this—and I did not believe what I did was all that wrong, so I felt no relief, no sense of having gotten away with something, as I walked out of the store with my ill-gotten bandage.

Nor did I anticipate the head of ShopKo security's hand coming down on my shoulder about ten steps from my car. "I need you to come with me, son," he said, his hand clamping tightly. He was large, far larger than me, but even a much smaller man would have shocked me into compliance. Dutifully, I walked with him back into ShopKo leaving a little trail of blood drops, my grip on the shop rags having come loose.

Upon reentering the store, he steered me right, away from the show floor and toward the store's administrative offices. It was a frightening turn. Much like the Service Center earlier, this was a space not made for the general public, and again, I struggled to focus in an unfamiliar environment. Mr. X, the head of security, explained my transgression to me. I denied it. He showed me footage of my transgression. I insisted I paid for the bandages. He explained that did not matter, that what I had done still qualified as theft. Looking down at my still bleeding hand, I pondered my next move. For years, I told people that there was dried blood on the confession he made me sign, but I admit here that I do not know if that is true. However, I do remember asking him whether I could use one of the bandages I "stole" or whether they needed to be preserved as evidence. "They're yours now," he said, so I put the big Band-Aid brand bandage on my still bleeding wound. I know that happened.

In any of its many retellings, I have never been able to capture the weirdness of that day. The way one of the police officers who were summoned to issue me a citation laughed maniacally at my predicament, his laughter propelling him into flimsy pirouettes. The way Mr. Varley, not sure how to interpret my story, offered me half his sandwich before rescinding the offer, telling me it would be better if I went home for lunch and stayed there the rest of the day. The way my father, frustrated that I did not make note of the man's name, insisted on taking me back to ShopKo so I could point out the head of security to him. The way the head of security stormed out of his own office when my father unexpectedly pulled out a micro cassette recorder, reciting official sounding legal observations into it. Sometimes I don't know what actually happened and what is dream. But that's what makes it Lynchian. As his biographer, Kristine McKenna, put it, Lynch films are "felt and experienced rather than understood" (McKenna and Lynch 2018, 5).

The experience stuck with me, and not in the wholesome, character-building sense my parents intended when they arranged this job for me. A year later, I was arrested again, this time for being a minor in possession of alcohol, despite the fact that I had no alcohol in my possession or in my system. In fact, the only alcohol I had tasted at that point in my life was from a tumbler of scotch that my grandma had switched with my tumbler of 7UP, thinking it would be funny to see my reaction. But my habitual sobriety was irrelevant. Being in the same car as other people with alcohol was sufficient to charge me with being a minor in possession. When I went to court to plead my case, the judge refused to hear my excuses, demanding instead that I explain "this shoplifting charge" to him.

"It's quite a funny story, your honor," I said. He did not find the story funny. In the end, I was fined $250 and instructed to straighten up and play by the rules before it was too late for me. Incidentally, he didn't give my friend—the one whose beer it was and whom the police had instructed to "drive straight home" even though his blood alcohol level was above the legal limit—a similar warning because that friend wasn't in court. His lawyer had his case transferred to juvenile court where the charges were dismissed. So much for being a man, I guess.

I sometimes wonder if this bandage heist is what separates me from Ragged Dick. Throughout that novel, Alger emphasizes that Dick never stole—not before he committed himself to being respectable and certainly not after. He has the opportunity to steal from a client who overpaid him, but Dick goes out of his way to get the man his change. It only took one day in the real world for me to turn to a life of crime. My fall from grace was immediate. I never stood a chance in 'spectable America. More forgivingly, I learned early that America is an absurd place and its justice corrupt. Please realize I am aware that if I weren't white, the charges never would have been dropped by the district attorney like they were. The cops likely wouldn't have laughed when they arrived; they would have probably arrested me for real, taking me to a station to have my parents pick me up so they, too, could be shamed. But realizing these things doesn't make me grateful. It makes me angrier about the absurdity because people's lives have been ruined, their lives corrupted by stupid little incidents, like stealing a bandage.

I told the story of my first day as a professional lawn boy a few years ago to a bunch of students as we were making our way home from a place-based, service-learning excursion to Bancroft, Nebraska, home of John G. Neihardt. I arrange such trips from time to time. On them, students read a work connected to a place in Nebraska—in this case, they read *Black Elk Speaks* (1932)—and then discuss the work in that place while serving the local community. For this excursion, the Neihardt Foundation, not having any pressing needs, arranged for us to conduct our service on the nearby Winnebago Reservation.

The address the Neihardt Foundation gave me led to the trailer-based headquarters of a local missionary group, which was less than ideal. I mentally rehearsed explanations for a quick departure as the pastor, a wide-smiling white fellow in cutoffs and a freebie T-shirt, approached our van when we pulled in. Perhaps seeing the discomfort on my face, he immediately explained that we were not going to be doing any missionary work, that we were going to be helping elder residents of the reservation with yard work. This was good news. I had, in fact, been counting on yard work all along. After all, on every previous trip, our service had been doing yard work for literary sites, like the Pavelka farmstead outside Red Cloud or Wright Morris's childhood home in Central City. Yard work makes for good pictures, certainly better ones than pictures of paperwork.

We started our service at a ranch home on the edge of the reservation. It was a large property owned by a retired stand-up comedian who kept the students laughing throughout. The work was hard but rewarding. It was the highlight of the day, and I wish we spent all our time there, but there were other homes on our list. Curiously, as we moved through the list of addresses that I was given, the residents, whose expressions of appreciation became increasingly tepid, grew younger and younger, until they were almost as young as the students themselves. Meanwhile, the grass was getting taller and taller, a few of the lawns having gone to seed. At what would turn out to be the last house we served, a twenty-something fellow walked out of the house, a basketball under his arm. "Thanks, dudes," he muttered before getting in his car and driving away.

A student who had been pulling weeds walked over to me and said in a hushed tone, "This is kind of fucked up. Did they know we were coming?"

"I do not think they knew we were coming." I checked my phone, and it was late enough to call it a day. "Let's finish here, and head home. We'll talk about it then."

However unwittingly, we were sent to the reservation that day to enforce mainstream American lawn customs. Apparently, the peer pressure that keeps my neighborhoods' lawns closely cropped wasn't working on the reservation, so they—and I'm not entirely sure who all to include in "they"—sought to reinforce the importance of good lawn care through some gentle colonialism. When we first pulled up to the church, I was worried we were going to be asked to do some missionary work, and that's just what we ended up doing. We shared with the residents of the Winnebago Reservation the gospel of a fresh cut lawn.

But I was also extraordinarily proud of the students for sniffing out the vile ideology undergirding our charity work that day. On the drive back to Omaha, they expressed outrage about the situation, which eventually turned into laughter as they worked their way through the absurdity of it all. We realized we were drawn into a five-hundred-year-old project of terraforming North America so that it better reflects the dominant culture's values. And it was then that I told them about my experience being arrested for stealing a single bandage on my first day of work as a lawn boy, which they thought was utterly hilarious. And then they sang songs from *Frozen* (2013) the rest of the way home.

References

Alger, Horatio. [1868] 2014. *Ragged Dick: Or, Street Life in New York with the Boot Blacks.* New York: Signet.

Anderson, E. N. 1972. "On the Folk Art of Landscaping." *Western Folklore* 31, no. 3: 179–88.

Baker, Sarah. 2015. "My Township Calls My Lawn 'a Nuisance.' But I Still Refuse to Mow It." *Washington Post*, August 3, 2015. https://www.washingtonpost.com/posteverything/wp/2015/08/03/my-town-calls-my-lawn-a-nuisance-but-i-still-refuse-to-mow-it/.

Dundes, Alan. 1971. "Folk Ideas as Units of Worldview." *Journal of American Folklore* 84, no. 331: 93–103.

Jenkins, Virginia Scott. 1994. *The Lawn: A History of an American Obsession.* Washington DC: Smithsonian Books.

Lynch, David. n.d. *David Lynch Theater.* YouTube channel. Accessed September 5, 2022. www.youtube.com/c/davidlynchtheater.

Lynch, David, director. 1986. *Blue Velvet.* Los Angeles, CA: De Laurentiis Entertainment Group. DVD.

Lynch, David, director. 1999. *The Straight Story.* Los Angeles, CA: Asymmetrical Productions. DVD.

McKenna, Kristina, and David Lynch. 2018. *Room to Dream.* New York: Random House.

Milesi, Christina, Steven W. Running, Christopher D. Elvidge, John B. Dietz, Benjamin T. Tuttle, and Ramakrishna R. Nemani. 2005. "Mapping and Modeling the Biogeochemical Cycling of Turf Grasses in the United States." *Environmental Management* 36, no. 3: 426–38.

Moyer, Justin Wm. 2017. "Trump Lets an 11-Year-Old Boy Mow the White House Lawn." *Washington Post*, September 15, 2017. https://www.washingtonpost.com/news/local/wp/2017/09/15/trump-lets-an-11-year-old-boy-mow-the-white-house-lawn/.

Pollan, Michael. 1989. "Why Mow? The Case Against Lawns." *New York Times*, May 28, 1989, section 6, 22.

Robbins, Paul. 2007. *Lawn People: How Grasses, Weeds, and Chemicals Make Us Who We Are.* Philadelphia: Temple University Press.

Chapter 15

CANNING FOR THE APOCALYPSE

Climate Change, Zombies, and the
Early Twenty-First-Century Canning Renaissance

Claire Schmidt

Weather is a decisive factor in food availability and consumption. In October 2019, the Pew Research Center found that 62 percent of Americans said climate change was affecting their local community "a great deal" or "some," whether from long periods of heat, erratic and severe storms, fires, or floods (Funk and Kennedy 2020, n.p.). Climate change already affects the way major crops are harvested and distributed (Lockwood 2016, 80). It is clear that American foodways are and will continue to be tied to weather and climate. One expression of that relationship is the twenty-first-century canning renaissance embraced by new millennial and Gen Z canners and continued by older practitioners. I am one of those. When new acquaintants ask why I spend so much of my time canning when I don't "need" to, I never say "because I love it" or "because my mother and grandmother and great-grandmother and great-great grandmother ad nauseum canned" or "because June means strawberry jam and August means tomato sauce" or "because I want to be as self-sufficient as possible" or "because I am compulsive and anxious and canning brings me peace." These are some of the reasons why I can. But I usually say something like: "'cause I want to survive the zombie apocalypse. Look, I have two chest freezers!" I'm not particularly interested in zombie books, films, or TV shows, and I'm not really worried about zombies, so why is this how I and others answer questions about canning work? This chapter explores the role changing weather patterns play in how twenty-first-century Americans talk about, think about, and practice home canning. I suggest that canning as a folk response to climate change is fear management through in-home means that don't resist capitalism but do allow the idea of populist resistance to capitalism and that canning for the zombie apocalypse is an aesthetic and communicative act that exerts local and imaginative control over a future that, largely because of the weather, remains out of our hands.

While weather is, colloquially, safe conversational territory, climate change is politicized. That politicization calls upon shared ideas of the end of the world, deserved or not. As Erik Swyngedouw (2018, 216) observes, images of the apocalypse accompany and fuel public debate about climate change. Talking about the weather now means talking about fear. Swyngedouw (2018, 216) points out, "our ecological predicament is sutured by millennial fears, sustained by an apocalyptic rhetoric and representational tactics, and by a series of performative gestures signaling an overwhelming, mind-boggling danger, one that threatens to undermine the very coordinates of our everyday lives and routines, and may shake up the foundations of all we took and take for granted." While Swyngedouw is concerned with Jacques Derrida's 1980s vision of a nuclear apocalypse without redemption, I find that twenty-first-century canning practices are firmly about a belief of survival.

Canning, Weather, and Time

Weather pressures have always dictated traditional foodways. Canning falls into the category of food preservation alongside other techniques like drying, fermenting, or freezing. While home fermentation is on the rise thanks in part to the now-established home-brewing movement and the more recent fermented-foods boom personified in print by Sandor Katz (2012) and more recently online by Brad Leone (2016–2022), this project focuses exclusively on canning that produces a shelf-stable, hermetically sealed product in a glass jar with a ring-and-lid cap rather than the perhaps more traditional and often more nutritious fermenting traditions that include kimchi, sauerkraut, kefir, or sour pickles.

As a technique and process supported by virtually every supermarket and chain home furnishing store in the United States, canning is still presented as the purview of an archetypal past. Canning publications nearly always associate canning knowledge and experience with roots in a bygone age. A 1992 *Mother Earth News* article cheerfully encourages budding canners to look for old jars and an old mentor at the same time: "A good way to collect low-cost jars is to put a few ads up here and there. You'll be surprised at the response. Not only will you find jars, but sometimes you can find a good canner. If you're lucky, you may find an 'old-time' canner who is willing to share his knowledge and experience with you" (Clay 1992, n.p.). While this suggests visions of grizzled old prospectors, the grandmother remains the most frequently referenced link to handed-down knowledge. The Amazon (n.d.a, n.p.) description of *Canning and Preserving for Beginners* (2013) states: "Canning and preserving is a great solution to always having local, seasonal, and budget-friendly food in your kitchen. But unless you learned how to do it in your grandmother's kitchen, it can be difficult—not to mention dangerous—to know where to start." This imagined grandmother is also called into service by America's Test Kitchen: "The art of preserving produce by canning and preserving has come full circle from grandmother's kitchen to a whole new generation now eager to learn it" (Amazon, n.d.b, n.p.). In 1977, Charles Taylor (1977, 26) of *The Saturday Evening Post* announced, "You're not with it these days unless you know how to out-do your own grandmother in home canning and preserving." Even Michelle Shocked (1992, n.p.) sings, "Saturday morning found me itching / To get on over to my grandma's kitchen / Where the sweetest little berries was cooking up right / And then we'd put them in a canning jar and seal them up tight."

Books and blogs promise to free new canners of the need for a matriarch (see, e.g., Rebecca Lindamood's [2016] *Not Your Mama's Canning Book*). The children and teenagers who grew up in the shadow of *An Inconvenient Truth* (2006) became part of a new canning movement. Cooking books, television shows, blogs, and vlogs featuring canning and DIY food preservation proliferated between 2008 and 2016. Michelle Obama planted a garden and launched historic improvement to American school meals. Young adults were buying and using canning supplies and books in record numbers in the first two decades of the twenty-first century (Campoy 2009). These products have been marketed aggressively to young Americans (see, e.g., Coyne and Knutzen 2010; Krissoff 2016; Lindamood 2016; McClellan 2012a; McClellan 2012b; Meredith 2014). As advertising researcher Rupal Parekh (2013, 1) notes, "The recession fueled a resurgence in home canning and DIY projects, while Americans' focus on homemade and artisan foods made with fresh ingredients has been a boon for Ball. And as a heritage brand, it's riding the throwback trend—when not used for actual canning, the jars serve as centerpieces at weddings or as glasses at comfort-food restaurants. Peruse Pinterest and the fandom is evident." As Danille Elise Christensen (2018, 114) observes, canning is tied to expanded and functional resource networks for seniors, and the same seems to be true for young adult canners both past and present (McClellan 2012a; McClellan 2012b; Parsons 1915, 608; Roger 2014). The juxtaposition of the family, the community, the corporate, and the state all combined to launch a new generation of canners whose priorities included taking action to slow climate change. However, they cannot unhitch canning from the experience of time itself.

The discourse surrounding canning invokes the past and the future because the practice of canning brings together both the past and the future in a particular space. One season determines what can be eaten in the next season. Take strawberry jam, perhaps the most celebrated home-canned product of the twentieth century. A deep freeze in January can kill plants (Darrow 1966, 323). A warm and wet March can encourage plants to leaf out early; the more leaves, the greater the number of flower clusters and thus fruit (Darrow 1966, 327). A late frost can kill flowers. A lack of pollinators results in nubbins or strangely shaped berries (Darrow 1966, 340). Drought in August can stress established plants, and because the next spring's buds are set in the fall, this results in fewer berries the next year. Too much rain and the berries are insipid and bland and can rot on the ground. A home canner who grows their own fruit ties the past year's dormant and growing seasons to the present year's jam making and then plans for the future.

Part of the purpose of strawberry jam is time travel. While the pastel pink foam (skimmed from the surface of the strawberry jam after the final boil) should be eaten within hours or days of cooking, the strawberry jam in the jars is eaten long after the strawberry season has ended. One eats fresh strawberries in June; one saves the experience of strawberry jam for a time of absence—when fresh, local strawberries are unavailable. The jam is "put up" or "put by"; it is out of reach, a kind of liminal storage, until one needs to experience the taste of June again. A dedicated canner's pantry may contain jars from multiple years of strawberry jam (discouraged in theory but common in practice) and comparing dates on jar lids calls attention to shifting or unpredictable seasons. Discovering that this year's strawberries ripened three weeks earlier than those two seasons ago immediately invokes thoughts of climate change and unfamiliar weather behavior.

Canning narratives also invoke a dangerous future as a reason for canning. In her introduction to home canning written for *Mother Earth News*, Jackie Clay (1992, n.p.) writes: "We keep a huge stock of canned goods to safeguard against hard times. And it's lucky we do. This past winter, a neighbor was laid off from his job with no warning. His family shared our bounty, and we were thrilled to give it. This year, his wife wants to begin canning. A 'lost' art is being revived." Like Clay, Melissa Click and Ronit Ridberg's (2010, 315) participants emphasize the sense of comfort that having a stock of home-canned food put by gave and emphasize job loss or loss of food systems we take for granted as threats that this stock defends against. That comfort extends to the absence of refrigeration or freezers; as one canner notes, "I can survive when the electric's out" (quoted in Christensen 2018, 111). Recent food-preservation initiatives in British Columbia by the First Nations Health Authority focuses on peer-to-peer canning education to ensure continuation of traditional foodways and food security; one result of the initiative has been the inclusion of food preservation and food security into tribal emergency-response planning, as many First Nations have been disproportionately affected by climate-change impacts (Yung and Neathway 2019, 52).

Canning, Climate Change, and the Post-Apocalyptic Imagination

Just as canning books, YouTube channels, blogs, and products surged in the post-2008 recession, our narratives of collapse and survival have fixed on the zombie apocalypse as a way of talking about our fears, whether of climate change or other mass disaster. While the popular-culture zombie phenomenon of the early twenty-first century has been interpreted as a response to global terrorism, xenophobia, and alienated modernity, it can also be understood as a response to climate change. Michael E. Webber (2013, 32) notes, "The fictional zombie invasion in World War Z can be interpreted as a symbolic stand-in for real-life climate change." *An Inconvenient Truth* (both book—see Gore 2006—and film—see Guggenheim 2006) was released in 2006 and its impact on millennials has been great (Funk and Hefferon 2019). With cannibals but without proper zombies, Octavia Butler's ([1993] 2010) novel *The Parable of the Sower* firmly establishes literary apocalypse as a weather-based, climate-change force.

Changing cultural concerns, like climate change, causes the figure of the on-screen zombie to shift over time in response (Scott 2020). Penny Crofts and Anthea Vogl (2019, 29) connect early twenty-first-century pop-culture zombies to "transgression of borders, as well as the failure of containment." Crofts and Vogl are concerned with the dehumanization of refugees expressed through a fictional zombie apocalypse, and any discussion of twenty-first-century refugees must consider the increasing role weather and climate play in movement of people. As the United Nations Advisory Group on Climate Change warns, an estimated two billion people could be displaced from their homes by the end of this century due to rising ocean levels; in recent years, entire island populations have been evacuated due to severe weather events, like Hurricane Irma and Cyclone Pam (United Nations Refugee Agency n.d.). As Shea Flanagan (2019, 2) observes, hundreds of thousands of the poorest humans on the globe are expected to become "climate refugees" by 2050. The question of where those refugees will find welcome remains open. These anxieties about permeable borders and insufficient containment can be extended to the impulse to can. Canning is, of course, a process of containment, and while people, the undead, and COVID-19 germs cannot be contained, we derive solace and reassurance from our visibly and audibly sealed jars. Colson Whitehead argues that fear of the zombie is fear of the future self (Fassler 2011). We expect that future self to change as the weather and the climate changes. I suggest that home canning in response to that fear of the future self is an aesthetic expression of hope for the future self.

What we become depends on the weather. In my lifetime, the zucchini has ceased to function as a viable symbol of Midwestern agrarian identity. When I was a child, my parents' garden was plagued by the overnight appearance of massive overgrown zucchini. If you don't pick that little tender squash today, by tomorrow it'll be the size of a fat baseball bat. The only possible use for this vegetable is to club your sister with it while playing caveman or wrap it in a baby blanket, carve a face on it, and play dolls. My mother's recipe files from the 1970s contain dozens of zucchini-based recipes to use up the harvest from ubiquitous "tomato-zucchini combo" to chocolate zucchini cake to canned ersatz hot-dog relish made from zucchini instead of cucumber. My father-in-law repeatedly tells an apocryphal Garrison Keillor-esque narrative of the need to lock car doors at church during zucchini season lest neighbors gift you with unwanted grocery sacks of those green logs. We were the people of the zucchini.

I am no longer the Midwesterner for whom an abundance of monstrous zucchini is a shared identity; instead, zucchini (like the weather) has become a divisive topic. Warmer winters mean rising overwintered populations of squash borers (*Melittia cucurbitae*) and squash bugs (*Anasa tristis*), both of which suck the life out of my squash plants even before the blossoms are pollinated. I find it almost impossible now to grow zucchini without pesticides, and so whether I choose to spray my plants becomes a much more valent feature of division and identification into folk groups. My father-in-law insistently repeats warnings about stealth squash deliveries despite the fact that this narrative has no basis in our present-day lived experience. His prediction has become inapplicable. We no longer have zucchini coming out of our ears, and thus, we no longer have that identity of a surplus of life and energy from the earth without a liberal dousing of poison.

Canning gives at least an illusion that a person can learn and practice skills that kept ancestors alive and in good health in a dangerous past and can again do so in a frightening future. P. Suzanne Pennington's (2013, 163, 180) unpublished 2013 dissertation clearly articulates the relationship between home canning as "post-apocalyptic skill set" and climate-related events (she terms it "weathering the storm"), and she grounds the practice in a contemporary assumption of environmental and ecological responsibility. Canning becomes a good and logical step in the face of an environmental apocalypse. Unlike freezing raw or blanched food, which is often felt to be easier, faster, and fresher than pressure or water-bath canning, sealed jars don't require electricity, thus reducing energy use. The jars can be reused year after year and create less waste. They can withstand significant weather events and require little equipment and technology. Canning is historically a response to excess and abundance in place, so the foods a person grew up canning are likely to be foods that continue to be viable even in a changing climate.

Canning is also part of a system of exchange. As Christensen (2015; 2016; 2018), Elizabeth L. Andress and Susan F. Clark (2015), Click and Ridberg (2010), and others observe, canners have skills and goods to sell and trade. If an apocalypse leads to local currency collapse, strawberry jam will continue to command value. Those who have enough to spare have power to help, cheer, or manipulate. How and what we eat greatly affects the health of the planet (O'Hara 2011, 25; Reinhardt 2020, 1–2). Nearly any twenty-first-century canning cookbook or blog touts the environmental benefits of eating locally, eating plants, and reducing food and energy waste. That canning reduces waste is not news; canning clubs in the early twentieth century were pressed into service as a solution to waste (United States Bureau of Labor Statistics 1917). My grandmother still hoards canning jars, though she has quit canning because jar shortages during the 1970s energy crisis were so disruptive (Taylor 1977, 26). While scholars, like Michael Newbury, point out the perceived naiveté and privilege of nouvelle sustainable eating espoused by Barbara Kingsolver or Alice Waters, food activists, like Dick Gregory, have been advocating sustainable plant-based diets for many decades (McQuirter 2010, xxii). To homogenize and oversimplify participation in food-based environmental activism reifies existing power structures. Christensen (2016, 43–44) explores the complexity of political positions held and performed within the broad DIY community online and finds that individuals take everyday action for environmental sustainability through DIY for many reasons and in many ways.

Twenty-First-Century Canning Artistic Communication

It is clear that canning is both artistic and communicative in addition to functional and sustainable. Pennington (2013) pays careful attention in her study to the aesthetics of canning—the textures, sounds (particularly the auditory element of the pop that comes from a sealing jar), smells, and visual appearance of the jars. The latter is clearly evident throughout commercial canning books, canning blogs, and canning social media posts. Between putting food into jars and taking it out again, the necessary step of photographing and sharing the evidence of this productive work is a key part of canning performance in the early twenty-first century. Christensen questions the accusations of hedonism and "rococo" in twenty-first-century canning displays (Dickerson 2010), arguing that American canners have always emphasized visual display, whether at public state-fair exhibitions or in-home storage and usage (Christensen 2015, 33–34; Winship 1918, 32). Participants in Click and Ridberg's (2010, 314–15) study similarly emphasize the sensory aesthetics that go far beyond the consumption of the food, specifically visual display and the auditory signal of the *pop*. The attention paid to beauty in the jar, on the shelf, or on the plate clearly signals that food is worth a person's time and personal labor. Aesthetic display, whether from food coloring of sweet pickles or crab apples, carefully arranged pantries, or color-coordinated plating demands a response. I cannot deny that I feel compelled to justify and defend my choice to spend hours in the kitchen, and the aesthetics of canning is one dimension of my defense. Another dimension is, of course, sustainability and reducing the carbon footprint of the food I buy.

The aesthetics of canning thus become the aesthetics of weather-disaster preparedness. Most academic studies of disaster aesthetics focus on aesthetic opportunities *after* the disaster and art as a response to things that have already happened (Jimenez 2018; Nygaard-Christensen 2011; Rivera-Santana 2020; Thorson 2012). These same opportunities for interpretation, advocacy, and self-expression are being seized in advance of a disaster. Preparation allows for artistic communication in advance, during, *and* after a disastrous weather event. It seems clear that individual weather-disaster preparation comes with a performative visual and narrative aesthetic dimension, clearly apparent on social media platforms through memes, gifs, images, and stories. In the canning world in particular, this preparation aesthetic is particularly apparent in recent wider choices in shape, size, color, and embellishment of jars and lids (see, e.g., the limited "collector's edition vintage" aqua-blue Ball Perfect jar). Not only do we believe we will survive, but we also believe we will survive in a style that we have designed and performed in advance.

Home canning communicates three clear messages regarding impending global catastrophe, whether due to zombies or climate change. First, a warning: prepare or die. When someone visits my basement and comments on the hundreds of full and empty jars, I say, "Yeah, I'm just waiting for the zombies," and my visitor usually says, "Wow, I'll come over to your house when they arrive, then." I've essentially told my visitor that I am canning in fear for my life, and they should fear for their own too. My suggestion of zombie apocalypse says, "Hey, you should consider getting prepared." This warning acts both to make preparation logical and necessary but also can be understood as "you too can reduce your carbon footprint and eat locally and sustainably, and if we don't do this kind of thing as a world, we're going to be in terrible, terrible trouble." Canning is the foodways version of the proverb "plan for the worst, hope for the best." Sharing, serving, displaying, or talking about home-canned food demonstrates that it is both possible and wise to buy local, seasonal food and take responsibility for one's own nourishment, despite the natural and institutional systems to which we must bow or break our hearts and minds fighting.

Thus, the second message is hope: I intend to be here to eat this food. To spend the money, effort, storage space, and time is a bargain with one's future self or, at least, future community. The intent to be there to eat food in the next season or next year put by today is the intent to be one of those in the paradise/wasteland of "after the end," as James Berger (1999, xii) puts it. It also means continuing the knowledge and skill of food production and self-sufficiency out on the far side of disaster. Just as Alexandria grows and cans food in Season 7 of the zombie show *The Walking Dead*, the canner communicates the intention to inhabit a future world, kettle and funnel in hand.

The third message is that we may better bear the existential crisis of climate change with its increasing extinction rates, loss of human life, and uncertainty if we treat it as a game, allowing players to win at life by canning, gardening, or reducing waste as much as possible. We might understand this through Roberto Benigni's *La Vita e Bella*, the 1997 tragicomedy set in World War II concentration camps. To turn tragedy into a game is not to deny the tragedy but to manage emotional and actional responses to that tragedy. Scholars have explored the role gamification can play in disaster management and emergency-response planning (Kankanamge et al. 2020). Video-game franchises like *The Last of Us*, *Resident Evil*, *State of Decay*, and *Dead Rising* allow players to enact an active, imaginatively violent response to a zombie apocalypse, while canning allows participants to enact a real-life preparation scenario. Many Americans (including myself) treated the 2020 COVID-19 lockdowns and food (and toilet paper) shortages as problem solving games. What can I cook with these five disparate ingredients? How long can I go without leaving my house? These were and are competitive games that we play against ourselves and others. The end result is that we follow instructions and mitigate disease spread, but we do it under our own terms and make our own game rules. I'm not worried about climate change—I'm playing a zombie game in my head!

The COVID-19 pandemic brought reassurance to many home canners that their work was needed. The spiced pears and canned peaches I made last summer sat uneaten all winter, but we ate them all in March and April because they helped us limit our grocery trips while keeping fruit on the table. I cannot deny the comfort and solace I took in having prepared a significant stock of food last summer during the weeks this spring when grocery shelves remained empty and when bags of King Arthur flour resold online for ten times the original price. And I also cannot deny my self-righteous satisfaction nor my rants about those people who suddenly discovered baking, gardening, or canning and bought up the supplies I myself took for granted. When lids and rings disappeared from grocery store shelves in August and September and were never restocked, I became incensed. I had been planting, baking, and preserving since childhood! I always baked my own bread! I always kept a garden! Surely, I had ownership of these sustainable food traditions and the moral high ground to look down on these Johnny-come-latelies. Scorning those who do not "put by" or plan ahead is not charming behavior but is undeniably quite common on blogs and in conversations among canners and preppers.

However, I don't think home canners really believe that their basements and pantries full of jars are going to protect them, whether from zombies or climate change. Few Americans are willing to completely divorce themselves from participation in capitalist modernity. Planting a garden and canning will unquestionably make an economic and environmental impact, but they will not topple corporate food giants, nor will they dismantle the complex systems that put food on grocery store shelves. Weather can and does disrupt food systems of processing, packing, and shipping, and canning gives us a (largely unrealistic) sense that we have the power to beat the weather. As Newbury (2012, 91) points out, the zombie-apocalypse vision yearns for a reintegration of humans in the environment but "zombie films insist in their imagery of apocalypse on the problematic provisionality of any such reference to the 'natural,' offering instead a world and food that are always and inescapably made by cultural and economic power." Twenty-first-century canners choose to continue family- or community-canning traditions; they find pleasure in the physical acts of canning food; they find both nourishment and cultural capitol in their jars. This work may be done individually, but it is never done in a vacuum. Michael Owen Jones (2007, 151) makes it clear that individual food choices are inextricably linked to tradition, group membership, and aspiration. While there are certainly vacuous, pretentious, and absurd home canners, being annoying on Pinterest is hardly the most important limiting factor of the quantifiable environmental, economic, social, or health impacts of home canning.

While a zombie-apocalypse scenario may be understood as a stand-in for the weather disasters of climate change, the question of who survives and why will not be determined by who has the best stocked pantry or the most productive backyard garden and chicken coop. In the United States, we may wait for our world to collapse, but for many around the world, that day has already come and gone. On April 27, 2020, United Nations (2020, n.p.) Secretary General António Guterres tweeted, "#COVID19 has exposed the fragility of our societies to global shocks, such as disease or the climate crisis." People are already dying from climate change not because they are improvident but rather because resources and risks are increasingly unevenly divided both in the United States and around the world. This is becoming clearer in light of the shared but highly unequal global experience of COVID-19, which "is revealing deep societal inequities and also demonstrating the interconnectedness of health, climate and sustainability issues" (Montez 2020, n.p.). While *World War Z* (2013) and *Parable of the Sower* both show fictional populations fleeing the global south for a safer, more resource-secure north, real-world climate refugees are as real today as they were during Woody Guthrie's weather-driven dustbowl. As Butler ([1993] 2010) clearly demonstrates in *The Parable of the Sower*, individual actions to educate oneself and prepare for weather disaster can be constrained and negated by larger structures of power, injustice, and control. But, as she also makes clear, self-sufficiency and existential self-determination are essential skills in adaptation and survival. Canning, like the zombie, is a folk response to the fear of climate change. We have to be able to imagine a world worth living in before we can fight for it. When we can food, we imagine ourselves and our loved ones alive and well fed on the other side. Knowledge leads to skill, skill leads to agency, and agency leads to hope.

Acknowledgements

Many thanks to the Missouri Valley College Murrell Memorial Library staff, Janet Gilmore, Lisa Higgins, Debbie Olson, and Daniel Ripley for their invaluable assistance obtaining publications during the spring 2020 COVID-19 pandemic and again to Daniel Ripley as well as Willow Mullins and Laurel Schmidt for their insight on our shared enthusiasm for food preservation and end-of-the-world-as-we-know-it conversations.

References

Amazon. n.d.a. "Canning and Preserving for Beginners: The Essential Canning Recipes and Canning Supplies Guide. Paperback—June 16, 2013." Accessed September 22, 2022. https://www.amazon.com/Canning-Preserving-Beginners-Essential-Supplies/dp/162315183X.

Amazon. n.d.b. "Foolproof Preserving: A Guide to Small Batch Jams, Jellies, Pickles, Condiments & More. Paperback—Illustrated, April 26, 2016." Accessed September 22, 2022. https://www.amazon.com/Foolproof-Preserving-Jellies-Pickles-Condiments/dp/1940352517/ref=sr_1_1?crid=1R9PMO1DB33LT&keywords=Foolproof+Preserving%3A+A+Guide+to+Small+Batch+Jams%2C+Jellies%2C+Pickles%2C+Condiments+and+More&qid=1663884748&s=books&sprefix=foolproof+preserving+a+guide+to+small+batch+jams%2C+jellies%2C+pickles%2C+condiments+and+more%2Cstripbooks%2C122&sr=1-1.

America's Test Kitchen. 2016. *Foolproof Preserving: A Guide to Small Batch Jams, Jellies, Pickles, Condiments and More.* Boston, MA: America's Test Kitchen.

Andress, Elizabeth L., and Susan F. Clark. 2015. "Our Own Food: From Canning Clubs to Community Gardens." *Remaking Home Economics: Resourcefulness and Innovation in Changing Times.* Athens: University of Georgia Press.

Berger, James. 1999. *After the End: Representations of Post-Apocalypse.* Minneapolis: University of Minnesota Press.

Butler, Octavia. [1993] 2010. *The Parable of the Sower.* Updated ed. New York: Grand Central Publishing.

Campoy, Ana. 2009. "Putting Up Produce: Yes, You Can." *Wall Street Journal*, October 15, 2009. https://www.wsj.com/articles/SB10001424052748703787204574449160079437536.

Christensen, Danille Elise. 2015. "Simply Necessity? Agency and Aesthetics in Southern Home Canning." *Southern Cultures* 31, no. 1: 15–42.

Christensen, Danille Elise. 2016. "Provident Living: Ethnography, Material Culture, and the Performance of Mormonism in Everyday Life." *Mormon Studies Review* 3, no.1: 37–52.

Christensen, Danille Elise. 2018. "Still Working: Performing Productivity through Gardening and Home Canning." In *The Expressive Lives of Elders: Folklore, Art, and Aging,* edited by Jon Kay, 106–37. Bloomington: Indiana University Press.

Clay, Jackie. 1992. "Home Canning for Beginners: How to Can Your Food Year-Round." *Mother Earth News,* August 1, 1992. https://www.motherearthnews.com/real-food/home-canning-for-beginners-zmaz92aszshe.

Click, Melissa, and Ronit Ridberg. 2010. "Saving Food: Food Preservation as Alternative Food Activism." *Environmental Communication* 4, no. 3: 301–17.

Coyne, Kelly, and Erik Knutzen. 2010. *Making It: Radical Home Ec for a Post-Consumer World.* New York: Rodale.

Crofts, Penny, and Anthea Vogl. 2019. "Dehumanized and Demonized Refugees, Zombies and World War Z." *Law & Humanities* 13, no. 1 (June): 29–51.

Darrow, George M. 1966. *The Strawberry: History, Breeding, and Physiology.* New York: Holt, Rinehart, and Winston.

Dickerson, Sara. 2010. "Can It: At-Home Preserving Is Ridiculously Trendy." *Slate,* March 10, 2010. http://www.slate.com/id/2246148/.

Fassler, Joe. 2011. "Colson Whitehead on Zombies, 'Zone One,' and His Love of the VCR." *The Atlantic,* October 18, 2011. https://www.theatlantic.com/entertainment/archive/2011/10/colson-whitehead-on-zombies-zone-one-and-his-love-of-the-vcr/246855/.

Flanagan, Shea. 2019. "'Give Me Your Tired, Your Poor, Your Huddled Masses': The Case to Reform U.S. Asylum Law to Protect Climate Change Refugees." *DePaul Journal for Social Justice* 13, no. 1: 1–33.

Funk, Cary, and Meg Hefferon. 2019. "Millennial and Gen Z Republicans Stand Out from Their Elders on Climate and Energy Issues." Pew Research Center, November 25, 2019. https://www.pewresearch.org/fact-tank/2019/11/25/younger-republicans-differ-with-older-party-members-on-climate-change-and-energy-issues/.

Funk, Cary, and Brian Kennedy. 2020. "How Americans See Climate Change and the Environment in Seven Charts." Pew Research Center, April 29, 2020. https://www.pewresearch.org/fact-tank/2020/04/21/how-americans-see-climate-change-and-the-environment-in-7-charts/.

Gore, Al. 2006. *An Inconvenient Truth: The Planetary Emergency of Global Warming and What We Can Do About It.* New York: Rodale.

Guggenheim, Davis, director. 2006. *An Inconvenient Truth.* Beverly Hills, CA: Lawrence Bender Productions. DVD.

Jimenez, Chris D. 2018. "Nuclear Disaster and Global Aesthetics in Gerald Vizenor's *Hiroshima Bugi: Atomu 57* and Ruth Ozeki's *A Tale for the Time Being*." *Comparative Literature Studies* 55, no. 2: 262–84.

Jones, Michael Owen. 2007. "Food Choice, Symbolism, and Identity: Bread-and-Butter Issues for Folkloristics and Nutrition Studies." *Journal of American Folklore* 120, no. 476: 129–77.

Kankanamge, Nayomi, Tan Yigitkanlar, Ashantha Goonetilleke, and Md. Kamruzzaman. 2020. "How Can Gamification Be Incorporated into Disaster Emergency Planning? A Systematic Review of the Literature." *International Journal of Disaster Resilience in the Built Environment* 11, no. 4: 481–506. https://eprints.qut.edu.au/199587/1/QUT_e_prints.pdf.

Katz, Sandor. 2012. *The Art of Fermentation*. White River Junction, VT: Chelsea Green Publishing.

Krissoff, Liana. 2016. *Canning for a New Generation: Bold, Fresh Flavors for the Modern Pantry*. New York: Harry N. Abrams.

Leone, Brad. 2016–2022. *It's Alive! with Brad Leone*. October 21, 2016–April 4, 2022. Bon Appetit video series. https://www.bonappetit.com/video/series/it-s-alive-with-brad.

Lindamood, Rebecca. 2016. *Not Your Mama's Canning Book: Modern Canned Goods and What to Make with Them*. Salem, MA: Page Street.

Lockwood, Allan H. 2016. *Heat Advisory: Protecting Health on a Warming Planet*. Cambridge, MA: MIT Press.

McClellan, Marisa. 2012a. "A Beginner's Guide to Canning." Serious Eats, February 29, 2012. https://www.seriouseats.com/2012/02/how-to-can-canning-pickling-preserving-ball-jars-materials-siphoning-recipes.html.

McClellan, Marisa. 2012b. *Food in Jars: Preserving Small Batches Year-Round*. Philadelphia, PA: Running Press.

McQuirter, Tracye. 2010. *By Any Greens Necessary*. Chicago: Lawrence Hill Books.

Meredith, Leda. 2014. *Preserve Everything: Can, Culture, Pickle, Freeze, Ferment, Dehydrate, Salt, Smoke, and Store Fruits, Vegetables, Meat, Milk, and More*. Woodstock, VT: Countryman Press.

Montez, Roqua. 2020. "Climate Change and Covid-19: Can This Crisis Shift the Paradigm?" Berkeley News, April 27, 2020. https://news.berkeley.edu/2020/04/27/climate-change-and-covid-19-can-this-crisis-shift-the-paradigm/.

Newbury, Michael. 2012. "Fast Zombie/Slow Zombie: Food Writing, Horror Movies and Agribusiness Apocalypse." *American Literary History* 28, no. 1: 87–114.

Nygaard-Christensen, Maj. 2011. "Building from Scratch: The Aesthetics of Post-Disaster Reconstruction." *Anthropology Today* 27, no. 6: 8–10.

O'Hara, Jeffrey K. 2011. *Market Forces: Creating Jobs through Public Investment in Local and Regional Food Systems*. Cambridge, MA: Union of Concerned Scientists.

Parekh, Rupal. 2013. "Ball Gets Better with Age, Thanks to Pinterest, DIY." *Advertising Age* 84, no. 15 (April): 1.

Parsons, Eloise. 1915. "How I Made My Crop." *Journal of Education* 81, no. 22: 608.

Pennington, P. Suzanne. 2013. "Making Sense of Mason Jars: A Qualitative Exploration of Contemporary Home Canning." PhD diss. Albany: SUNY. ProQuest.

Reinhardt, Sarah. 2020. "In Support of Sustainable Eating: Why the United States Dietary Guidelines Should Prioritize Healthy People and a Healthy Planet." Cambridge, MA: Union of Concerned Scientists.

Rivera-Santana, Carlos. 2020. "Aesthetics of Disaster as Decolonial Aesthetics: Making Sense of the Effects of Hurricane María through Puerto Rican Contemporary Art." *Cultural Studies* 34, no. 3: 341–62.

Roger, Ellen. 2014. "What Will I Eat During the Zombie Apocalypse . . . and Why Canning Might Not Be the Perfect Solution." YWCA Niagara Region, September 10, 2014. https://www.ywcaniagararegion.ca/canning/.

Scott, Paul. 2020. "From Contagion to Cogitation: The Evolving Television Zombie." *Science Fiction Studies* 47, no. 1: 93–110.

Shocked, Michelle. 1992. "Strawberry Jam." Track 9 on *Arkansas Traveler*. Mercury, 512 101-2, CD.

Swyngedouw, Erik. 2018: *Promises of the Political: Insurgent Cities in a Post-Political Environment.* Cambridge, MA: MIT Press.

Taylor, Charles. 1977. "The Canning Renaissance." *Indianapolis (IN) Saturday Evening Post,* October 1, 1977.

Thorson, Esther. 2012. "The Quality of Disaster News: Frames, Disaster Stages, and a Public Health Focus." In *Reporting Disaster on Deadline,* edited by Lee Wilkins, Martha Steffens, Esther Thorson, Greeley Kyle, Kent Collins, and Fred Vultee, 69–80. New York: Routledge.

United Nations. 2020. "Parallel Threats of COVID-19, Climate Change, Require 'Brave, Visionary and Collaborative Leadership': UN Chief." United Nations News, April 28, 2020. https://news.un.org/en/story/2020/04/1062752.

United Nations Refugee Agency. n.d. "Climate Change and Disaster Displacement." Accessed September 22, 2022. https://www.unhcr.org/en-us/climate-change-and-disasters.html.

United States Bureau of Labor Statistics. 1917. "Waste in Food Distribution in New York." *Monthly Review of the US Bureau of Labor Statistics* 5, no. 3 (September): 95–98.

Webber, Michael E. 2013. "World War G: Zombies, Energy and the Geosciences." *Earth* 58, no. 12 (December): 32–33.

Winship, A. E. 1918. "Looking About." *Journal of Education* 87, no. 2: 32–33.

Yung, Kathleen, and Casey Neathway. 2019. "Community Champions for Safe, Sustainable, Traditional Food Systems." *Current Developments in Nutrition* 4, S1: 49–52.

ABOUT THE CONTRIBUTORS

Emma Frances Bloomfield is associate professor of Communication Studies at the University of Nevada, Las Vegas. She studies scientific controversies and environmental communication and how they relate to identity, ideologies, and storytelling. Her work has explored the intersections of science and religion, the environment and economics, and the technical and public spheres. She is author of *Communication Strategies for Engaging Climate Skeptics: Religion and the Environment*, which was published in 2019 as part of Routledge's Advances in Climate Change Research series. Her work appears in journals such as *Rhetoric Society Quarterly*, *Science Communication*, and *Environmental Communication*.

Sheila Bock is associate professor in the Department of Interdisciplinary, Gender, and Ethnic Studies at the University of Nevada, Las Vegas. Her research interests include the contested domains of illness experience, material/digital enactments of personal and community identities, the intersections between folklore and popular culture, and the values and challenges of cross-disciplinary collaboration. Her work has been published in the *Journal of American Folklore*, the *Journal of Folklore Research*, *Western Folklore*, the *Journal of Folklore and Education*, the *Western Journal of Black Studies*, the *Journal of Medical Humanities*, *Diagnosing Folklore: Perspectives on Disability, Health, and Trauma*, and *The Oxford Handbook of American Folklore and Folklife Studies*.

Kristen Bradley is an English professor at Nashville State Community College. She holds a doctorate in English with a concentration in folklore from the University of Louisiana at Lafayette. Her other publications include articles about southern foodways and folklore. She currently resides in New Johnsonville, Tennessee, with her dog and two cats. Bradley would like to invite everyone to attend St. Patrick's Irish Picnic in McEwen, Tennessee, the last Saturday in July.

Hannah Chapple is the dean of interdisciplinary studies at the Weber School in Atlanta, Georgia, where she works to enrich the curriculum through interdisciplinary connections and student-led research. She holds a doctorate in English from the University of Louisiana at Lafayette with a concentration in creative writing. A writer and educator, Chapple's recent poetry appears in *Sooth Swarm Journal*, *Ibis Head*, and *DMQ Review*.

James I. Deutsch is a curator and editor at the Smithsonian Institution's Center for Folklife and Cultural Heritage, where he has helped plan and develop public programs on California, China, Hungary, Peace Corps, Apollo Theater, Circus Arts, National Aeronautics and Space Administration, Mekong River, US Forest Service, World War II, Silk Road, and White House workers. In addition, he serves as an adjunct professor—teaching courses on American film history and folklore—in the American Studies Department at George Washington University. Deutsch has also taught American studies classes at universities in Armenia, Belarus, Bulgaria, Germany, Kyrgyzstan, Norway, Poland, and Turkey. He has earned academic degrees from Williams College, University of Minnesota, Emory University, and George Washington University and has published numerous articles and encyclopedia entries on a wide range of topics relating to film and folklore.

Rev. Máirt Hanley is from a small fishing village in County Wexford and went to University College Cork to attain a BA in philosophy and archaeology. Having worked as an archaeologist and musician, he started working in the Church of Ireland, first as a groundskeeper in Cork's St. Finbarr's Cathedral, then as a youth worker and religious education teacher. He then studied for the priesthood in Dublin gaining a BTh from Trinity College. After ordination, he served in several parishes in Kerry before becoming rector of Baltinglass in County Wicklow. He was a founding member of the Community of Brendan the Navigator, Cumann Breandán Naofa, and is currently serving as vicar/chair. He has been a regular contributor to several church and local magazines and radio stations and has had articles in national church publications.

Christine Hoffmann is associate professor and assistant chair of the English department at West Virginia University. She writes and teaches about early modern English literature and culture; the rhetoric and ethics of social media; and the transhistorical connections between literary and curatorial practices of collecting. Her essays have appeared or are forthcoming in cultural studies and literary theory journals, including *SubStance*, *PMLA*, *College Literature*, and *Rhizomes*. Her book *Stupid Humanism: Folly as Competence in Early Modern and Twenty-First-Century Culture* was published in 2017 as part of Palgrave Macmillan's Early Modern Cultural Studies series.

Kate Parker Horigan is associate professor in the Department of Folk Studies and Anthropology at Western Kentucky University. Her research focuses on expressive culture in communities affected by conflict and disaster. She is the author of *Consuming Katrina: Public Disaster and Personal Narrative* (2018). Her current work explores memory and narration of war and genocide in Bosnia (1992–1995), especially among Bosnian refugees in Bowling Green, Kentucky.

Shelley Ingram is associate professor of English at the University of Louisiana at Lafayette. Her research focuses primarily on the relationship between folklore and literature, including its connections to ethnography and race, folk narrative, food, and place. She has written on the literature of Shirley Jackson, Ishmael Reed, James Hannaham, and Tana French, among others, with essays appearing in edited collections and in journals, such as *African American Review*, *Journal of Folklore Research*, and *Food & Foodways*. Her cowritten book *Implied Nowhere: Absence in Folklore Studies* was published by the University Press of Mississippi in 2019.

John Laudun is professor of English at the University of Louisiana at Lafayette, where his research focuses on vernacular discourses, especially narratives, and how they cascade through socio-cultural networks both online and off. In addition to his work on folk narrative and the intellectual history of folklore studies, he has published a monograph on material folk culture, *The Amazing Crawfish Boat* (2016), one of the first folklore studies to make a case for the application of network theory to creativity and tradition. His work has appeared in a variety of academic journals and edited volumes as well as cited in newspapers and documentaries. He has been supported by a Jacob K. Javits Fellowship, a MacArthur Scholarship, the EVIA Digital Archive, the NEH Institute on Network Studies in the Humanities, and UCLA's Institute for Pure and Applied Mathematics. For more information, please see http://johnlaudun.net/.

Jordan Lovejoy is an American Council of Learned Societies Emerging Voices fellow, Southern Futures assistant director, and visiting assistant professor of American studies at the University of North Carolina at Chapel Hill. As a folklorist of Appalachia, she explores reciprocal ethnographic engagement and environmental artistic expression. Her research examines floods in Appalachian literature and storytelling and the cultural and environmental memory of the 2001 flooding in southern West Virginia.

Lena Marander-Eklund is professor of folkloristics in the Department of Folklore at Åbo Akademi University in Turku, Finland. Her research interests are narratives and narration, methodological aspects on interviews and questionnaires, life stories, emotions, weather folklore, and folk medicine. Her dissertation dealt with childbirth stories. She has also studied homemakers in Finland—the way women talk about themselves being housewives in the 1950s. She has been part of the research project Vardagens rum (Everyday Life and Space) funded by Society of Swedish Literature in Finland.

Jennifer Morrison is assistant professor of English at Xavier University of Louisiana. Her research focuses on the nuances and specifics of African American culture in American literary and cultural studies. Her current project looks at the way Black southern writers have articulated a cultural and geographic landscape, the Gulf South, in contemporary literature. Her essay "Three Women in a System: Black Domestics and the Complications of Race and Gender in Ernest Gaines's *Of Love and Dust*" appeared in the 2016 issue of *Louisiana Folklore Miscellany*.

Willow G. Mullins is a lecturer in the School of Celtic and Scottish Studies at the University of Edinburgh (Scotland). She is interested in material culture and in how people navigate the inevitable—-death, birth, and the weather. She has published on Shirley Jackson and Tana French and has three books, *The Folklorist in the Marketplace* (2019), *Implied Nowhere: Absence in Folklore Studies* (2019), and *Felt* (2009).

Anne Pryor, PhD, is a folklorist based in Madison, Wisconsin. Her continuing interest in weatherlore is primarily occupational, with a focus on tornado chasers and weather workers. She was the codirector of the award-winning K–12 curriculum project Wisconsin Weather Stories. She is married to atmospheric scientist Steve Ackerman.

Todd Richardson is a professor in the Goodrich Scholarship Program at the University of Nebraska at Omaha, where he holds the James R. Schumacher Chair of Ethics. His writing has appeared in a variety of popular and academic publications, including the *Writer's Chronicle*, *The Reader*, and the *Journal of American Folklore*.

Claire Schmidt is associate professor of English at Missouri Valley College. She teaches courses in British and world literature and writing. Her research interests include contemporary experiences of the everyday—specifically occupational humor, foodways, and race—as well as early medieval British literature (Anglo Saxon hagiography and oral tradition), Chaucer, postcolonial British fiction, and global mystery fiction. She is the author of *If You Don't Laugh You'll Cry: The Occupational Humor of White Wisconsin Prison Workers* (2017) and has published her research in the journals *Western Folklore, Digest: A Journal of Foodways and Culture,* and *Oral Tradition* as well as in collections, including *Reading Mystery Science Theatre: Critical Approaches* (2013) and *Folklore in the United States and Canada: An Institutional History* (2020).

INDEX

Ackerman, Steve A., 7, 59–63, 67–71, 73–74
Ahmed, Sara, 183–84, 187
"All Along the Watchtower" (Dylan), 166
allegory, 31, 34–36
All Quiet on the Western Front (Remarque), 172
Anderson, E. N., 266
Anthropocene, 125, 207–8; Anthropocentric, 10–17, 20–23, 106, 124
apocalypse, 276–77, 280–85; beliefs, 22–24; biblical, 166, 256; destruction, 34; post-apocalyptic, 281; rhetoric, 277
Appalachia, 109–11, 118, 203, 205–23
Aristotle, *Meteorology* (*Meteorologia*), 46, 72, 128, 256, 259
Armstrong, Henry, 36–37

Bailey, Holly, *The Mercy of the Sky*, 51–52, 100, 101
barbecue, 82, 94, 248–53, 258
Bible, the: language, 7, 35, 39; motifs, 7, 29, 154, 230; stories, 93, 154; teachings, 10–11, 19; text, 23, 28
bioregionalism, 106–9, 119
blizzards, 7, 43–44, 50, 54, 162, 177, 202
"Blowin' in the Wind" (Dylan), 163–64
Boudreaux and Thibodeaux. *See* numbskulls
Britton, Joan, 36
Brooks, Max, *World War Z*, 280, 285
Burnett, T-Bone, 168
Butler, Octavia, *The Parable of the Sower*, 280, 285

Cantore, Jim, 51
"Caribbean Wind" (Dylan), 168–69
Catholicism (Roman Catholic), 50, 80, 82–83, 85, 248–49, 254–56, 258; Diocese of Nashville, 249
chemtrails, 7, 64–74
Christianity, 7, 10–23, 32, 80, 94
Christian weatherlore, 23
Church of Ireland, 78, 93, 95; prayer book, 79, 83–85, 87
climate change, xiv–xv, xvii–xx, 5, 7, 10–12, 16, 19, 20–23, 38, 66, 70, 90, 101–2, 105–13, 115, 117, 119, 177–78, 203–4, 206–13, 217, 219, 276–80, 283–86; anxiety, xviii, 74; denial, 14, 110, 115–17, 119, 120; skepticism, 10–12, 22, 24
clouds, xvi–xvii, 3–4, 17, 19, 46, 48, 56, 67, 70–71, 75, 176, 179, 189, 227
conspiracy narratives, 73
conspiracy theories (or beliefs), xiii, xix, 56, 65, 69, 71–73
contrails, 7–8, 65, 67–75
Cooley, Peter, 28, 29
Cornwall Alliance, 13–14
COVID-19, 66, 200, 207, 280, 284–85
Creation Care, 11–13, 21

Darenbourg, Charles A., 29, 39
deluge, 29–32, 35–37, 114; Motif A1021, 29, 36
"Dignity" (Dylan), 170–71
divine retribution/punishment, 28, 31, 35, 38, 91
dominion verse, 10, 13, 18

Duluth, Minnesota, 160–64
Dundes, Alan, 29–30, 40, 264; *The Flood Myth*, 29
Dylan, Bob, 102, 160–72; Bob Dylan Archive, 165, 167, 169, 171

earthquakes, 7, 160, 191; Lisbon (1755), 37–38; Los Angeles (1971), xvii, 4
Eco Congregation, 92
"11 Outlined Epitaphs" (Bob Dylan), 161, 164
EmiSunshine, 215
England, Gary, 51–52
environmental ideologies, 11, 12, 20
environmentalism, 11, 14–15, 20, 23
Evangelical Environmental Network, 13

Faerie Queene, The (Spenser), 133–37
fake news, 73–74
"Farewell" (Dylan), 165–67
Faulkner, William, *The Old Man*, 227
fear, 22, 67, 140, 148, 177, 180, 183–84, 186–92, 200, 258, 276–77, 280, 283
fire, 7, 34, 54, 66, 168, 171, 187, 276
first-person experience, 179
Fletcher's Grove, 213–15
Flood Myth, The. See Dundes, Alan
floods, 7, 19, 32–33, 203, 205–8, 212–24, 228, 230, 234–37, 243, 258; Louisiana (2005, Hurricane Katrina), 29, 31–37, 39, 154; Louisiana (2016), 236; motif (metaphor), 7, 29–36, 38, 39–40, 206; Old Testament (Noah), 14, 29–31, 34, 39, 94, 154, 230, 257–58; stories, narratives, and legends, xv, 29–34, 235; thousand-year, 19, 213–16; West Virginia (2001), 216–18
folklore: "folkloristic," xiv, 20, 100, 164, 264; groups, 110, 116–17, 281; ideas, 227–28, 239; knowledge, 47, 49, 54, 112, 115, 145; music, 162, 213–16, 218; science, 5, 6
food, 21, 34, 153, 199–204, 245, 276–79, 281–86; canned or canning, 279, 283, 285–86; ferial, 201–3; festal, 201–3; foodways, 202–3, 276–77, 279, 283
forecasting, 43–45, 55, 63, 251; history of, 45–48; as magic, 45, 48

fracking, 210–11
Frankenstein (Shelley), 99–101

Galloway Young, Katharine, 183
Genesis, Book of, 11, 14, 18, 29, 30, 34
geoengineering, 65, 69–71
"Girl from the North Country" (Dylan), 102, 166–67
Glassie, Henry, 106
global warming, 19, 38, 66, 70, 86, 207, 208, 210
God (gods), xvi, 11, 14–18, 21, 28, 30–38, 40, 46, 48, 52, 82–95, 109, 146, 148, 151, 180, 184, 186, 187, 241, 256–58
Goldstein, Diane, 190, 191
Gore, Al, *An Inconvenient Truth*, 116, 278, 280
Greenwich Village, 162
Gulf South, 7, 144–45, 151–52, 155, 157

Hamilton, Edith, *Mythology*, 146
Harris, Lester, 34–35
Harvey, Todd, 163
Hasted, Nick, 167–68
Heylin, Clinton, 163
Hibbing, Minnesota, 161–62, 164, 171
Homer, 127; *Odyssey*, 172
Horigan, Kate Parker, xiv, 217
Howard, Mark, 165
hurricanes, xiii, xiv, xvii–xix, 4, 7, 16, 19, 28, 32–33, 40, 48, 50, 54, 102, 123, 144–46, 148, 152–58, 177, 180, 199, 201–2, 208, 226, 230–32, 238, 241, 243, 256; Dorian (2019), 230–32; Elena (1985), 147; Frederic (1979), 200; Galveston (1900), 226, 243; Georges (1998), 147; Harvey (2017), 7, 18, 19, 22, 226; Hazel (1954), 230; Irma (2017), 280; Ivan (2004), 153; Katrina (2005), xiv, 7, 28–40, 102, 144–55, 157–58, 177, 202–3, 238; Rita (2005), 238
Hurston, Zora Neale: *Mules and Men*, 230; *Their Eyes Were Watching God*, 227

"Idiot Wind" (Dylan), 166–68
"I'm Cold" (Dylan), 167–68
Inconvenient Truth, An (Gore), 278, 280

Junger, Sebastian, *The Perfect Storm*, 99

Kees, Weldon, 167
Khalidi, Rashid, 167
Kverndokk, Kyrre, 37–38, 177

Labov, William, *Language in the Inner City*, 179, 181–85, 191, 240
Larry Meiller Show, 59–60, 74
Larry X, 34
Laymon, Kiese, *Long Division*, 102, 144, 149, 151, 155, 158
legends, xiv–xv, 56, 67, 73–74, 190, 203, 227–29, 233, 236, 238–42, 245
Lindahl, Carl, xiv, 32, 37, 39–40, 238–42
Little Ice Age, 99, 137
Liturgical Advisory Committee, 83
Long Division (Laymon), 102, 144–45, 149–58
Louisiana, xx, 3, 6–8, 153–55, 229, 233–36, 238, 240, 243–44
"Love Minus Zero/No Limit" (Dylan), 164, 167
Lynch, David, 261–62, 272; *Blue Velvet*, 261–62; *The Straight Story*, 264

M, Patrice, 35
Maravich, "Pistol" Pete, 170
Martin, Craig, 256, 258
Martin, Jonathan E., 8, 59–66, 68–69, 71–75
McCain, John, 167
McEwen, Tennessee, 248–49, 251, 253–56
Medea (mythology), 146–49
Meiller, Larry, 59–65, 67–69, 70, 74
Mercy of the Sky, The (Bailey), 100, 101
meteorologists, xiv, xvi, 3, 6, 43–45, 47–52, 54–56, 161
meteorology, xvi, 49, 53, 56, 100, 179
Meteorology/Meteorologia (Aristotle), 256, 259
Meyer, Travis, 50–51
Mintz, Elliott, 165
Mississippi Gulf Coast, 145, 147, 149
Moby-Dick (Melville), 172
Motif-Index of Folk-Literature (Thompson), 29
motifs, 7, 29–38, 110; Motif A1010, Deluge, 29, 36; Motif A1018, Flood as punishment, 29, 32; Motif A1020, Escape from, 29, 36; Motif A1021, Deluge: escape in boat, 29, 36
Mother Earth News, 277, 279
Mules and Men (Hurston), 230
Myrtle Beach Jeep, 232, 239, 244
Mythology (Hamilton), 146

naming storms, 180
narrative, xiv–xv, 28–36, 38, 65–67, 69, 74, 100–101, 108–10, 122, 126, 129, 131–33, 135, 137–39, 146–52, 154, 157–58, 177, 179–92, 203, 209, 219, 234–35, 237–39, 241, 248, 252–54, 257–58, 264, 279–81, 283; allegorical, 33; "almost" stories, 177, 189, 192; belief, 7; chemtrail, 67, 69, 71, 74; complete, 181–83, 189; folk, 29–31, 145–46; micronarratives, 177, 184–85, 189–90; mythological, 29; narrative strategy, 177, 184–85, 189–90; Noah, 94; personal, xix, 7, 29–31, 61–62, 74, 102; stylistic device, 185; of survivors, 29, 33, 40; untellability, 190
National Oceanic and Atmospheric Administration (NOAA), 43, 213, 233
New Orleans, Louisiana, 7, 28–29, 31–35, 40, 153, 155, 202, 214, 249
New York City, New York, 104–5, 162, 206
"Nice Weather Today!" questionnaire, 177–79
Noah, xix, 29–30, 86, 92–94, 230, 257–58
Nobel Prize in Literature, 172
nostalgia, 4, 200, 263, 266
numbskulls, 229, 243

Obama, Barack, 167
Obama, Michelle, 278
Odyssey (Homer), 172
Old Man, The (Faulkner), 227
Old Testament, 19, 29, 34, 39, 257–58
"One More Night" (Dylan), 166
Oration on the Dignity of Man (Pico), 130
Orleans (Smith), 102, 144–45, 152–54, 156, 158

Parable of the Sower (Butler), 280, 285
"Percy's Song" (Dylan), 166
Perfect Storm, The (Junger), 99

phatic communication, 177
Pico, Giovanni, *Oration on the Dignity of Man*, 130
Plutarch, 126, 160
Porché, Bernie, 33–36
Primiano, Leonard Norman, 10

Quart Livre, Le (Rabelais), 102, 127–30

Raeben, Norman, 167
Rabelais, François, *Le Quart Livre*, 102, 127–30
rain, 4, 6, 22, 44, 48, 53, 55–56, 68, 70, 79, 80, 82–84, 87, 90, 93–94, 111–16, 156, 160, 165–66, 180, 199, 212, 214–15, 224, 227, 229–30, 250–55, 257–58, 265, 279
Remarque, Erich Maria, *All Quiet on the Western Front*, 172
rescue, 31, 36–37
Revelation, Book of, 34
Robbins, Paul, 266
Robbins, Terry, 164
Rudd, Mark, 164

saints, 248, 255–56; All Saints Day, 155–56; saints' days, 255
Salvage the Bones (Ward), 102, 144–46, 156–57
San Francisco, California, xx, 118, 169, 230
Shelley, Mary, *Frankenstein*, 99–100, 101
"Shot of Love" (Dylan), 169
Shuman, Amy, 31, 181, 190
sky, xv, xvii, 3–4, 6–7, 34, 46, 51, 66–67, 74, 81, 100–101, 151, 158, 171, 261
slow violence, 208, 211–14, 216–17, 219
Smith, Sherri L., *Orleans*, 102, 144–45, 152–54, 156, 158
snow, xiii, 4–5, 7, 22, 52, 62, 64, 114, 126, 131, 133–35, 160, 162, 199–200; as insult, 102, 122–26, 132–33, 138–40; snowflake, xviii, xix, 123–24, 136
"Snow Child, The" (fable), 130–32, 135
Soles, Steven, 167–68
Sounes, Howard, 168
South Carolina, 7–8, 230, 232
Spenser, Edmund, *The Faerie Queene*, 102, 133–36

Stattin, Jochum, 184
stewardship, 15–16
Stewart, George R., *Storm*, 100
Storm (Stewart), 100, 102–3
storms, xiii, xv, xvii–xviii, 3–4, 6, 18–19, 28–30, 32, 34–35, 39, 44, 47, 50–51, 62, 86, 99, 100, 102–3, 111, 116, 131, 144–45, 147–48, 150–54, 158, 166, 176–92, 199–203, 207, 212, 226–27, 231, 233–34, 239, 242–44, 256–57, 281
Straight Story, The (Lynch), 264
"Subterranean Homesick Blues" (Dylan), 164–65
sun, xviii–xix, 16, 34, 44, 46, 66, 68, 111, 131, 146, 148, 160, 162, 168–69, 180, 214, 224, 251, 255
Surviving Katrina and Rita in Houston (SKRH), 28
sustainability, 118, 140, 263, 265, 282–83, 285

Tennessean (*Nashville Tennessean*), 248, 250–51, 253–55, 257
Tennessee Crossroads, 257
Their Eyes Were Watching God (Hurston), 227
thunderstorms, xiii, 4, 102, 177–81, 184–85, 187–92, 202, 213, 234
tornadoes, xiii, xv–xvi, xix, 3–7, 50–51, 54, 100–101, 208; Oklahoma City (2013), 51, 100; Waynoka (1898), xv–xvi
tricksters, 229, 233
trucks, 235–38; abandoned, 203, 227, 235, 237–38, 240–41; and individualism, 242; and masculinity, 240
Turner, Gil, 163

"Under the Red Sky" (Dylan), 171
University of Minnesota, 162

Van Ronk, Dave, 162

Ward, Jesmyn, *Salvage the Bones*, 144–47, 154–55, 158
Weather Guys, 8, 60–61, 65–67, 72, 74–75
weatherlore, xiii–xx, 3, 7–8, 20, 23, 46, 101, 160, 164, 167, 170, 203, 248, 251–52, 254–56, 258
Weathermen (Weather Underground), 164

weathermen, 49–50, 52, 55, 164
West Virginia, 209–10, 213–17, 220–24
Williams, David, 232
Williams, Paul, 163
windlore, 102, 160, 163–64, 168, 172
Wisconsin, 8, 59–60, 62–63, 74–75, 264–65
Wisconsin Idea, 60
World War Z (Brooks), 280, 285

Yale Project on Climate Change, 12

Zimmerman, Robert Allen, 160. *See also* Dylan, Bob
Zollo, Paul, 167
zombies, 203, 276, 280, 283–84

Made in United States
North Haven, CT
02 June 2023